Springer Series on Atomic, Optical, and Plasma Physics

Volume 114

The Springer Series on Atomic, Optical, and Plasma Physics covers in a comprehensive manner theory and experiment in the entire field of atoms and molecules and their interaction with electromagnetic radiation. Books in the series provide a rich source of new ideas and techniques with wide applications in fields such as chemistry, materials science, astrophysics, surface science, plasma technology, advanced optics, aeronomy, and engineering. Laser physics is a particular connecting theme that has provided much of the continuing impetus for new developments in the field, such as quantum computation and Bose-Einstein condensation. The purpose of the series is to cover the gap between standard undergraduate textbooks and the research literature with emphasis on the fundamental ideas, methods, techniques, and results in the field.

More information about this series at http://www.springer.com/series/411

Krzysztof Sacha

Time Crystals

Springer

Krzysztof Sacha
Jagiellonian University in Krakow
Krakow, Poland

ISSN 1615-5653 ISSN 2197-6791 (electronic)
Springer Series on Atomic, Optical, and Plasma Physics
ISBN 978-3-030-52522-4 ISBN 978-3-030-52523-1 (eBook)
https://doi.org/10.1007/978-3-030-52523-1

This Springer imprint is published by the registered company Springer Nature Switzerland AG
The registered company address is: Gewerbestrasse 11, 6330 Cham, Switzerland

To Ewa, Ola, and Wojciech

Preface

What are time crystals? I have been asked this question many times. Even if I did my best to explain the phenomena related to the time crystal research, there were further questions. Where can we read about them? Original scientific publications and even review articles are often written for specialists, is there a book on that subject? It was the main motivation why I decided to write this monograph.

The field of time crystals, despite being in its infancy, is developing quite rapidly and it is impossible to describe all interesting results related to the time crystal area. If I tried to do so, I would never finish writing this book. I have made a selection and the selection is obviously subjective. I concentrate on quantum aspects of time crystals and try to describe how the research initiated by the seminal articles by Alfred Shapere and Frank Wilczek is developing and in which directions.

Basic knowledge of the reader on quantum mechanics is assumed. If I feel, subjectively, that certain formalism which I have to use is more advanced or a considered problem needs an introduction, background information is provided. I am aware that this monograph is far from being perfect and try to excuse myself by saying that the primary purpose of the book is to support the field in its infancy and to introduce researchers, especially young scientists, to the new ideas: the ideas where the time dimension is as good to host crystalline structures as the space dimensions do.

Kraków, Poland
March 2020

Krzysztof Sacha

Acknowledgements

I am deeply grateful to: Bryan Dalton, Alexandre Dauphin, Dominique Delande, Krzysztof Giergiel, Weronika Golletz, Peter Hannaford, Arkadiusz Kosior, Arkadiusz Kuroś, Maciej Lewenstein, Paweł Matus, Bruno Mera, Marcin Mierzejewski, Florian Mintert, Artur Miroszewski, Luis Morales-Molina, Rick Mukherjee, Yasser Omar, Frederic Sauvage, Andrzej Syrwid, Jia Wang, and Jakub Zakrzewski for the collaboration on the time crystal research. It was my great pleasure.

I would also like to thank Jakub Zakrzewski for coming to my office in 2012 with the seminal papers on time crystals by Alfred Shapere and Frank Wilczek and asking for my opinion. It was the key moment when I got attracted to the time crystal problems which I did not have any idea about.

I am extremely grateful to Peter Hannaford for very close and intensive collaboration. His enthusiasm for time crystals and many difficult questions encouraged me to put a lot of effort into the research and allowed me to understand what I was actually doing.

A new research area needs a lot of support; otherwise, it will die out. I thank very much Maciej Lewenstein for his openness to novel ideas and his support of the time crystal field.

I am grateful to Arkadiusz Kosior for the scientific cooperation and for his invaluable help in organizing the workshop on time crystals which brought to Kraków enthusiasts of time crystallization in 2019.

I thank all my students and young collaborators, especially Krzysztof Giergiel without whom the progress of our research on time crystals, if any, would be much slower.

I acknowledge the support of the National Science Centre, Poland via Project No. 2018/31/B/ST2/00349.

Contents

1 Introduction 1
References 4

2 Spontaneous Breaking of Space Translation Symmetry 7
2.1 Spontaneous Formation of Space Crystals 7
2.1.1 Consequences of Continuous Space Translation Symmetry .. 7
2.1.2 Spontaneous Emergence of Crystals in Space 9
2.2 Spontaneous Breaking of Space Translation Symmetry in a Non-interacting System 14
References 16

3 Spontaneous Breaking of Continuous Time Translation Symmetry ... 19
3.1 Time Crystal Idea 19
3.1.1 Wilczek Model 21
3.1.2 Excited Eigenstates and Time Crystals Formation 25
3.1.3 Lack of Time Crystal Behavior of Systems with Two-Body Interactions in the Ground State and in the Equilibrium State 27
3.1.4 Time Crystal Behavior of Systems with Multi-Body Interactions 28
3.1.5 Classical Time Crystals 31
3.2 Josephson Effect and Superfluid Phase Oscillations 33
3.2.1 Josephson Effect 33
3.2.2 Superfluid Phase Oscillations 35
References 36

4 Discrete Time Crystals and Related Phenomena 39
4.1 Floquet Formalism and Discrete Time Translation Symmetry 39
4.1.1 Floquet Formalism 39
4.1.2 Discrete Time Translation Symmetry of Periodically Driven Systems 46

4.2 Discrete Time Crystals in Ultra-Cold Atoms Bouncing on an Oscillating Atom Mirror ... 46
4.2.1 A Particle Bouncing Resonantly on an Oscillating Mirror: Classical and Quantum Description ... 47
4.2.2 Formation of Discrete Time Crystals ... 56
4.2.3 Phase Diagram ... 63
4.2.4 Dramatic Breaking of the Discrete Time Translation Symmetry and Formation of n-Tupling Discrete Time Crystals ... 72
4.2.5 Stability of Discrete Time Crystals in Ultra-Cold Atoms Bouncing on an Oscillating Mirror ... 77
4.3 Discrete Time Crystals in Periodically Driven Lattice Systems ... 81
4.3.1 Discrete Time Crystals in Spin Systems with Disorder ... 81
4.3.2 Clean Discrete Time Crystals and Prethermal Character of Discrete Time Crystals in Lattice Systems ... 87
4.4 Fractional Time Crystals ... 99
4.4.1 Fractional Time Crystals in Ultra-Cold Atoms Bouncing on a Mirror ... 100
4.4.2 Fractional Time Crystal in Spin Systems ... 104
4.5 Discrete Time Quasi-Crystals ... 108
4.6 Experimental Realizations ... 115
4.6.1 Periodically Driven Chain of Ions ... 116
4.6.2 Periodically Driven Nitrogen-Vacancy Centers in Diamond . 119
4.6.3 Periodically Driven Nuclear Magnetic Moments ... 123
4.6.4 Space-Time Crystals in an Atomic Bose–Einstein Condensate ... 127
4.7 Dissipative Systems ... 133
4.7.1 Dicke Time Crystals in Driven Dissipative Systems ... 134
4.7.2 Discrete Time Crystals in Dissipative Systems in the Absence of Manifest Symmetries ... 140
4.7.3 Analysis of Discrete Time Crystals in Open Systems ... 147
4.7.4 Breaking of Continuous Time Translation Symmetry in Dissipative Systems ... 151
4.7.5 Imaginary Time Crystals ... 155
4.8 Further Reading ... 161
References ... 162

5 Condensed Matter Physics in the Time Dimension ... 173
5.1 Platform for Condensed Matter Physics in the Time Dimension ... 173
5.2 Anderson Localization in the Time Dimension ... 179
5.2.1 Anderson Localization in Time Crystals ... 181
5.2.2 Anderson Localization in the Time Dimension Without Crystalline Structure in Time ... 187
5.2.3 Anderson Localized-Delocalized Transition in the Time Dimension ... 192

5.3 Single-Particle Condensed Matter Phenomena in the Time Dimension ... 200
5.3.1 Topological Time Crystals ... 200
5.3.2 Engineering of Effective Potentials: Quasi-Crystalline Structure in the Time Dimension ... 210
5.4 Many-Body Condensed Matter Physics in the Time Dimension ... 212
5.4.1 Mott Insulator Phase in the Time Dimension ... 212
5.4.2 Many-Body Localization with Temporal Disorder ... 216
5.4.3 Engineering of Long-Range Interactions in Time Crystals ... 219
5.4.4 Time Lattices with Properties of Multi-Dimensional Space Crystals ... 223
5.5 Time Engineering: Anderson Molecules ... 226
References ... 229

6 Phase Space Crystals ... 237
6.1 Idea of Phase Space Crystals ... 237
6.2 Topological Phase Space Crystals ... 240
6.3 Many-Body Physics in Phase Space Crystals ... 245
References ... 248

7 Photonic Time Crystals ... 251
7.1 Photonic Crystals in Space ... 251
7.1.1 Introduction to Photonic Crystals in Space ... 251
7.1.2 Topological Properties of Photonic Crystals in Space ... 254
7.2 Photonic Crystals in Time ... 258
7.2.1 Introduction to Photonic Time Crystals ... 259
7.2.2 Topological Properties of Photonic Time Crystals ... 262
References ... 264

Index ... 267

Chapter 1
Introduction

Periodic motion can be observed everywhere ranging from atomic systems and a clock on a wall to the Solar System and beyond the Milky Way. Periodic evolution in time can be considered as a temporal counterpart of periodic behavior in the space dimensions which is also common because there are solid state crystals. Research of periodic motion of systems as crystalline structures in the time domain is a novel field initiated by Alfred Shapere and Frank Wilczek in the classical systems [21] and by Wilczek himself in the quantum case [25].

Space crystals form as a result of self-organization of many-body systems in a regular way in space due to interactions between particles. In the field of time crystals we expect an analogous self-organization process but in the time dimension. In other words, an interacting many-body system should be able to self-organize its time evolution and start moving in a periodic way in time. Research of synchronization of classical many-body systems, like in the seminal Kuramoto model [1, 13, 14, 26], is an extensively developing field. However, we do not describe the classical synchronization problems in the present monograph because we focus on quantum systems.

In the formation of solid state crystals quantum effects are involved. Especially, space crystals break space translation symmetry. In quantum mechanics, symmetries of the Hamiltonian of a system are very important things and must be obeyed by system eigenstates. Interactions between electrons and between electrons and nuclei do not favor any location of a crystal in space but in a real experiment we observe a crystalline structure anchored to a certain position. The translation symmetry is broken and whether it can happen when a solid state system is in the lowest energy state is a non-trivial question. While in condensed matter physics we know it is possible and even we understand quite well why, the crystallization in the time domain is less obvious especially when we are ambitious and would like to see the crystallization, and thus periodic motion of a system, in the lowest energy state.

K. Sacha, *Time Crystals*, Springer Series on Atomic, Optical, and Plasma Physics 114, https://doi.org/10.1007/978-3-030-52523-1_1

Usually, a classical particle possessing the lowest possible energy is at rest. In order to perform motion a particle needs additional energy which can be converted into the kinetic energy. This intuition is not always correct because there are classical systems where the energy is minimal if a particle is moving periodically. These systems are classical time crystals proposed by Shapere and Wilczek [21] which will be shortly described in the present book.

If a classical particle can minimize its energy by performing periodic motion, quantum many-body systems should be able to do so as well. That was the idea of quantum time crystals proposed by Wilczek in 2012 [25]. However, it turns out that in isolated time-independent quantum many-body systems it is not so easy. The model proposed by Wilczek does not reveal periodic motion in its ground state which was shown soon enough [3]. Formation of crystals (both space and time crystals) in quantum many-body systems is related to spontaneous breaking of a continuous translation symmetry of the Hamiltonian into a discrete translation symmetry. A time-independent system possesses the Hamiltonian which does not change in time but we expect that it prefers to move with a certain period and the phase it starts periodic evolution is chosen spontaneously. The question if a quantum many-body system can form periodic motion in the lowest energy state can be rephrased into a question whether the system in the ground state reveals spontaneous breaking of the continuous time translation symmetry into a discrete one? For a class of systems with two-body interactions, Watanabe and Oshikawa showed that it is not possible [23, 24]. Going beyond this class of systems, the question is open. Recently, in Ref. [12], Kozin and Kyriienko constructed a system which fulfills the definition of time crystals given in [23]. If we relax the criteria and allow for spontaneous breaking of the symmetry not in the ground state but in the lowest energy eigenstate among states of a certain symmetry, then time crystal behavior can be observed. Such a symmetry-protected time crystal can be observed also in the original Wilczek model [22].

The original Wilczek work on quantum time crystals [25] was inspiring and triggered new thinking about periodically evolving quantum systems. If there are difficulties with creating time crystals in time-independent systems why not look for time crystal behavior in isolated but driven systems? Quantum systems driven with a period T possess already the discrete time translation symmetry because we may shift the time coordinate only by an integer multiple of T if we want to leave the Hamiltonian unchanged. Crystallization in time in this kind of systems would mean that a system prefers moving with a period which is different from the period dictated by the drive [19]. In other words, a discrete time translation symmetry of the Hamiltonian should be spontaneously broken into another discrete time translation symmetry.

When we consider a classical particle driven with a period T, it is not surprising if there exists a stable periodic solution of the Hamilton equations that evolves with a period $2T$ for instance. The same particle but in the quantum mechanical description cannot perform stationary motion with the period $2T$ because the discrete time translation symmetry forces all periodic solutions of the Schrödinegr equation to evolve with the driving period T. Thus, in the classical case stable periodic evolution of a particle which does not respect the discrete time translation symmetry of the Hamiltonian is possible while the corresponding quantum system must obey the symmetry.

Discrete time translation symmetry must be respected by a quantum periodically driven particle. Is it possible that a many-body system of interacting particles is able to break this symmetry spontaneously and start evolving with a period different than the driving period? This question was addressed in Ref. [19] and it was shown that indeed an isolated and periodically driven system of ultra-cold atoms can break spontaneously the discrete time translation symmetry if atom–atom interactions are sufficiently strong. Subsequent publications proposed the same phenomenon in systems of spins [6, 11] and the research of the so-called discrete time crystals began [20]. There are different aspects of this research area. The phenomenon is interesting on its own and it is also important because discrete time crystals can break ergodicity. A generic many-body system is expected to be heated to the infinite temperature by time-periodic drive but it can be not the case for a discrete time crystal. The field is also stimulated by experimental research—the first experiments were performed soon after the publications of the theoretical predictions [4, 27].

Spontaneous breaking of time translation symmetry is the realization in the time domain of one of the important properties of solid state systems. Drawing inspiration from the Wilczek work [25] one may ask whether other condensed matter phenomena can be also observed in the time dimension? Guo and coworkers showed that a particle driven periodically can reveal energy structure (more precisely quasi-energy structure) which resembles a band structure of a particle in a spatially periodic potential if the driving is resonant [8]. Similar band structures can be also observed in propagation of electromagnetic waves in media if the refractive index is modulated periodically in time [2, 5, 9, 10, 17, 28]. The latter systems are called photonic time crystals which are temporal counterparts of photonic crystals in space. It turns out that many different condensed matter phenomena like Anderson or many-body localization in the time dimension or topological time crystals can be realized in periodically driven systems [7, 15, 16, 18, 20]. While all these phenomena are externally induced, that is, they are not emergent and no spontaneous symmetry breaking is involved, they reproduce in the time dimension what is observed in condensed matter physics in the space dimensions.

The monograph is organized as follows. Before we switch to time crystals, we shortly describe spontaneous breaking of space translation symmetry and formation of space crystals in Chap. 2. Spontaneous breaking of continuous time translation symmetry and possibilities of formation of time crystals in time-independent systems are discussed in Chap. 3. Chapter 4 describes discrete time crystals. It contains theoretical analyses of spontaneous breaking of discrete time translation symmetry by periodically driven isolated systems, description of experimental demonstrations of discrete time crystals and also a section where dissipative systems are considered. In Chap. 5 we start considerations of time crystal phenomena that are induced externally by periodic drive and that are not a result of spontaneous breaking of time translation symmetry. We show that various phases known in condensed matter physics can be observed in the time dimension too. Periodically driven systems can also reveal crystalline structures in the phase space which are discussed in Chap. 6. Finally, in Chap. 7, we describe photonic time crystals.

As we mention in the Preface it is impossible to describe all aspects of the time crystal field. There are many interesting works which we do not report in the present monograph and we encourage a reader to refer to the scientific literature.

References

1. Acebrón, J.A., Bonilla, L.L., Pérez Vicente, C.J., Ritort, F., Spigler, R.: The Kuramoto model: a simple paradigm for synchronization phenomena. Rev. Mod. Phys. **77**, 137–185 (2005). https://doi.org/10.1103/RevModPhys.77.137
2. Biancalana, F., Amann, A., Uskov, A.V., O'Reilly, E.P.: Dynamics of light propagation in spatiotemporal dielectric structures. Phys. Rev. E **75**, 046607 (2007). https://doi.org/10.1103/PhysRevE.75.046607
3. Bruno, P.: Comment on "space-time crystals of trapped ions". Phys. Rev. Lett. **111**, 029301 (2013). https://doi.org/10.1103/PhysRevLett.111.029301
4. Choi, S., Choi, J., Landig, R., Kucsko, G., Zhou, H., Isoya, J., Jelezko, F., Onoda, S., Sumiya, H., Khemani, V., von Keyserlingk, C., Yao, N.Y., Demler, E., Lukin, M.D.: Observation of discrete time-crystalline order in a disordered dipolar many-body system. Nature **543**(7644), 221–225 (2017). Letter. https://doi.org/10.1038/nature21426
5. Elachi, C.: Waves in active and passive periodic structures: a review. Proc. IEEE **64**(12), 1666–1698 (1976). https://doi.org/10.1109/PROC.1976.10409
6. Else, D.V., Bauer, B., Nayak, C.: Floquet time crystals. Phys. Rev. Lett. **117**, 090402 (2016). https://doi.org/10.1103/PhysRevLett.117.090402
7. Giergiel, K., Dauphin, A., Lewenstein, M., Zakrzewski, J., Sacha, K.: Topological time crystals. New J. Phys. **21**(5), 052003 (2019). https://doi.org/10.1088/1367-2630/ab1e5f
8. Guo, L., Marthaler, M., Schön, G.: Phase space crystals: a new way to create a quasienergy band structure. Phys. Rev. Lett. **111**, 205303 (2013). https://doi.org/10.1103/PhysRevLett.111.205303
9. Harfoush, F., Taflove, A.: Scattering of electromagnetic waves by a material half-space with a time-varying conductivity. IEEE Trans. Antenn. Propag. **39**(7), 898–906 (1991). https://doi.org/10.1109/8.86907
10. Holberg, D., Kunz, K.: Parametric properties of fields in a slab of time-varying permittivity. IEEE Trans. Antenn. Propag. **14**(2), 183–194 (1966). https://doi.org/10.1109/TAP.1966.1138637
11. Khemani, V., Lazarides, A., Moessner, R., Sondhi, S.L.: Phase structure of driven quantum systems. Phys. Rev. Lett. **116**, 250401 (2016). https://doi.org/10.1103/PhysRevLett.116.250401
12. Kozin, V.K., Kyriienko, O.: Quantum time crystals from Hamiltonians with long-range interactions. Phys. Rev. Lett. **123**, 210602 (2019). https://doi.org/10.1103/PhysRevLett.123.210602
13. Kuramoto, Y.: Self-entrainment of a population of coupled non-linear oscillators. In: Araki, H. (ed.) International Symposium on Mathematical Problems in Theoretical Physics, pp. 420–422. Springer, Berlin (1975)
14. Kuramoto, Y., Nishikawa, I.: Statistical macrodynamics of large dynamical systems. case of a phase transition in oscillator communities. J. Statist. Phys. **49**(3), 569–605 (1987). https://doi.org/10.1007/BF01009349
15. Lustig, E., Sharabi, Y., Segev, M.: Topological aspects of photonic time crystals. Optica **5**(11), 1390–1395 (2018). https://doi.org/10.1364/OPTICA.5.001390. http://www.osapublishing.org/optica/abstract.cfm?URI=optica-5-11-1390

16. Mierzejewski, M., Giergiel, K., Sacha, K.: Many-body localization caused by temporal disorder. Phys. Rev. B **96**, 140201 (2017). https://doi.org/10.1103/PhysRevB.96.140201
17. Morgenthaler, F.R.: Velocity modulation of electromagnetic waves. IRE Trans. Microw. Theory Tech. **6**(2), 167–172 (1958). https://doi.org/10.1109/TMTT.1958.1124533
18. Sacha, K.: Anderson localization and Mott insulator phase in the time domain. Sci. Rep. **5**, 10787 (2015). https://doi.org/http://dx.doi.org/10.1038/srep1078710.1038/srep10787. https://www.nature.com/articles/srep10787
19. Sacha, K.: Modeling spontaneous breaking of time-translation symmetry. Phys. Rev. A **91**, 033617 (2015). https://doi.org/10.1103/PhysRevA.91.033617
20. Sacha, K., Zakrzewski, J.: Time crystals: a review. Rep. Prog. Phys. **81**(1), 016401 (2017). https://doi.org/10.1088/1361-6633/aa8b38
21. Shapere, A., Wilczek, F.: Classical time crystals. Phys. Rev. Lett. **109**, 160402 (2012). https://doi.org/10.1103/PhysRevLett.109.160402
22. Syrwid, A., Zakrzewski, J., Sacha, K.: Time crystal behavior of excited eigenstates. Phys. Rev. Lett. **119**, 250602 (2017). https://doi.org/10.1103/PhysRevLett.119.250602
23. Watanabe, H., Oshikawa, M.: Absence of quantum time crystals. Phys. Rev. Lett. **114**, 251603 (2015). https://doi.org/10.1103/PhysRevLett.114.251603
24. Watanabe, H., Oshikawa, M., Koma, T.: Proof of the absence of long-range temporal orders in Gibbs states. J. Stat. Phys. **178**(4), 926–935 (2020). https://doi.org/10.1007/s10955-019-02471-5
25. Wilczek, F.: Quantum time crystals. Phys. Rev. Lett. **109**, 160401 (2012). https://doi.org/10.1103/PhysRevLett.109.160401
26. Witthaut, D., Timme, M.: Kuramoto dynamics in Hamiltonian systems. Phys. Rev. E **90**, 032917 (2014). https://doi.org/10.1103/PhysRevE.90.032917
27. Zhang, J., Hess, P.W., Kyprianidis, A., Becker, P., Lee, A., Smith, J., Pagano, G., Potirniche, I.D., Potter, A.C., Vishwanath, A., Yao, N.Y., Monroe, C.: Observation of a discrete time crystal. Nature **543**(7644), 217–220 (2017). Letter. https://doi.org/10.1038/nature21413
28. Zurita-Sánchez, J.R., Halevi, P., Cervantes-González, J.C.: Reflection and transmission of a wave incident on a slab with a time-periodic dielectric function $\epsilon(t)$. Phys. Rev. A **79**, 053821 (2009). https://doi.org/10.1103/PhysRevA.79.053821

Chapter 2
Spontaneous Breaking of Space Translation Symmetry

Abstract Spontaneous symmetry breaking is related to a situation where equations that describe a system possess a symmetry but stable solutions of these equations break this symmetry. In quantum mechanics if the Hamiltonian commutes with a certain unitary (symmetry) operator, its eigenfunctions have to be eigenstates of this operator. In condensed matter physics, interactions between particles depend on relative distances between them and therefore the Hamiltonian does not change if all particles are translated by the same arbitrary vector. Consequently the ground state, or any other eigenstate, of the system corresponds to the single-particle probability density uniform in space. However, the symmetry can be spontaneously broken if an eigenstate is extremely fragile to a perturbation. In this chapter, the continuous space translation symmetry of condensed matter systems is considered and the approach that allows one to identify spontaneous breaking of this symmetry into a discrete space translation symmetry is introduced.

2.1 Spontaneous Formation of Space Crystals

2.1.1 Consequences of Continuous Space Translation Symmetry

It is often forgotten that formation of crystals in space requires breaking of continuous space translation symmetry into discrete space translation symmetry. Standard approach in condensed matter physics assumes that crystals are already formed and their properties are analyzed with the help of models where a spatially periodic potential is inserted by hand. One of the aims of the present monograph is to analyze spontaneous formation of crystalline structures in time but first we have to remind ourselves what quantum mechanics tells us about the formation of conventional space crystals.

Formation of space crystals is related to self-organization of a quantum many-body system in a periodic way in space. In the *classical world* crystalline structures can also form, however, on the atomic scale it is not possible to avoid a quantum approach because both the formation of crystals and their properties are determined

K. Sacha, *Time Crystals*, Springer Series on Atomic, Optical, and Plasma Physics 114, https://doi.org/10.1007/978-3-030-52523-1_2

by quantum mechanics. Solid state systems consist of atoms which in turn are built with electrons and nuclei. The Hamiltonian of such systems can be written in a general form

$$H = \sum_{i=1}^{N} \frac{\mathbf{p}_i^2}{2m_i} + \frac{1}{2}\sum_{i \neq j}^{N} U_{ij}(\mathbf{r}_i - \mathbf{r}_j), \tag{2.1}$$

where N is the total number of electrons and nuclei, m_i's stand for masses of the particles, and U_{ij} for Coulomb interactions between electrons and between electrons and nuclei. We assume that N particles are in space of volume V with periodic boundary conditions. Looking at the Hamiltonian (2.1) we do not see any crystalline structure in space but we hope that when we calculate its eigenstates, a space crystal emerges. The hope is slightly threatened by the fact that the Hamiltonian is translationally invariant and so should its eigenstates. When we translate all particles by the same vector, $\mathbf{r}_i \to \mathbf{r}_i + \mathbf{R}$, the Hamiltonian does not change. In other words, H commutes with the unitary operator $\mathcal{T}_{\mathbf{R}}$ that translates all particles by an arbitrary vector $\mathbf{R}$. Thus, an eigenstate ψ of H is also an eigenstate of $\mathcal{T}_{\mathbf{R}}$,[1]

$$\begin{aligned} \psi(\mathbf{r}_1 + \mathbf{R}, \ldots, \mathbf{r}_N + \mathbf{R}) &= \mathcal{T}_{\mathbf{R}}\, \psi(\mathbf{r}_1, \ldots, \mathbf{r}_N) \\ &= e^{i\varphi} \psi(\mathbf{r}_1, \ldots, \mathbf{r}_N). \end{aligned} \tag{2.2}$$

Equation (2.2) shows that if we shift the many-body system, prepared in the ground state (or in any other eigenstate), by any arbitrary vector $\mathbf{R}$, the N-particle probability density does not change. Actually, it is something that should be expected because the position of the center of mass of particles decouples from the degrees of freedom for relative positions. In the center of mass coordinate frame, the Hamiltonian takes the form

$$H = \frac{\mathbf{P}^2}{2\left(\sum_i m_i\right)} + \text{relative degrees of freedom}, \tag{2.3}$$

where $\mathbf{P}$ is the total momentum which is a good quantum number. Eigenstates of H are characterized by the well defined momentum $\mathbf{P}$ and consequently the center of mass position is totally delocalized. Hence, even if a space crystal exists we do not know where it is located because its center of mass is delocalized [10].

[1] If some energy levels of the Hamiltonian (2.1) are degenerate, then one has to choose proper combinations of eigenstates belonging to a degenerate subspace in order to obtain states which are also eigenvectors of $\mathcal{T}_{\mathbf{R}}$.

2.1.2 Spontaneous Emergence of Crystals in Space

The translational invariance of the N-particle probability density, cf. (2.2), does not mean necessarily that a crystalline structure cannot emerge when we start measuring positions of particles because there can be strong correlations between particles.

It is easy to show that the single-particle probability density, i.e. the probability density for a measurement of a single electron or a single nucleus, is uniform in space if the system is prepared in an eigenstate,

$$\rho(\mathbf{r}_1+\mathbf{R}) = \int d^3(\mathbf{r}_2+\mathbf{R})\ldots d^3(\mathbf{r}_N+\mathbf{R})\,|\psi(\mathbf{r}_1+\mathbf{R},\ldots,\mathbf{r}_N+\mathbf{R})|^2$$
$$= \int d^3\mathbf{r}_2\ldots d^3\mathbf{r}_N\,|\psi(\mathbf{r}_1,\ldots,\mathbf{r}_N)|^2 = \rho(\mathbf{r}_1), \tag{2.4}$$

where we have taken into account the property (2.2). Thus, when we ask where one can observe a single particle, there is no preferable location and it can happen anywhere.

If the single-particle probability does not reveal any crystalline structure, let us analyze a two-particle probability density which is actually the second order correlation function G_2,

$$\rho_2(\mathbf{r}_1,\mathbf{r}_2) = \int d^3\mathbf{r}_3\ldots d^3\mathbf{r}_N\,|\psi(\mathbf{r}_1,\mathbf{r}_2,\mathbf{r}_3\ldots,\mathbf{r}_N)|^2 \propto G_2(\mathbf{r}_1,\mathbf{r}_2). \tag{2.5}$$

We have seen that the translation symmetry requires that the single-particle probability density must be uniform in space. In the case of the two-particle probability ρ_2, symmetry consequences are not so dramatic. That is, ρ_2 need not be uniform but it can change only with a change of relative position of two particles, $\rho_2(\mathbf{r}_1,\mathbf{r}_2) = \rho_2(\mathbf{r}_1-\mathbf{r}_2) = \rho_2(\mathbf{r})$. If $\rho_2(\mathbf{r})$ reveals a periodic structure in space, then distances between particles are regular and a crystalline structure is present in the system. We can interpret such a behavior of $\rho_2(\mathbf{r}_1,\mathbf{r}_2)$ as follows. Measurement of a particle at $\mathbf{r}_1$ results in probability density for a measurement of a next particle which shows a crystalline structure in space. One can say that due to the interactions between particles, a condensed matter system would prefer to reveal periodic structure in space but due to the symmetry requirement it forms a crystal whose center of mass is totally delocalized in space. The measurement of the position of a particle is able to localize the center of mass position and uncover a crystalline structure.

The measurement of the first particle breaks the continuous translation symmetry because the probability density for a next particle is no longer uniform in space. The first particle can be detected at any point in space and consequently we do not know where a crystalline structure localizes. It is a spontaneous process and therefore formation of space crystals is related to spontaneous breaking of the continuous space translation symmetry into a discrete space translation symmetry.

Another important issue is the lifetime of a space crystal described by a symmetry broken state. Once the continuous space translation symmetry is broken and a crystal emerges, one can ask how quickly it can decay because a state of the system is no longer an eigenstate? We may expect that in the thermodynamic limit, i.e. when N and V tend to infinite values but the particle density N/V = constant, the lifetime increases to infinity because quantum spreading of the center of mass position of a massive object takes infinitely long time. Let us carry out a simple analysis assuming that the center of mass is described by a wavepacket localized on a length scale $\sigma \approx 10^{-11}$ m. The corresponding kinetic energy is $E_k = h^2/(2\sigma^2 \sum_i m_i)$. The quantum evolution of the center of mass leads to spreading of the wavepacket and consequently to delocalization of a space crystal. However, in order to observe that the crystalline structure is smeared, the center of mass must be delocalized on a length scale of the order of the lattice constant of a crystal, i.e. $a \approx 10^{-10}$ m, that takes time of the order of $t = a/[h/(\sigma \sum_i m_i)] \approx 5 \cdot 10^4$ years for $\sum_{i=1}^{N} m_i = 1$ kg [8]. This estimation of time needed to blur a crystal assumes an isolated system. If during this period we, or an environment, perform a measurement of the center of mass position, a crystal gets localized again and the spreading of the wavepacket starts from the very beginning.

Let us also estimate how much energy is needed in order to break the continuous space translation symmetry into a discrete one. The kinetic energy E_k of the localized center of mass of a crystal, for the parameters we have chosen, is extremely close to the ground state energy and about 10^{-26} smaller than energy of an optical photon. It means that absorption of a single optical photon provides by far enough energy to localize a crystal. In the thermodynamic limit an infinitesimally small perturbation is able to break the continuous space translation symmetry and therefore it occurs spontaneously [8].

Analysis of spontaneous symmetry breaking can be also performed in a different way. First a symmetry breaking perturbation is applied and the ground state of a finite system is calculated. Then, the thermodynamic limit is taken and finally the perturbation is turned off which results in a symmetry broken state [1, 5, 6].

The phenomenon analyzed here belongs to a wide class of spontaneous symmetry breaking problems where equations describing a system possess a symmetry but a system prefers to choose a symmetry broken solution because symmetry preserving solutions are not stable [9].

Second Quantization Formalism

It is useful to rewrite single- and two-particle probability densities and describe spontaneous breaking of space translation symmetry also in the second quantization formalism [3, 7]. The formalism is a convenient technique for indistinguishable particles because antisymmetrization (or symmetrization) of many-body wavefunctions is ensured automatically.[2]

[2]The name "second quantization" is commonly used but it can be misleading because it suggests that the nonrelativistic quantum mechanics is quantized again which is obviously not the case. The advantage of the formalism is that we do not label particles and therefore we do not have to worry about the symmetrization or antisymmetrization of many-body states with respect to an exchange of particles.

Let us introduce the field operator $\hat{\psi}(\mathbf{r})$ which annihilates a particle at position $\mathbf{r}$ [3]. In the case of identical fermions (bosons), the field operator fulfills anticommutation (commutation) relations,

$$\left[\hat{\psi}(\mathbf{r}), \hat{\psi}(\mathbf{r}')\right]_{\pm} = 0, \quad \left[\hat{\psi}(\mathbf{r}), \hat{\psi}^{\dagger}(\mathbf{r}')\right]_{\pm} = \delta(\mathbf{r} - \mathbf{r}'), \tag{2.6}$$

where $[\hat{A}, \hat{B}]_{\pm} = \hat{A}\hat{B} \pm \hat{B}\hat{A}$ and $\delta(\mathbf{r})$ is the Dirac-delta distribution—the plus (minus) sign corresponds to the anticommutation (commutation) relation. The operator can be written as a sum,

$$\hat{\psi}(\mathbf{r}) = \sum_{i=0}^{\infty} \phi_i(\mathbf{r})\, \hat{a}_i, \tag{2.7}$$

where the operators $\hat{a}_i$ fulfill the relations,

$$\left[\hat{a}_i, \hat{a}_j\right]_{\pm} = 0, \quad \left[\hat{a}_i, \hat{a}_j^{\dagger}\right]_{\pm} = \delta_{ij}, \tag{2.8}$$

and annihilate a particle in the wavefunctions $\phi_i(\mathbf{r})$ which form a compete orthonormal single-particle basis. For a many-body system we can define the Fock state basis, $|n_0, n_1, \ldots\rangle$, where n_i particles occupy a given single-particle mode ϕ_i [3, 7]. For bosons the action of the annihilation and creation operators reads

$$\begin{aligned} \hat{a}_i|\ldots, n_i, \ldots\rangle &= \sqrt{n_i}|\ldots, n_i - 1, \ldots\rangle, \\ \hat{a}_i^{\dagger}|\ldots, n_i, \ldots\rangle &= \sqrt{n_i + 1}|\ldots, n_i + 1, \ldots\rangle. \end{aligned} \tag{2.9}$$

For fermions it is important if we annihilate (or create) a particle first in a mode j and then in a mode i or vice versa because the operators anticommute $\hat{a}_i\hat{a}_j = -\hat{a}_j\hat{a}_i$.[3] Therefore, once introduced, the order of the modes in the Fock states has to be kept the same for all calculations. The action of the operators in the case of a Fermi system reads

$$\begin{aligned} \hat{a}_i|\ldots, n_i, \ldots\rangle &= n_i\theta_i|\ldots, 0_i, \ldots\rangle, \\ \hat{a}_i^{\dagger}|\ldots, n_i, \ldots\rangle &= (1 - n_i)\theta_i|\ldots, 1_i, \ldots\rangle, \end{aligned} \tag{2.10}$$

(continued)

[3]Note that the anticommutation relations imply $\hat{a}_i^{\dagger}\hat{a}_i^{\dagger} = -\hat{a}_i^{\dagger}\hat{a}_i^{\dagger} = 0$ and it is not possible to create two identical fermions in the same state.

where $\theta_i = \sum_{j=0}^{i-1} n_j$. Equations (2.10) show that the order of modes in Fock states is important, i.e. the action of the operators $\hat{a}_i$ and $\hat{a}_i^\dagger$ depends if the modes with $j < i$ are occupied or not. Both in the Bose and Fermi cases the operators $\hat{n}_i = \hat{a}_i^\dagger \hat{a}_i$ are the particle number operators,

$$\hat{n}_i|\ldots, n_i, \ldots\rangle = n_i|\ldots, n_i, \ldots\rangle. \tag{2.11}$$

Any many-body state of a Fermi or Bose system can be written in the Fock state basis,

$$|\psi\rangle = \sum_{n_0, n_1, \ldots} c_{\{n_0, n_1, \ldots\}}|n_0, n_1, \ldots\rangle. \tag{2.12}$$

To complete basic information about the second quantization formalism, let us show how a many-body Hamiltonian of identical particles looks like in this formalism. Assume that in the position representation, a many-body Hamiltonian that describes N identical particles which interact via a two-body interaction potential $V(\mathbf{r}_i - \mathbf{r}_j)$ reads

$$\hat{H} = \sum_{i=1}^{N} \left[\frac{\mathbf{p}_i^2}{2m} + U(\mathbf{r}_i) \right] + \sum_{i<j}^{N} V(\mathbf{r}_i - \mathbf{r}_j), \tag{2.13}$$

where m is the mass of particles and $U(\mathbf{r})$ is an external potential which in general can be present, e.g. a trapping potential for ultra-cold atoms. The same Hamiltonian expressed by means of the field operators (2.7) takes the form [3, 7]

$$\begin{aligned} \hat{H} &= \int d^3r\, \hat{\psi}^\dagger(\mathbf{r}) \left[\frac{\mathbf{p}^2}{2m} + U(\mathbf{r}) \right] \hat{\psi}(\mathbf{r}) \\ &+ \frac{1}{2} \int d^3r \int d^3r'\, \hat{\psi}^\dagger(\mathbf{r})\, \hat{\psi}^\dagger(\mathbf{r}')\, V(\mathbf{r} - \mathbf{r}')\, \hat{\psi}(\mathbf{r}')\, \hat{\psi}(\mathbf{r}). \end{aligned} \tag{2.14}$$

For non-interacting particles the equivalence of (2.13) and (2.14) can be easily shown if as the wavefunctions $\phi_i(\mathbf{r})$ in the expansion for the field operator (2.7) we choose the single-particle eigenfunctions, $[\mathbf{p}^2/2m + U(\mathbf{r})]\phi_i(\mathbf{r}) = E_i\phi_i(\mathbf{r})$. Then, substituting the expansion (2.7) in (2.14) we obtain the non-interacting part of the Hamiltonian as the sum of the single-particle eigenenergies multiplied by numbers of particles that occupy the energy levels, i.e. $\sum_{i=0}^{\infty} E_i \hat{n}_i$. It is in agreement with what we expect from the non-interacting part of the Hamiltonian (2.13).

(continued)

The equivalence of the interaction parts of the energies can be proven by comparing the corresponding matrix elements in the Fock state basis. For example for the Bose system, the average value of the interactions energy in the Hamiltonian (2.14) in a Fock state $|\ldots, 0_{i-1}, N, 0_{i+1}, \ldots\rangle$ where N particles occupy a mode ϕ_i is

$$\frac{N(N-1)}{2}\int d^3r \int d^3r' \phi_i^*(\mathbf{r})\phi_i^*(\mathbf{r}')V(\mathbf{r}-\mathbf{r}')\phi_i(\mathbf{r}')\phi_i(\mathbf{r}). \qquad (2.15)$$

The same Fock state in the particle representation reads $\phi_i(\mathbf{r}_1)\phi_i(\mathbf{r}_2)\ldots\phi_i(\mathbf{r}_N)$ and the average value of the interaction energy in the Hamiltonian (2.13) in this state is exactly the same as (2.15) because all particles are in the same mode and there are $N(N-1)/2$ identical terms of the interaction energy in (2.13).

Now let us return to the analysis of spontaneous breaking of the space translation symmetry. In the second quantization formalism, the single- and two-particle probability densities read

$$\rho_1(\mathbf{r}) \propto \langle\psi|\, \hat{\psi}^\dagger(\mathbf{r})\, \hat{\psi}(\mathbf{r})\, |\psi\rangle, \qquad (2.16)$$

$$\rho_2(\mathbf{r}_1, \mathbf{r}_2) \propto \langle\psi|\, \hat{\psi}^\dagger(\mathbf{r}_1)\, \hat{\psi}^\dagger(\mathbf{r}_2)\, \hat{\psi}(\mathbf{r}_2)\, \hat{\psi}(\mathbf{r}_1)\, |\psi\rangle. \qquad (2.17)$$

If a system, which possesses the continuous space translation symmetry (which requires that the external potential $U(\mathbf{r}) = 0$ in the Hamiltonians (2.13) and (2.14)), is prepared in the ground state $|\psi\rangle$ (or any other eigenstate), the single-particle probability density is uniform, $\rho_1(\mathbf{r})$ =constant. However, the continuous translation symmetry can be spontaneously broken into a discrete translation symmetry if the two-particle probability density $\rho_2(\mathbf{r}_1, \mathbf{r}_2) = \rho_2(\mathbf{r}_1 - \mathbf{r}_2) = \rho_2(\mathbf{r})$ reveals periodic behavior in space. Let us assume that it is the case and illustrate the spontaneous symmetry breaking by means of the following experiment. The single-particle probability density corresponding to an eigenstate of the system is uniform but when we perform measurement of the position of a single particle and the measurement is destructive, i.e. a particle is annihilated at the position $\mathbf{r}_1$ where it is measured, the state of the remaining $N-1$ particles after the measurement,

$$|\psi_{SB}\rangle \propto \hat{\psi}(\mathbf{r}_1)\, |\psi\rangle, \qquad (2.18)$$

need not correspond to the uniform single-particle probability density. In the considered example, the single-particle probability density of the $N-1$ particle system reveals periodic structure in space because it reduces to the two-particle probability density $\rho_2(\mathbf{r}_1, \mathbf{r}_2)$ of the initial eigenstate and we have

(continued)

assumed that $\rho_2(\mathbf{r}_1, \mathbf{r}_2) = \rho_2(\mathbf{r}_1 - \mathbf{r}_2)$ is periodic in space,

$$\begin{aligned} \langle \psi_{SB}| \, \hat{\psi}^\dagger(\mathbf{r}_2) \, \hat{\psi}(\mathbf{r}_2) \, |\psi_{SB}\rangle &\propto \langle \psi| \, \hat{\psi}^\dagger(\mathbf{r}_1) \, \hat{\psi}^\dagger(\mathbf{r}_2) \, \hat{\psi}(\mathbf{r}_2) \, \hat{\psi}(\mathbf{r}_1) \, |\psi\rangle \\ &\propto \rho_2(\mathbf{r}_1 - \mathbf{r}_2). \end{aligned} \tag{2.19}$$

Thus, according to such an experimental interpretation of the two-particle probability density $\rho_2(\mathbf{r}_1, \mathbf{r}_2)$, a measurement of a particle position is responsible for the spontaneous symmetry breaking where the system spontaneously switches from a symmetry preserving eigenstate to a symmetry broken state,

$$|\psi\rangle \rightarrow |\psi_{SB}\rangle. \tag{2.20}$$

The measurement can be intentionally performed by an experimentalist or it can be *done by the environment* due to a loss of a particle.

2.2 Spontaneous Breaking of Space Translation Symmetry in a Non-interacting System

Space crystals are condensed matter systems that, due to interactions between atoms, are able to form spontaneously periodic structures in space. Such a self-organization of many particles can be quantitatively described within the quantum mechanics only and it is related to spontaneous breaking of the continuous space translation symmetry of the many-body Hamiltonian into a discrete space translation symmetry.

Spontaneous breaking of the continuous space translation symmetry of the Hamiltonian into a discrete one can be also observed in systems which have little in common with space crystals. In the present section we will see that even non-interacting systems can reveal spontaneous breaking of the continuous translation symmetry and form periodic structures in space. However, such a phenomenon is not induced by interactions between particles like in the solid state case—it takes place due to the fact that particles are indistinguishable.

Let us consider N non-interacting bosons in a one-dimensional ring of the unit circumference which are described by the Hamiltonian $H = \sum_{j=1}^{N} p_j^2/2$. We assume unit mass of bosons and $\hbar = 1$. The Hamiltonian does not change when we translate positions of particles by an arbitrary vector and consequently its eigenstates can be chosen as eigenstates of the corresponding space translation operator, see Sect. 2.1.1. As an example let us consider an eigenstate $|\psi\rangle$ where $N/2$ bosons are moving with the momentum k and the remaining $N/2$ with the opposite momentum $-k$ [4]. In the second quantization formalism (see background information in Sect. 2.1.2) it reads

$$|\psi\rangle = |N/2, N/2\rangle, \tag{2.21}$$

where $N/2$ particles occupy the single-particle state (mode) $\phi_1(x) = e^{ikx}$ and also $N/2$ particles are in the state $\phi_2(x) = e^{-ikx}$. The continuous space translation symmetry is reflected by the fact that the single-particle probability density is uniform in space,

$$\rho_1(x) \propto \langle\psi|\, \hat{\psi}^\dagger(x)\, \hat{\psi}(x)\, |\psi\rangle = N, \tag{2.22}$$

where the bosonic field operator

$$\hat{\psi}(x) = e^{ikx}\hat{a}_1 + e^{-ikx}\hat{a}_2, \tag{2.23}$$

with $\hat{a}_{1,2}$ annihilation operators that annihilate bosons in the modes $\phi_{1,2}$. However, if we calculate the two-particle probability density $\rho_2(x_1, x_2)$ we observe periodic behavior as a function of $x_1 - x_2$,

$$\begin{aligned}\rho_2(x_1, x_2) &\propto \langle\psi|\, \hat{\psi}^\dagger(x_1)\, \hat{\psi}^\dagger(x_2)\, \hat{\psi}(x_2)\, \hat{\psi}(x_1)\, |\psi\rangle \\ &= N(N/2-1) + N^2\cos^2[k(x_1 - x_2)].\end{aligned} \tag{2.24}$$

The measurement of the position of the first particle can happen equally likely at any place because $\rho_1(x) =$ constant. When it happens at x_1, then the probability density for the measurement of the second boson reveals periodic behavior as a function of the distance from the location of the first particle because $\rho_2(x_1, x_2) = \rho_2(x_1 - x_2)$ is periodic function of $x_1 - x_2$ [4].

With the help of computer simulations one can continue the analysis of measurements of positions of particles. Knowing a state of $N - j$ particles after the measurements of j particles we can calculate the conditional probability for the measurement of the position of the $(j+1)$th particle provided the previous particles have been measured at positions $\{x_1, \ldots, x_j\}$. At the end we obtain N positions and we can prepare a histogram [4]—an example is plotted in Fig. 2.1 where one can see periodic distribution of particles in space.

Actually, the problem we consider is related to the interference of two independent Bose–Einstein condensates [2, 4]. If two Bose–Einstein condensates are prepared independently in two different laboratories and they are pushed towards each other, then one can ask if interference fringes are observed or not. The initial state of the system can be modeled by a single Fock state like (2.21). There is no initial phase relation between two Bose–Einstein condensates. However, the phase relation is being established if we keep measuring positions of particles because bosons are indistinguishable and we do not know from which single-particle state, ϕ_1 or ϕ_2, a particle measured at a certain position comes from. At the end we obtain a distribution of bosons which clearly shows an interference pattern like in Fig. 2.1. Formation of the periodic distribution in space is a spontaneous symmetry breaking phenomenon. Before we perform the measurements we do not know at

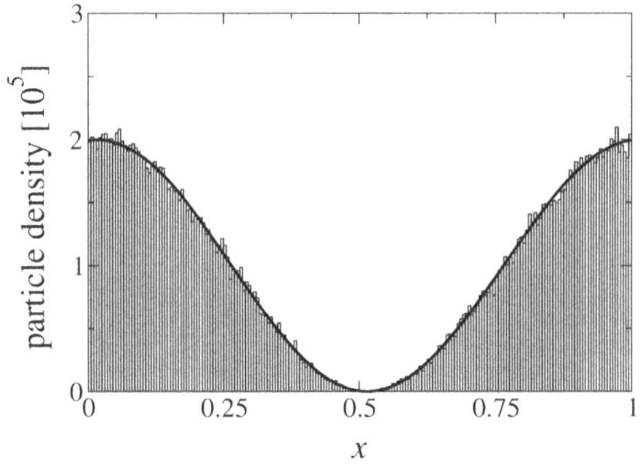

Fig. 2.1 Particle density corresponding to a single realization of the measurement of positions of all bosons prepared initially in the Fock state (2.21) where two modes, $\phi_1(x) = e^{ikx}$ and $\phi_2(x) = e^{-ikx}$, are occupied by $N/2$ bosons. The space is restricted to $0 \le x \le 1$, $k = \pi$ and $N = 10^5$. The figure shows an interference pattern. If we simulate the measurement many times, we will always observe similar patterns but the position of its maximum will be different in different realizations of the same experiment

which position the interference fringes will localize. They can be located equally likely at any point due to the continuous space translation symmetry of the system.

It should be stressed once again that the fact that particles are indistinguishable is responsible for the spontaneous breaking of the continuous space translation symmetry into a discrete one in the system that we have just described. There is no interaction between particles and the symmetry breaking phenomenon is different as compared to the spontaneous formation of space crystals.

References

1. Anderson, P.W.: Basic Notions of Condensed Matter Physics. Westview Press, Boulder (1997)
2. Andrews, M.R., Townsend, C.G., Miesner, H.J., Durfee, D.S., Kurn, D.M., Ketterle, W.: Observation of interference between two Bose condensates. Science **275**(5300), 637–641 (1997). https://doi.org/10.1126/science.275.5300.637. http://science.sciencemag.org/content/275/5300/637
3. Fetter, A.L., Walecka, J.D.: Quantum Theory of Many-Particle Systems. McGraw-Hill, Boston (1971)
4. Javanainen, J., Yoo, S.M.: Quantum phase of a Bose–Einstein condensate with an arbitrary number of atoms. Phys. Rev. Lett. **76**, 161–164 (1996). https://link.aps.org/doi/10.1103/PhysRevLett.76.161
5. Kaplan, T., Horsch, P., von der Linden W.: Order parameter in quantum antiferromagnets. J. Phys. Soc. Jpn. **58**, 3894 (1989)
6. Koma, T., Tasaki, H.: Symmetry breaking in Heisenberg antiferromagnets. Commun. Math. Phys. **158**, 191 (1993)

7. Pethick, C., Smith, H.: Bose–Einstein condensation in dilute gases. Cambridge University Press, Cambridge (2002)
8. Sacha, K., Zakrzewski, J.: Time crystals: a review. Rep. Prog. Phys. **81**(1), 016401 (2017). https://doi.org/10.1088/1361-6633/aa8b38
9. Strocchi, F.: Symmetry Breaking. The Lecture Notes in Physics. Springer, Berlin (2005, 2008)
10. van Wezel, J., van den Brink, J.: Spontaneous symmetry breaking in quantum mechanics. Am. J. Phys. **75**, 160401 (2007)

Chapter 3
Spontaneous Breaking of Continuous Time Translation Symmetry

Abstract Time-independent systems possess a continuous time translation symmetry which says that eigenstates of the Hamiltonian are also eigenstates of the time translation operator. In the present chapter we analyze if this symmetry can be spontaneously broken into a discrete time translation symmetry in a quantum many-body system that would correspond to spontaneous emergence of a periodic motion of a system and thus formation of a time crystal. We also present the idea of classical time crystals.

3.1 Time Crystal Idea

Formation of space crystals is related to spontaneous breaking of continuous space translation symmetry of the condensed matter Hamiltonian into a discrete space translation symmetry. In 2012 Frank Wilczek came out with the idea of quantum time crystals—systems that possess continuous time translation symmetry but this symmetry can be spontaneously broken into a discrete time translation symmetry [49]. Basically such systems are expected to form spontaneously periodic structures in time in a similar way as solid state systems do it in space. Put it differently time crystals are systems which are able to switch to periodic motion under any, even infinitesimally weak, perturbation. Wilczek paper triggered new thinking and initiated novel research area: condensed matter behavior in the time domain [34].

Let us consider what are the consequences of the continuous time translation symmetry of a quantum system. Assume we have a time-independent N-particle system. The Hamiltonian H of such a system commutes with the time translation operator which is simply the evolution operator $\mathcal{T}_t = e^{-iHt}$, where we have chosen the units where $\hbar = 1$. An eigenstate ψ of a system corresponding to energy E is necessarily also an eigenstate of $\mathcal{T}_t$,

$$\mathcal{T}_t\, \psi(\mathbf{r}_1, \ldots, \mathbf{r}_N, 0) = \psi(\mathbf{r}_1, \ldots, \mathbf{r}_N, t) = e^{-iEt}\, \psi(\mathbf{r}_1, \ldots, \mathbf{r}_N, 0). \quad (3.1)$$

It means that the corresponding probability density does not change in time, $|\psi(t)|^2 = |\psi(0)|^2$, which is of course a trivial conclusion because eigenstates of

K. Sacha, *Time Crystals*, Springer Series on Atomic, Optical, and Plasma Physics 114, https://doi.org/10.1007/978-3-030-52523-1_3

time-independent systems are stationary states. Wilczek asked the question if, due to interactions between particles, such a continuous time translation symmetry can be spontaneously broken into a discrete one by a many-body system prepared in the ground state?

Before we present the Wilczek model let us consider what would be the signatures of the spontaneous formation of time crystals. If a time-independent N-body system is prepared in an eigenstate ψ, the corresponding single-particle probability density does not depend on time. For identical particles it can be easily shown if we employ the second quantization formalism, see background information in Sect. 2.1.2, that is

$$\rho_1(\mathbf{r},t) \propto \langle\psi|\ \hat{\psi}^\dagger(\mathbf{r},t)\ \hat{\psi}(\mathbf{r},t)\ |\psi\rangle = \langle\psi|\ \hat{\psi}^\dagger(\mathbf{r},0)\ \hat{\psi}(\mathbf{r},0)\ |\psi\rangle \propto \rho_1(\mathbf{r},0), \tag{3.2}$$

where we have used the Heisenberg picture, $\hat{\psi}(\mathbf{r},t) = e^{iHt}\hat{\psi}(\mathbf{r},0)e^{-iHt}$, and the fact that we consider an eigenstate, $e^{-iHt}|\psi\rangle = e^{-iEt}|\psi\rangle$. However, when we switch to the two-time correlation function,

$$\begin{aligned} G_2(t',t) &= \langle\psi|\ \hat{\psi}^\dagger(\mathbf{r},t)\ \hat{\psi}^\dagger(\mathbf{r},t')\ \hat{\psi}(\mathbf{r},t')\ \hat{\psi}(\mathbf{r},t)\ |\psi\rangle \\ &= \langle\psi|\ \hat{\psi}^\dagger(\mathbf{r},0)\ e^{iH(t'-t)}\ \hat{\psi}^\dagger(\mathbf{r},0)\ \hat{\psi}(\mathbf{r},0)\ e^{-iH(t'-t)}\ \hat{\psi}(\mathbf{r},0)\ |\psi\rangle \\ &= G_2(t'-t), \end{aligned} \tag{3.3}$$

the continuous time translation symmetry implies only that G_2 depends on the time interval $t'-t$. The time translation symmetry does not forbid $G_2(t'-t)$ to change periodically as a function of $t'-t$. If it is the case, then we can observe spontaneous breaking of the continuous time translation symmetry into a discrete time translation symmetry [47]. That is, when one locates a detector at a certain point $\mathbf{r}$ in the space and ask how the probability density for the detection of a single particle at this space point changes in time if a system is prepared in an eigenstate, it turns out it must be constant in time because $\rho_1(\mathbf{r},t) = \rho_1(\mathbf{r},0)$. However, if the detector clicks at t, then the probability of the second click of the detector changes periodically versus t' if $G_2(t'-t)$ is a periodic function of $t'-t$—compare similar discussion in the case of spontaneous formation of space crystals in Sect. 2.1.2.

We will see that in the Wilczek model spontaneous breaking of the continuous time translation symmetry into a discrete one is not possible in the ground state but it may happen if the system is prepared in specific excited eigenstates [41]. There are publications proving that, under certain conditions, spontaneous breaking of the continuous time translation symmetry is not possible if a system is prepared in the ground state or in the equilibrium state [5, 47, 48]. We will discuss these results in Sect. 3.1.3.

At the end of the present section it is worth comparing space and time crystals. In space crystals we ask if the probability for the detection of particles changes periodically in the configuration space at a certain moment of time, i.e. at the moment we perform the detection of particle positions. When we switch from space to time crystals, the role of space and time is exchanged. In the time crystal case we fix position in the space and ask if the probability for the detection of particles at this fixed space point changes periodically versus time [24, 34].

3.1.1 Wilczek Model

Frank Wilczek considered N bosons moving on a ring of unit length in the presence of *magnetic* flux α [49], that is the problem which is often dubbed the Aharonov–Bohm ring [1]. Bosons are assumed to possess unit mass and interact via the attractive contact Dirac-delta potential $V(x) = g_0\delta(x)$. We assume also $\hbar = 1$. The many-body Hamiltonian of the system reads

$$H = \sum_{i=1}^{N} \frac{(p_i - \alpha)^2}{2} + g_0 \sum_{i<j}^{N} \delta(x_i - x_j), \tag{3.4}$$

where $g_0 < 0$ determines the strength of the attractive interactions between particles.

If only one particle is placed on the ring, the probability current along the ring may not vanish even if a particle is prepared in the ground state. Periodic boundary conditions in the ring problem lead to quantization of the momentum of a particle, $p_n = 2\pi n$ where n is an integer number, and consequently energy eigenvalues of a particle are $E_n = (p_n - \alpha)^2/2$. Even if a particle is in the ground state, the probability current,

$$\frac{dE_n}{dp_n} = 2\pi n - \alpha, \tag{3.5}$$

does not vanish provided α is not an integer multiple of 2π. Thus, for an appropriate choice of the *magnetic* flux α, we can get a *flow* of a particle along the ring in the lowest energy state. When we switch to N non-interacting bosons on the ring, then the ground state energy is NE_n and we deal with a Bose–Einstein condensate where all particles occupy the same single-particle eigenstate. There is no correlation between particles—the result of the measurement of a particle position does not depend if any other particles have been previously measured. Consequently, even if there is non-vanishing probability current along the ring, the single-particle probability density (3.2) and the two-time correlation function (3.3) are time independent and do not reveal any periodic behavior. This analysis indicates that no spontaneous breaking of time translation symmetry can be observed in the ground state of the non-interacting bosons on the ring.

Let us now switch to the interacting particles but first consider the case with no *magnetic* flux, i.e. $\alpha = 0$. In the mean-field description of the system it is assumed that bosons form a Bose–Einstein condensate, i.e. the many-body state is a product state,

$$\psi(x_1, \ldots, x_N) = \phi(x_1)\,\phi(x_2)\ldots\phi(x_N), \tag{3.6}$$

where all particles occupy the same single-particle wavefunction ϕ. It is a good approximation for the ground state and collectively excited eigenstates of weakly interacting bosons [30]. If we look for an eigenstate of the system in the form of (3.6), the wavefunction ϕ is a solution of the time-independent Gross–Pitaevskii equation [30] (see the background information at the end of the present section)

$$-\frac{1}{2}\frac{\partial^2\phi}{\partial x^2}+g_0(N-1)|\phi|^2\phi=\mu\phi. \tag{3.7}$$

In the case of bosons with the attractive contact interactions ($g_0<0$) moving in the one-dimension ring, the mean-field approximation for the ground state results in a bright soliton solution which, for $2/|g_0(N-1)|\ll 1$, reduces to

$$\phi(x)\approx\cosh^{-1}\left[\frac{g_0(N-1)}{2}(x-X)\right], \tag{3.8}$$

where X is the position of the center of mass of the particles. The wavefunction (3.8) describes a localized wavepacket and the bright soliton name comes from the nonlinear optics [19]. Within the full many-body description, that goes beyond the mean-field approximation, it is easy to show that (3.6) with (3.8) cannot be the ground state of the system because it is not an eigenstate of the space translation operator, see Sect. 2.1.1. The many-body ground state is extremely fragile to any perturbation. For example, it is sufficient to measure a position of one particle and the system collapses to a state with the probability density localized around a certain position on the ring. In the experiment with a large number N of bosons one never deals with the many-body ground state but with a state which is very well described by the mean-field bright soliton solution, i.e. (3.6) with (3.8). The center of mass position X around which particles localize is determined during preparation of the experiment. If we start with the many-body ground state and perform measurement of particle positions, X is determined randomly and reflects spontaneous breaking of the continuous space translation symmetry of the Hamiltonian (3.4), compare Sect. 2.2.

We have presented the discussion that in the absence of the *magnetic* flux α, N interacting bosons prepared in the many-body ground state can spontaneously break the continuous space translation symmetry of the Hamiltonian and form a bright soliton. We have mentioned also that if non-interacting particles are prepared in the ground state, there is non-zero probability current along the ring for a proper choice of the *magnetic* flux α. Wilczek combined these two phenomena together aiming that when interacting particles form a bright soliton, a proper choice of the *magnetic* flux α causes that the spontaneous breaking of the continuous space translation symmetry will be also accompanied with the breaking of the continuous time translation symmetry into the discrete one because the soliton will be moving periodically along the ring. It turned out it was not possible that was quickly pointed out by Bruno [4, 5, 51]. Probably the simplest way to show that the Wilczek idea does not work is to analyze the center of mass degree of freedom of the system [41]. The interaction between particles in (3.4) depends on relative distances between them and consequently the total momentum P (which is conjugate to the center of mass position X) is a good quantum number and in an eigenstate of the system it takes quantized value $P_n = 2\pi n$ with integer n. Switching to the center of mass coordinate frame the Hamiltonian of the system takes the form,

$$H = \frac{(P - N\alpha)^2}{2N} + \text{relative degrees of freedom.} \tag{3.9}$$

In order to get the ground state energy we have to minimize the kinetic energy of the center of mass degree of freedom. However, when $N \to \infty$ the corresponding probability flux of the center of mass degree of freedom,

$$\frac{\partial H}{\partial P_n} = 2\pi \frac{n}{N} - \alpha, \tag{3.10}$$

vanishes because it is always possible to find such n that $2\pi n/N$ is arbitrary close to α regardless what α is chosen. The large N limit in the Bose system is defined so that $N \to \infty$ and $g_0 \to 0$ but $g_0 N$ =constant. It ensures that the mean-field predictions are the same for any N because the Gross–Pitaevskii equation (3.7) depends on the product of $g_0 N$ only (more precisely it depends on $g_0(N-1)$ but for a large particle number $N \approx N - 1$). Thus, even if $\alpha \neq 0$ and in the ground state of the system a bright soliton is spontaneously formed, the probability current vanishes and there is no motion of a bright soliton when $N \to \infty$ (see also a debate in Refs. [28, 29, 42, 43]). The large N limit is necessary if we want the bright soliton to live forever and not being destroyed by quantum many-body effects that are beyond the mean-field approximation [23, 39–41].

The initial Wilczek proposal turned out to be impossible for realization but his idea triggered new thinking and opened a novel research area.

Mean-Field Approach and Gross–Pitaevskii Equation

Assume that we have N bosons in the three-dimensional space interacting via the contact Dirac-delta potential $V(\mathbf{r}) = g_0\delta(\mathbf{r})$ in the presence of an external potential $U(\mathbf{r})$.[1] The many-body Hamiltonian reads

$$\hat{H} = \sum_{i=1}^{N} \left[\frac{\mathbf{p}_i^2}{2m} + U(\mathbf{r}_i) \right] + g_0 \sum_{i<j}^{N} \delta(\mathbf{r}_i - \mathbf{r}_j), \tag{3.11}$$

where m is the mass of particles. The same Hamiltonian in the second quantization formalism (see the background information in Sect. 2.1.2) takes the form

(continued)

[1]Strictly speaking in the three-dimensional space the Dirac-delta potential needs regularization which can be done by employing a pseudo potential $V(\mathbf{r}) = g_0\delta(\mathbf{r})\partial_r(r\cdot)$ [8]. In the one-dimensional space no regularization is needed.

$$\hat{H} = \int d^3r\, \hat{\psi}^\dagger(\mathbf{r}) \left[\frac{\mathbf{p}^2}{2m} + U(\mathbf{r}) + \frac{g_0}{2}\hat{\psi}^\dagger(\mathbf{r})\, \hat{\psi}(\mathbf{r}) \right] \hat{\psi}(\mathbf{r}). \quad (3.12)$$

If the interactions between bosons are weak, we can expect that the ground state of the many-body system is not very different from the non-interacting case, i.e. it is a Bose–Einstein condensate state where all bosons occupy the same single-particle state $\phi(\mathbf{r})$ [8, 23, 30]. Thus, the many-body ground state can be approximated by a product state $\phi(\mathbf{r}_1)\phi(\mathbf{r}_2)\ldots\phi(\mathbf{r}_N)$ which in the second quantization formalism reads $|N\rangle$ meaning that all N particles occupy a mode ϕ. The average energy of the system in such a product state is

$$\begin{aligned} E &= \langle N|\hat{H}|N\rangle \\ &= N\int d^3r \left(\frac{\hbar^2}{2m}|\nabla\phi(\mathbf{r})|^2 + U(\mathbf{r})|\phi(\mathbf{r})|^2 + \frac{g_0(N-1)}{2}|\phi(\mathbf{r})|^4 \right). \end{aligned} \quad (3.13)$$

For non-interacting particles the energy E is minimal for ϕ being the ground state of a single particle in the presence of the external potential $U(\mathbf{r})$. When weak interactions are on we can expect that a product state can be still a good approximation for the ground state but we have to find an optimal single-particle wavefunction ϕ which can be done by minimization of the energy E. Extremal values of the functional E, with the condition that $\langle\phi|\phi\rangle = 1$, correspond to vanishing of the variational derivative which reduces to the Gross–Pitaevskii equation [8, 30]

$$-\frac{\hbar^2}{2m}\nabla^2\phi + U(\mathbf{r})\phi + g_0(N-1)|\phi(\mathbf{r}))|^2\phi = \mu\phi, \quad (3.14)$$

where μ is a Lagrange multiplier which is actually the chemical potential of the system. The Gross–Pitaevskii equation allows us to find many-body states corresponding to extremal values of the energy within the subspace of the many-body product states. The form of the equation indicates that in the mean field sense, a particle experiences an external potential $U(\mathbf{r})$ but also an additional *potential* proportional to a mean density of particles $g_0(N-1)|\phi(\mathbf{r})|^2$.

There is also a time-dependent version of the Gross–Pitaevskii equation,

$$i\hbar\frac{\partial\phi}{\partial t} = -\frac{\hbar^2}{2m}\nabla^2\phi + U\phi + g_0(N-1)|\phi|^2\phi, \quad (3.15)$$

which allows one to describe time evolution of a Bose–Einstein condensate, for example, when an external potential changes in time, $U(\mathbf{r}, t)$.

3.1.2 Excited Eigenstates and Time Crystals Formation

In the Wilczek model it is not possible to observe spontaneous breaking of the continuous time translation symmetry of the Hamiltonian into a discrete time translation symmetry if the system is prepared in the ground state. However, all eigenstates of a time-independent Hamiltonian H are also eigenstates of the time translation operator $\mathcal{T}_t = e^{-iHt}$. One can ask if spontaneous breaking of the time translation symmetry can occur if a system is prepared in an excited eigenstate? The question is not academic only because nowadays it is possible to prepare, control, and detect many-body systems not only in equilibrium states but also in highly excited states—it is especially the case in ultra-cold atomic gases [22].

Let us consider the Hamiltonian (3.4) from the Wilczek model but for $\alpha = 0$. The total momentum of the system is conserved and the Hilbert space can be divided into subspaces labeled by the total momentum quantum numbers. In each such subspace one can identify the lowest energy eigenstates which are called the *yrast* states [38].[2] If we choose the lowest energy eigenstate in the subspace corresponding to the total momentum $P_N = 2\pi N$, then the probability current

$$\frac{\partial H}{\partial P_N} = 2\pi, \tag{3.16}$$

which indicates that there is *flow* of particles along the ring with the period $T = 1/(2\pi)$. If spontaneous breaking of the space translation symmetry is possible, like in the case of the many-body ground state where a bright soliton state is formed, then we may expect that inhomogeneous profile of particles density will be moving with the period T demonstrating spontaneous breaking of the continuous time translation symmetry into the discrete time translation symmetry. The spontaneous breaking of the space translation symmetry is expected because the *yrast* state for $P_N = 2\pi N$ corresponds to the ground state wavefunction of the system multiplied by the factor $\exp(i2\pi NX)$ where $X = \sum_{j=1}^{N} x_j/N$ is the center of mass position. This reasoning shows that formation of a time crystal not in the ground state, as Wilczek expected, but in an excited eigenstate is possible. In order to demonstrate that it is true, one can switch to numerical simulations which allow also for the analysis of the lifetime of the symmetry broken state that describes the time crystal evolution.

In Fig. 3.1 we present the conditional probability density $\rho_2(x, t)$ for the detection of the second particle, provided the first one has been measured destructively at $t = 0$ at the position $x = 0$, i.e.

$$\rho_2(x, t) \propto \langle\psi|\, \hat{\psi}^\dagger(0, 0)\, \hat{\psi}^\dagger(x, t)\, \hat{\psi}(x, t)\, \hat{\psi}(0, 0)\, |\psi\rangle, \tag{3.17}$$

where $|\psi\rangle$ is the *yrast* state, i.e. the lowest energy eigenstate in the subspace corresponding to the total momentum $P_N = 2\pi N$. For $x = 0$, the probability

[2]The name *yrast* comes from the Swedish word *yr*, which means dizzy.

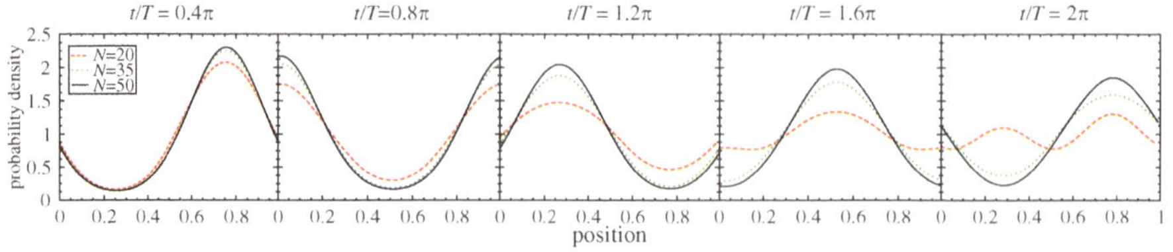

Fig. 3.1 Evolution of the conditional probability density $\rho_2(x, t)$, see (3.17), for the detection of the second particle provided the first one has been measured at $t = 0$ at the position $x = 0$ and annihilated. Different panels are related to different moments of time as indicated in the figure. Dashed, dotted, and solid lines correspond to different numbers of particles in the system, i.e. $N = 20$, 35, and 50, respectively. Note that with increasing N the initial shape of the profile of the probability density is preserved for a longer and longer period of time. The results are obtained for $g_0(N-1) = -15$. The figure reproduced from [41]

$\rho_2(x, t)$ reduces to the two-time correlation function (3.3). The measurement of the first particle breaks the continuous space translation symmetry, i.e. it localizes the center of mass of the system on the length scale comparable with the width of the bright soliton solution (3.8), i.e. $\sigma_X(0) \propto |g_0(N-1)|^{-1}$. The time evolution of the center of mass degree of freedom is described by a free particle-like Hamiltonian, see (3.9) with $\alpha = 0$. Assuming that after the first particle measurement the center of mass is described by the Gaussian wavepacket of the width $\sigma_X(0)$ we immediately get that $\sigma_X(t) \propto t/N$. Consequently the time t_c when $\sigma_X(t_c)$ becomes comparable with the circumference of the ring, i.e. when the bright soliton disappears, scales linearly with N that is confirmed in numerical simulations based on results presented in Fig. 3.1 [41]. Thus, in the limit where $N \to \infty$ but $g_0 \to 0$ so that $g_0(N-1)$ =constant, spontaneous breaking of the continuous space translation symmetry leads to a localized wavepacket that travels along the ring with the period $T = 1/(2\pi)$ without any decay for time period that increases linearly with N. In other words we observe spontaneous breaking of the continuous time translation symmetry into a discrete time translation symmetry, i.e. formation of a time crystal that lives forever in the limit of the large number of particles.

The described phenomenon can be realized in the ultra-cold atoms experiment. The entire experiment can look as follows. First N attractively interacting atoms should be prepared in the ground state in the presence of an artificial *magnetic* field corresponding to the flux $\alpha = 2\pi N$—in ultra-cold atoms artificial *magnetic* field can be realized in many different ways [15]. Afterwards the *magnetic* flux should be set to zero, i.e. $\alpha = 0$, which does not perturb the state of the system but makes it an excited eigenstate. Ideally, the prepared excited eigenstate would possess both space and time translation symmetry but because it is fragile to a perturbation these symmetries will be spontaneously broken (probably during the preparation of the experiment) and a bright soliton will form which will be moving periodically along the ring.

The time crystal evolution that we have just described belongs to the so-called symmetry-protected time crystals [18]. In the system considered here it means that the time crystal forms not in the ground state but in the lowest energy eigenstate in the subspace characterized by the total momentum $P \neq 0$ which is the constant of motion of the system.

3.1.3 Lack of Time Crystal Behavior of Systems with Two-Body Interactions in the Ground State and in the Equilibrium State

Wilczek paper on quantum time crystals was published in 2012 [49] and quickly his idea was questioned by Bruno [4, 5, 51], Nozières [27], and later by Watanabe and Oshikawa [47, 48] (see also [16]).

Bruno analyzed spontaneous breaking of the continuous time translation symmetry into a discrete time translation symmetry by considering N particles on a ring with two-body interactions and in the presence of a *magnetic* flux, i.e. a problem similar to the Wilczek model (3.4). He introduced a perturbing potential that rotated periodically along a ring with a frequency Ω [5]. Switching to the frame rotating with the frequency Ω one ends up with the time-independent Hamiltonian and in that frame it is possible to find energy eigenstates. In the rotating frame, when one calculates the ground state and takes the limit $N \to \infty$ first and then turns off the perturbation, spontaneous breaking of the continuous space translation symmetry, if it occurs, can be detected. Assuming that it is the case, Bruno argued that returning to the laboratory frame the energy of the system is minimal for $\Omega = 0$ and concluded that no spontaneous periodic motion is possible if the system is prepared in the ground state. He also analyzed a thermal equilibrium state by assuming that it is determined in the rotating frame where the Hamiltonian is time independent and obtained similar conclusion. Determination of thermal equilibrium in a rotating frame should be performed with caution because it implies actually that a system is in contact with a reservoir of rotating particles. A ground state in a rotating frame can correspond to an excited eigenstate in the laboratory frame—e.g. description of spontaneous emission of an atom in a rotating frame can correspond to a jump of an electron upwards in energy [11].

A different strategy was adopted by Watanabe and Oshikawa [47, 48] who did not assume the presence of a symmetry breaking time-periodic perturbation in order to analyze spontaneous breaking of the continuous time translation symmetry into a discrete one. They considered a class of time-independent systems with two-body interactions. In the ground state or in the equilibrium state, the average value of an observable in the Heisenberg picture, like, for example, the density operator $\hat{\rho}_1(\mathbf{r}, t) = \hat{\psi}^{\dagger}(\mathbf{r}, t)\hat{\psi}(\mathbf{r}, t)$, does not depend on t, see (3.2),

$$\langle\hat{\rho}_1(\mathbf{r}, t)\rangle = \langle\hat{\rho}_1(\mathbf{r}, 0)\rangle. \tag{3.18}$$

However, if in the thermodynamic limit (where the volume of the system $V \to \infty$), the density–density correlation function,

$$\lim_{V\to\infty} \langle \hat{\rho}_1(\mathbf{r}', t')\, \hat{\rho}_1(\mathbf{r}, t)\rangle = f(t' - t), \tag{3.19}$$

where $f(t' - t)$ is a periodic function of $t' - t$, spontaneous breaking of the continuous time translation symmetry into a discrete time translation symmetry occurs.[3] Such a strategy overcomes the difficulty in the definition of the lowest energy eigenstate of a system in the presence of time-periodic perturbation and thus there is no problem with the definition of the statistical ensemble based on energy eigenvalues. The authors show that in the limit of $V \to \infty$, both in the ground state and in the equilibrium state, the correlation function does not tend to $f(t' - t)$ that is a periodic function of $t' - t$. It indicates that no spontaneous breaking of the continuous time translation symmetry into a discrete one can occur.

In the Watanabe and Oshikawa paper taking the standard thermodynamic limit, i.e. the volume of a system $V \to \infty$, was essential. They also assumed that two-body interactions had to decay with the relative distance r between particles at least like $1/r^{\beta}$ with $\beta > 0$. Systems that fulfill such conditions cannot reveal spontaneous formation of a time crystal in its ground state or in the thermal equilibrium state.

In the Bruno paper the large V limit is not necessary [5]. He considers particles on a ring and assumes that $N \to \infty$ without further conditions. The Bruno approach is directly applicable to the Wilczek model where the size of the ring is kept fixed but $N \to \infty$, see Sects. 3.1.1 and 3.1.2.

3.1.4 Time Crystal Behavior of Systems with Multi-Body Interactions

In order to probe spontaneous breaking of the continuous time translation symmetry into a discrete time translation symmetry by a time-independent system in the ground state, one can analyze the auto-correlation function $f(t) = \langle\psi_0|O(t)O|\psi_0\rangle$ where $|\psi_0\rangle$ is the ground state (i.e. $H|\psi_0\rangle = E_0|\psi_0\rangle$) and O is an observable [47]. One expects that a time crystal forms if $f(t)$ does not vanish in the thermodynamic limit and is a periodic function of time t. If we allow for N-body observables, then such time crystal behavior is a common phenomenon. Indeed, for $O = |\psi_0\rangle\langle\psi_n| + |\psi_n\rangle\langle\psi_0|$, where $|\psi_n\rangle$ is an excited eigenstate of a system corresponding to eigenenergy $E_n > E_0$, the auto-correlation function reveals time-periodic behavior [18],

[3] For an extensive observable O, the auto-correlation function used in [47] to probe time crystal behavior reads $\langle O(t)O\rangle/V^2$ which may vanish in the absence of long range correlations in large volume limit, at all times. In order to capture measurable persistent oscillations in the absence of long range correlations, Medenjak et al. propose the following auto-correlation function $\langle O(t)O\rangle/\langle O^2\rangle$, see Ref. [25].

$$f(t) = \langle\psi_0|O(t)O|\psi_0\rangle = \langle\psi_0|e^{iHt}Oe^{-iHt}O|\psi_0\rangle = e^{-i(E_n-E_0)t}, \tag{3.20}$$

where $\hbar = 1$ is assumed. In the field of time crystals we are not interested in such, rather trivial, phenomena.

Kozin and Kyriienko proposed a strategy how to construct systems where the auto-correlation function of one-body observables can reveal periodic time evolution in the ground state [20]. The basic idea is to introduce multi-body interactions in a many-body system. Then, the assumptions of the Watanabe and Oshikawa theorem are not satisfied, its restrictions do not apply and an avenue towards realization of time crystals is open.

Kozin and Kyriienko considered N-qubit system (which can be a chain of N spins-1/2) and took as an observable O the total magnetization along the z axis, $\hat{M}_z = \sum_{i=1}^{N} \sigma_i^z/N$ where $\sigma_i^{x,y,z}$ are Pauli matrices and i is the index of a site of a chain of spins-1/2. In order to fulfill the Watanabe and Oshikawa criterion of the existence of a time crystal, the auto-correlation function,

$$f(t) = \lim_{N\to\infty} \langle\psi_0|\hat{M}_z(t)\hat{M}_z|\psi_0\rangle, \tag{3.21}$$

calculated in the system ground state $|\psi_0\rangle$ corresponding to the eigenenergy E_0, must be period in time. The construction of the desired system goes as follows. Let us define the normalized state

$$|\psi_1\rangle = \frac{\hat{M}_z|\psi_0\rangle}{\sqrt{\langle\psi_0|\hat{M}_z^2|\psi_0\rangle}}, \tag{3.22}$$

and require that $|\psi_0\rangle$ is chosen so that $\langle\psi_1|\psi_0\rangle \propto \langle\psi_0|\hat{M}_z|\psi_0\rangle = 0$. Next, let us assign the energy $E_1 > E_0$ to the state $|\psi_1\rangle$ and construct the following Hamiltonian

$$\hat{H} = E_0|\psi_0\rangle\langle\psi_0| + E_1|\psi_1\rangle\langle\psi_1| + \sum_{j>1} E_j|\psi_j\rangle\langle\psi_j|, \tag{3.23}$$

where $\langle\psi_j|\psi_0\rangle = 0$ and $\langle\psi_j|\psi_1\rangle = 0$ with $E_j > E_0$ for $j > 1$. Then, the correlation function (3.21) reads

$$\begin{aligned} f(t) &= \lim_{N\to\infty} \langle\psi_0|e^{i\hat{H}t}\hat{M}_z e^{-i\hat{H}t}\hat{M}_z|\psi_0\rangle \\ &= e^{-i(E_1-E_0)t} \lim_{N\to\infty} \langle\psi_0|\hat{M}_z^2|\psi_0\rangle. \end{aligned} \tag{3.24}$$

Thus, the constructed system fulfills the criteria for the presence of time crystal behavior in the ground state $|\psi_0\rangle$ provided $\langle\psi_0|\hat{M}_z|\psi_0\rangle = 0$ and $\langle\psi_0|\hat{M}_z^2|\psi_0\rangle \neq 0$ in the limit of $N \to \infty$. In order to show that a state $|\psi_0\rangle$ with such properties exists, let us expand $|\psi_0\rangle$ in the eigenbasis of the Hermitian operator $\hat{M}_z$,

$$|\psi_0\rangle = \sum_{k,d} c_{k,d}|m_k, d\rangle, \tag{3.25}$$

where $\hat{M}_z|m_k, d\rangle = m_k|m_k, d\rangle$ with $m_k = (N-2k)/N$ and $d = 1, 2, \ldots, C_N^k$ labels degenerate eigenstates that correspond to the same eigenvalue m_k. In this basis we obtain

$$\langle\psi_0|\hat{M}_z|\psi_0\rangle = \sum_{k,d} |c_{k,d}|^2 m_k, \qquad \langle\psi_0|\hat{M}_z^2|\psi_0\rangle = \sum_{k,d} |c_{k,d}|^2 m_k^2. \tag{3.26}$$

A state $|m_k, d\rangle$ corresponds to k spins pointing down and $N-k$ spins pointing up and when we exchange k and $N-k$ we get $m_{N-k} = -m_k$. Thus, any state (3.25) where $|c_{k,d}| = |c_{N-k,d}|$ fulfills $\langle\psi_0|\hat{M}_z|\psi_0\rangle = 0$ and $\langle\psi_0|\hat{M}_z^2|\psi_0\rangle \neq 0$ and consequently defines a many-body system (3.23) which reveals time crystal behavior in the ground state according to the definition of Watanabe and Oshikawa [47].

As an example we can choose $|\psi_0\rangle$ as one of the maximally entangled Greenberger–Horne–Zeilinger (GHZ) states [45],

$$|G_\pm\rangle = \frac{1}{\sqrt{2}}(|\uparrow\uparrow\ldots\uparrow\rangle \pm |\downarrow\downarrow\ldots\downarrow\rangle), \tag{3.27}$$

which fulfill $\hat{M}_z|G_{-,+}\rangle = |G_{+,-}\rangle$ and $\langle G_-|G_+\rangle = 0$. For instance, let us pick $|\psi_0\rangle = |G_-\rangle$, $E_0 = -1$ and the eigenenergies of all other $2^N - 1$ eigenstates of the system to be zero, $E_j = 0$. Then, the Hamiltonian of the constructed system reads

$$\hat{H} = -|G_-\rangle\langle G_-|. \tag{3.28}$$

The Hamiltonians (3.23) and (3.28) of the systems that reveal time crystal behavior are presented in terms of the many-body projectors which do not allow us to see that the interactions between spins have highly non-local character. An example of the Hamiltonian written in terms of the spin operators that reveals the time crystal behavior is provided in Ref. [20], i.e.

$$\hat{H} = -\frac{J}{N(N-1)} \sum_{1\le i<j\le N} \sigma_1^x\sigma_2^x\ldots\sigma_i^y\ldots\sigma_j^y\ldots\sigma_N^x, \tag{3.29}$$

which is a sum of N-length strings with exactly two operators σ^y and $N-2$ operators σ^x. The non-degenerate ground state of (3.29) is $|\psi_0\rangle = |G_-\rangle$ and corresponds to the energy $E_0 = -J/2$ and the highest excited (also non-degenerate) eigenstate is $|G_+\rangle$ with the energy $J/2$. Thus, the Hamiltonian (3.29) fulfills the time crystal criterion because $\langle\psi_0|\hat{M}_z(t)\hat{M}_z|\psi_0\rangle = e^{-iJt}$. Exactly the same behavior is revealed by the auto-correlation function $\langle\psi_0|\sigma_i^z(t)\sigma_j^z|\psi_0\rangle = e^{-iJt}$ [20]. The Hamiltonian (3.29) describes long range multi-spin interactions with N-

length strings. The most local Hamiltonian that possesses $|G_+\rangle$ as an eigenstate must contain at least $(N/2)$-length strings [13, 20].

Zero-temperature time crystals proposed by Kozin and Kyriienko are stable under a certain class of perturbations but according to their numerical simulations the crystals are melting when the temperature is non-zero [20].

In the case of spontaneous formation of space crystals, the ground state of a solid state system is highly degenerate in the thermodynamic limit and a space crystal can emerge out of the ground state manifold. In the case of time crystals considered in the present section, the Hamiltonians are time independent, the ground state is non-degenerate, and breaking of the continuous time translation symmetry and the resulting time crystal evolution is related to the gap between the ground state and excited states.

3.1.5 Classical Time Crystals

In 2012 Shapere and Wilczek introduced the concept of classical time crystals which are classical systems that are able to perform periodic motion despite the fact they have the lowest possible energy [36].

In the case of classical Hamiltonian systems in order to find the lowest energy of, e.g. a particle moving in one-dimensional space, we have to find the minimal value of the Hamiltonian and thus solve the following equations:

$$\frac{\partial H}{\partial p} = 0, \qquad \frac{\partial H}{\partial x} = 0. \tag{3.30}$$

The first of Eqs. (3.30) implies that an extremal value of the Hamiltonian does not allow for any motion of a particle,

$$\dot{x} = \frac{dx}{dt} = \frac{\partial H}{\partial p} = 0. \tag{3.31}$$

On the other hand when we consider a particle on a ring with energy given by

$$E = \frac{\dot{x}^4}{4} - \frac{\dot{x}^2}{2}, \tag{3.32}$$

we conclude that the lowest energy corresponds to the velocity $\dot{x} = \pm 1$ and consequently a particle is able to move periodically along a ring even though its energy is minimal what is in contradiction with (3.31). The puzzle can be solved when one tries to express the energy as a function of the particle momentum. Indeed, the Hamiltonian of a particle is a multi-valued function of the momentum with cusps corresponding precisely to the energy minima at $\dot{x} = \pm 1$. At the cusps the Hamilton equations are not defined and Eq. (3.31) cannot say anything about the velocity of a particle. The system described by the energy (3.32) is a classical time crystal which is moving periodically even if its energy is minimal. It does not mean that we deal with a perpetual motion because the system is in the lowest energy state and cannot perform any work.

Shapere and Wilczek considered also a more general problem where the quartic and quadratic kinetic energy terms are supplemented by a potential energy $V(x)$ [33, 37],

$$E = \frac{\dot{x}^4}{4} - \frac{\dot{x}^2}{2} + V(x), \tag{3.33}$$

where, e.g. $V(x) = x^2/2$. Then, there is competition between the minimization of the potential and kinetic energies. The static position $x = 0$ of a particle that minimizes the potential energy is in contradiction with the non-zero velocities $\dot{x} = \pm 1$ that make the kinetic energy minimal, see Fig. 3.2. One may wonder what is the lowest energy state of the system? The equation of motion for a particle,

$$(\dot{x}^2 - 1)\ddot{x} = -\frac{dV(x)}{dx}, \tag{3.34}$$

indicates that the quantity $\dot{x}^2 - 1$, which play the role of an effective mass, may vanish and thus time evolution of a particle can be ill-defined. In order to analyze

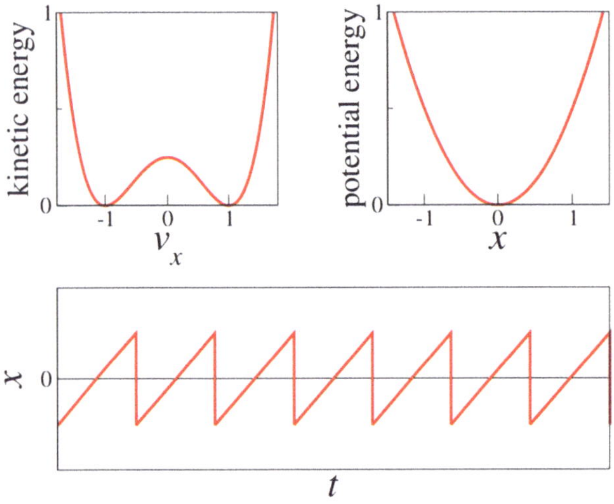

Fig. 3.2 Top panels show kinetic and potential energies of a particle. The former is minimized by particle's motion with the velocity $v_x = \dot{x} = 1$ (or $v_x = -1$), the latter by the particle localized at the position $x = 0$. In order to reconcile these two contradicting requirements, the particle performs Sisyphus motion depicted schematically in bottom panel. That is, a particle slowly climbs with a constant velocity $v_x = 1$ (or $v_x = -1$), crosses the position $x = 0$, and afterwards instantaneously rolls back and starts climbing again and so on

motion of a particle, Shapere and Wilczek show that such a system arises as an effective description of a charged particle in XY plane subject to the magnetic field along the z-direction, $B_z = y^2 - 1$, and the electric potential $y^4/4 - y^2/2 + V(x)$. The planar particle is assumed to have an anisotropic mass which vanishes along the x-direction and has a small value m_y along the y-direction. The mass m_y is actually a regularization parameter and when $m_y \to 0$, the equations of motion of the planar particle reduce to (3.34). For small m_y and for energy close to the minimal value, which for the planar particle is easy to find, the motion along the x-direction reflects a Sisyphus motion around $x = 0$. That is, starting with a small negative (or positive) value, the position $x(t)$ changes linearly with time with the velocity $\dot{x} = 1$ ($\dot{x} = -1$) but when it reaches a small positive (negative) value it becomes quickly reset back to the initial negative (positive) value, see Fig. 3.2. In other words the velocity is almost always positive (negative) except brief intervals when the position gets reset. Thus, the velocity nearly always minimizes the kinetic energy and the position is always close to $x = 0$ where the potential energy is minimal. If the energy of the planar particle tends to the minimal value, the amplitude of the Sisyphus oscillations decreases and their frequency increases but the approach to this limit is non-uniform [37].

An interesting application of the time crystal idea was presented by Das et al. [10] who considered a model of cycling cosmology as a classical time crystal (see also [9]). They showed that the scale factor $a(t)$ of a generalized Friedmann–Robertson–Walker metric is described by a mechanical model with the Hamiltonian that involved terms proportional to $\dot{a}^2$, $\dot{a}^4$, and even $\dot{a}^6$. Consequently classical time crystal structures turn out to be more complex than in the case of the system described by (3.33). The cyclic cosmological model presented by Das et al. is purely geometrical, that is, no matter field is introduced from the outside.

While the classical single-particle systems considered here can reveal periodic motion and at the same time minimize energy, their quantization and generalization to the many-body interacting systems are not straightforward [14, 35, 37].

3.2 Josephson Effect and Superfluid Phase Oscillations

3.2.1 Josephson Effect

In Sect. 2.2 we have considered spontaneous breaking of the continuous space translation symmetry into a discrete space translation symmetry. There two independently prepared Bose–Einstein condensates move in opposite directions and form spontaneously periodic interference pattern in space when measurement of positions of particles is performed [2, 17]. This kind of spontaneous process occurs even without interactions between particles and it is a consequence of the fact that particles are indistinguishable. In the present section we will show that spontaneous breaking of the continuous time translation symmetry into a discrete time translation symmetry can be observed in a similar Bose system and the resulting time-periodic oscillations are the so-called AC Josephson effect [47].

Similarly as in Sect. 2.2 we consider two Bose–Einstein condensates consisting of identical non-interacting bosons on the one-dimensional ring of unit circumference. One condensate is formed by $N/2$ particles that occupy the zero-momentum state $\phi_1(x) = 1$, while the other condensate consists of $N/2$ bosons but in a non-zero-momentum state $\phi_2(x) = e^{ikx}$. The state of the entire system reads (cf. Sect. 2.2)

$$|\psi\rangle = |N/2, N/2\rangle, \tag{3.35}$$

and it is an excited eigenstate of the system Hamiltonian $H = \sum_{j=1}^{N} p_j^2/2$,

$$H|\psi\rangle = \frac{Nk^2}{4}|\psi\rangle, \tag{3.36}$$

where we have assumed unit mass of particles and $\hbar = 1$.

The single-particle probability density corresponding to the state (3.35) does not reveal any non-trivial space or time dependence (cf. Sect. 2.1.2 or Sect. 2.2),

$$\rho_1(x,t) \propto \langle\psi|\hat{\psi}^\dagger(x,t)\,\hat{\psi}(x,t)|\psi\rangle = N, \tag{3.37}$$

where $\hat{\psi}(x,t) = \hat{a}_1 + e^{-ik^2t/2}e^{ikx}\hat{a}_2$ is the bosonic field operator in the Heisenberg picture and $\hat{a}_{1,2}$ are operators that annihilate bosons in the single-particle states $\phi_{1,2}$. However, when we calculate a two-time correlation function,

$$\begin{aligned} G_2(t',t) &= \langle\psi|\,\hat{\psi}^\dagger(x,t)\,\hat{\psi}^\dagger(x,t')\,\hat{\psi}(x,t')\,\hat{\psi}(x,t)\,|\psi\rangle \\ &= N(N-1) + (N^2/2)\cos\left[k^2(t'-t)/2\right], \end{aligned} \tag{3.38}$$

we conclude that the continuous time translation symmetry can be spontaneously broken into a discrete one. That is, a symmetry broken state emerges if, e.g. at time moment t a single particle is observed at the detector located at x. After the destructive detection of a particle (a particle is annihilated at the position x in the detection process) the state of the remaining $(N-1)$ particles reads

$$|\psi_{SB}\rangle \propto \hat{\psi}(x,t)|\psi\rangle, \tag{3.39}$$

which is a symmetry broken state and the corresponding single-particle density reveals periodic evolution in time,

$$\langle\psi_{SB}|\,\hat{\psi}^\dagger(x,t')\,\hat{\psi}(x,t')\,|\psi_{SB}\rangle = G_2(t',t). \tag{3.40}$$

The described phenomenon is the AC Josephson effect [47]. The time-periodic oscillations emerge spontaneously but the cause of their emergence, similarly as in the problem of the interference of two Bose–Einstein condensates analyzed in Sect. 2.2, is the indistinguishability of bosons and they occur even without interactions between particles—the interactions of course modify the oscillations [31]. It should be also stressed that the entire system must be prepared in an excited eigenstate. If all N bosons occupy the lowest energy zero-momentum state ϕ_1, no non-trivial time dependence is observed in the two-time correlation function $G_2(t', t)$.

The considered phenomenon demonstrates spontaneous breaking of the continuous time translation symmetry into a discrete time translation symmetry and can be identified with time crystal behavior [31, 32]—sometimes this kind of phenomena is referred to as *old* time crystals [52].

It is worth mentioning that interactions between particles are responsible for the spontaneous breaking of the space translation symmetry in solid state systems and for the formation of space crystals. In the example we have described in the current section no interactions are present and the spontaneous symmetry breaking results from the fact that particles are not distinguishable. We do not know from which condensates a detected particle comes from what establishes a phase relation between the two initially independent condensates that are described by the Fock state (3.35).

3.2.2 Superfluid Phase Oscillations

So far we have considered isolated systems which possess a well defined number of particles. However, Bose–Einstein condensates are often described in the framework of the ground canonical ensemble where the number of particles fluctuates and the average particle number N can be well defined only. Then, Bose–Einstein condensation is identified with breaking of the U(1) symmetry, i.e. the symmetry of the Hamiltonian related to the group of the unitary transformations $e^{iq\hat{N}}$ where $\hat{N}$ is the operator of the total number of particles and q is a real parameter. The condensate corresponds to a state of a Bose system which breaks the U(1) symmetry, i.e. the average of the bosonic field operator does not vanish,

$$\langle \hat{\psi}(\mathbf{r}, t) \rangle = \psi_0(\mathbf{r})\, e^{-i\mu t}. \tag{3.41}$$

The quantity $\langle \hat{\psi}(\mathbf{r}, t) \rangle$ is an order parameter (condensate wavefunction) and it reveals time-periodic oscillations with the frequency given by the chemical potential $\mu = dE/dN$ where E is energy of the system. The oscillations reflect the fact that we deal with the state of the system that breaks spontaneously time translation symmetry. They can be observed because our system is not isolated and can exchange particles with another system that plays a role of the reference frame with respect to which the oscillations can be measured.

Volovik considered a class of similar systems and identified two time scales: energy relaxation time scale τ_E and time scale τ_N which in our example is related to relaxation of the system to the true equilibrium state [46] (see also [7, 12, 26, 44, 50]). If τ_N is much larger than τ_E, the system characterized by an average number of particles N relaxes to the lowest energy state corresponding to a fixed N and afterwards slowly relaxes to the equilibrium state where no oscillations are observed. Thus, for time t fulfilling $\tau_E \ll t \ll \tau_N$, we can observe time translation symmetry breaking but a system is not strictly in the equilibrium state. While τ_N can tend to the infinity, the strict limit $\tau_N = \infty$ must not be reached because then there is no reference frame and the time-periodic oscillations are not observable.

Volovik analyzes a magnon Bose–Einstein condensate which is formed by bosonic magnetic quasi-particles (magnons) excited in the superfluid ^{3}He-B by means of the pulsed nuclear magnetic resonance [6, 46]. The application of the radio-frequency pulse deflects the magnetization of the superfluid ^{3}He-B but the resulting induction signal first decays due to inhomogeneity of the local field but afterwards spontaneously reappears, on a time scale which we denote by τ_E, with its own phase completely forgetting the phase of the radio-frequency pulse. Magnon–magnon interactions prevent dephasing and the induction signal decays slowly on a time scale denoted by τ_N. The magnon lifetime τ_N is much greater than τ_E so that they have enough time to form a Bose–Einstein condensate. Spontaneous precession of the magnetization in magnon condensates in the context of time crystals has been demonstrated experimentally in Refs. [3, 21].

References

1. Aharonov, Y., Bohm, D.: Significance of electromagnetic potentials in the quantum theory. Phys. Rev. **115**, 485–491 (1959). https://link.aps.org/doi/10.1103/PhysRev.115.485
2. Andrews, M.R., Townsend, C.G., Miesner, H.J., Durfee, D.S., Kurn, D.M., Ketterle, W.: Observation of interference between two Bose condensates. Science **275**(5300), 637–641 (1997). https://doi.org/10.1126/science.275.5300.637. http://science.sciencemag.org/content/275/5300/637
3. Autti, S., Eltsov, V.B., Volovik, G.E.: Observation of a time quasicrystal and its transition to a superfluid time crystal. Phys. Rev. Lett. **120**, 215301 (2018). https://link.aps.org/doi/10.1103/PhysRevLett.120.215301
4. Bruno, P.: Comment on "quantum time crystals". Phys. Rev. Lett. **110**, 118901 (2013). http://link.aps.org/doi/10.1103/PhysRevLett.110.118901
5. Bruno, P.: Impossibility of spontaneously rotating time crystals: a no-go theorem. Phys. Rev. Lett. **111**, 070402 (2013). http://link.aps.org/doi/10.1103/PhysRevLett.111.070402
6. Bunkov, Y.: Magnon BEC versus atomic BEC. J. Low Temp. Phys. **185**(5), 399–408 (2016). https://doi.org/10.1007/s10909-016-1583-z
7. Castillo, E., Koch, B., Palma, G.: On the dynamics of fluctuations in time crystals. arXiv e-prints, arXiv:1410.2261 (2014)
8. Castin, Y., Dum, R.: Low-temperature Bose-Einstein condensates in time-dependent traps: beyond the $u(1)$ symmetry-breaking approach. Phys. Rev. A **57**, 3008–3021 (1998). https://link.aps.org/doi/10.1103/PhysRevA.57.3008
9. Chakraborty, S., Ghosh, S.: Non-trivial ground state for gravitational perturbation in quadratic gravity. arXiv e-prints, arXiv:2001.04680 (2020)

10. Das, P., Pan, S., Ghosh, S., Pal, P.: Cosmological time crystal: cyclic universe with a small cosmological constant in a toy model approach. Phys. Rev. D **98**, 024004 (2018). https://link.aps.org/doi/10.1103/PhysRevD.98.024004
11. Delande, D., Zakrzewski, J.: Spontaneous emission of nondispersive Rydberg wave packets. Phys. Rev. A **58**, 466–477 (1998). https://link.aps.org/doi/10.1103/PhysRevA.58.466
12. Else, D.V., Bauer, B., Nayak, C.: Prethermal phases of matter protected by time-translation symmetry. Phys. Rev. X **7**, 011026 (2017). http://link.aps.org/doi/10.1103/PhysRevX.7.011026
13. Facchi, P., Florio, G., Pascazio, S., Pepe, F.V.: Greenberger-Horne-Zeilinger states and few-body Hamiltonians. Phys. Rev. Lett. **107**, 260502 (2011). https://link.aps.org/doi/10.1103/PhysRevLett.107.260502
14. Ghose-Choudhury, A., Guha, P.: Branched Hamiltonians and time translation symmetry breaking in equations of the Lienard type. arXiv e-prints, arXiv:1904.11225 (2019)
15. Goldman, N., Juzeliunas, G., Öhberg, P., Spielman, I.B.: Light-induced gauge fields for ultracold atoms. Rep. Prog. Phys. **77**(12), 126401 (2014). http://stacks.iop.org/0034-4885/77/i=12/a=126401
16. Huang, Y.: Absence of temporal order in states with spatial correlation decay. arXiv e-prints, arXiv:1912.01210 (2019)
17. Javanainen, J., Yoo, S.M.: Quantum phase of a Bose-Einstein condensate with an arbitrary number of atoms. Phys. Rev. Lett. **76**, 161–164 (1996). https://link.aps.org/doi/10.1103/PhysRevLett.76.161
18. Khemani, V., Moessner, R., Sondhi, S.L.: A brief history of time crystals. arXiv e-prints, arXiv:1910.10745 (2019)
19. Kivshar, Y.S., Agrawal, G.P.: Optical Solitons. Academic, Burlington (2003)
20. Kozin, V.K., Kyriienko, O.: Quantum time crystals from Hamiltonians with long-range interactions. Phys. Rev. Lett. **123**, 210602 (2019). https://link.aps.org/doi/10.1103/PhysRevLett.123.210602
21. Kreil, A.J.E., Musiienko-Shmarova, H.Y., Eggert, S., Serga, A.A., Hillebrands, B., Bozhko, D.A., Pomyalov, A., L'vov, V.S.: Tunable space-time crystal in room-temperature magnetodielectrics. Phys. Rev. B **100**, 020406 (2019). https://link.aps.org/doi/10.1103/PhysRevB.100.020406
22. Lewenstein, M., Sanpera, A., Ahufinger, V.: Ultracold Atoms in Optical Lattices: Simulating Quantum Many-Body Systems. OUP Oxford (2012). https://books.google.pl/books?id=Wpl91RDxV5IC
23. Lieb, E.H., Seiringer, R., Yngvason, J.: Bosons in a trap: a rigorous derivation of the Gross-Pitaevskii energy functional. Phys. Rev. A **61**, 043602 (2000). https://link.aps.org/doi/10.1103/PhysRevA.61.043602
24. Maccone, L., Sacha, K.: Quantum measurements of time. Phys. Rev. Lett. **124**, 110402 (2020). https://link.aps.org/doi/10.1103/PhysRevLett.124.110402
25. Medenjak, M., Buča, B., Jaksch, D.: Isolated Heisenberg magnet as a quantum time crystal. Phys. Rev. B. American Physical Society. **102**(4), 041117 (2020). https://link.aps.org/doi/10.1103/PhysRevB.102.041117
26. Nicolis, A., Piazza, F.: Spontaneous symmetry probing. J. High Energy Phys. **2012**(6), 25 (2012). http://dx.doi.org/10.1007/JHEP06(2012)025
27. Nozières, P.: Time crystals: can diamagnetic currents drive a charge density wave into rotation? Europhys. Lett. **103**(5), 57008 (2013). http://stacks.iop.org/0295-5075/103/i=5/a=57008
28. Öhberg, P., Wright, E.M.: Quantum time crystals and interacting gauge theories in atomic Bose-Einstein condensates. Phys. Rev. Lett. **123**, 250402 (2019). https://link.aps.org/doi/10.1103/PhysRevLett.123.250402
29. Öhberg, P., Wright, E.M.: Öhberg and Wright reply. Phys. Rev. Lett. American Physical Society. **124**(17), 178902 (2020). https://link.aps.org/doi/10.1103/PhysRevLett.124.178902
30. Pethick, C., Smith, H.: Bose–Einstein Condensation in Dilute Gases. Cambridge University Press, Cambridge (2002)
31. Prokof'ev, N.V., Svistunov, B.V.: Algebraic time crystallization in a two-dimensional superfluid. J. Exp. Theor. Phys. **127**(5), 860–864 (2018). https://doi.org/10.1134/S1063776118110092

32. Prokof'ev, N., Svistunov, B.: Space- and time-crystallization effects in multicomponent superfluids. Phys. Rev. B **101**, 020505 (2020). https://link.aps.org/doi/10.1103/PhysRevB.101.020505
33. Sacha, K., Hannaford, P.: Time crystal minimizes its energy by performing sisyphus motion. Proc. Natl. Acad. Sci. **116**(38), 18755–18756 (2019). https://doi.org/10.1073/pnas.1913075116. https://www.pnas.org/content/116/38/18755
34. Sacha, K., Zakrzewski, J.: Time crystals: a review. Rep. Prog. Phys. **81**(1), 016401 (2017). https://doi.org/10.1088/1361-6633/aa8b38
35. Shapere, A., Wilczek, F.: Branched quantization. Phys. Rev. Lett. **109**, 200402 (2012). http://link.aps.org/doi/10.1103/PhysRevLett.109.200402
36. Shapere, A., Wilczek, F.: Classical time crystals. Phys. Rev. Lett. **109**, 160402 (2012). http://link.aps.org/doi/10.1103/PhysRevLett.109.160402
37. Shapere, A.D., Wilczek, F.: Regularizations of time-crystal dynamics. Proc. Natl. Acad. Sci. **116**(38), 18772–18776 (2019). https://doi.org/10.1073/pnas.1908758116. https://www.pnas.org/content/116/38/18772
38. Siemens, P.J., Jensen, A.S.: Elements of Nuclei: Many-Body Physics With the Strong Interaction (Lecture Notes and Supplements in Physics). Addison-Wesley, Redwood City (1987)
39. Syrwid, A., Sacha, K.: Lieb-Liniger model: emergence of dark solitons in the course of measurements of particle positions. Phys. Rev. A **92**, 032110 (2015). https://link.aps.org/doi/10.1103/PhysRevA.92.032110
40. Syrwid, A., Brewczyk, M., Gajda, M., Sacha, K.: Single-shot simulations of dynamics of quantum dark solitons. Phys. Rev. A **94**, 023623 (2016). https://link.aps.org/doi/10.1103/PhysRevA.94.023623
41. Syrwid, A., Zakrzewski, J., Sacha, K.: Time crystal behavior of excited eigenstates. Phys. Rev. Lett. **119**, 250602 (2017). https://link.aps.org/doi/10.1103/PhysRevLett.119.250602
42. Syrwid, A., Kosior, A., Sacha, K.: Comment on "Quantum Time Crystals and Interacting Gauge Theories in Atomic Bose-Einstein Condensates". Phys. Rev. Lett. American Physical Society. **124**(17), 178901 (2020). https://link.aps.org/doi/10.1103/PhysRevLett.124.178901 Syrwid, A., Kosior, A., Sacha, K.: Comment on quantum time crystals and interacting gauge theories in atomic Bose-Einstein condensates. Phys. Rev. Lett. American Physical Society. **124**(17), 178901 (2020). https://link.aps.org/doi/10.1103/PhysRevLett.124.178901
43. Syrwid, A., Kosior, A., Sacha, K.: Lack of a genuine time crystal in a chiral soliton model. Phys. Rev. Research. American Physical Society. **2**(3), 032038 (2020). https://link.aps.org/doi/10.1103/PhysRevResearch.2.032038
44. Thies, M.: Semiclassical time crystal in the chiral Gross-Neveu model. arXiv e-prints, arXiv:1411.4236 (2014)
45. Tóth, G., Gühne, O.: Entanglement detection in the stabilizer formalism. Phys. Rev. A **72**, 022340 (2005). https://link.aps.org/doi/10.1103/PhysRevA.72.022340
46. Volovik, G.E.: On the broken time translation symmetry in macroscopic systems: precessing states and off-diagonal long-range order. J. Exp. Theor. Phys. Lett. **98**, 491 (2013)
47. Watanabe, H., Oshikawa, M.: Absence of quantum time crystals. Phys. Rev. Lett. **114**, 251603 (2015). http://link.aps.org/doi/10.1103/PhysRevLett.114.251603
48. Watanabe, H., Oshikawa, M., Koma, T.: Proof of the absence of long-range temporal orders in Gibbs states. J. Stat. Phys. **178**(4), 926–935 (2020). https://doi.org/10.1007/s10955-019-02471-5
49. Wilczek, F.: Quantum time crystals. Phys. Rev. Lett. **109**, 160401 (2012). http://link.aps.org/doi/10.1103/PhysRevLett.109.160401
50. Wilczek, F.: Superfluidity and space-time translation symmetry breaking. Phys. Rev. Lett. **111**, 250402 (2013). http://link.aps.org/doi/10.1103/PhysRevLett.111.250402
51. Wilczek, F.: Wilczek reply:. Phys. Rev. Lett. **110**, 118902 (2013). http://link.aps.org/doi/10.1103/PhysRevLett.110.118902
52. Wilczek, F.: The exquisite precision of time crystals. Sci. Am. **321**(5), 28–35 (2019)

Chapter 4
Discrete Time Crystals and Related Phenomena

Abstract In the present chapter we concentrate on periodically driven systems. Although such systems possess already discrete time translation symmetry, we will see that this symmetry can be spontaneously broken into another discrete time translation symmetry and a system is able to spontaneously reorganize its motion and move with a period which is different from the period of the driving. These phenomena are dubbed discrete (or Floquet) time crystals because discrete time translation symmetry becomes spontaneously violated. Discrete time crystals attract considerable attention: they are examples of quantum many-body systems that break ergodicity and they are believed to be resistant to decoherence. Moreover, discrete time crystals have been already demonstrated in the laboratory and further experiments are being planned.

4.1 Floquet Formalism and Discrete Time Translation Symmetry

4.1.1 Floquet Formalism

Let us consider a single particle which is described by a time-periodic Hamiltonian $H(t+T) = H(t)$ where $T = 2\pi/\omega$. It is not possible to find energy eigenstates of the system because the energy is not conserved. However, due to the fact that the Hamiltonian is time periodic we can still look for a kind of stationary states which are periodically evolving wavefunctions. The Floquet theorem (see [27, 183, 208]) says that we can expand any solution of the time-dependent Schrödinger equation in the form

$$\psi(\mathbf{r}, t) = \sum_n c_n e^{-iE_n t} u_n(\mathbf{r}, t), \tag{4.1}$$

K. Sacha, *Time Crystals*, Springer Series on Atomic, Optical, and Plasma Physics 114, https://doi.org/10.1007/978-3-030-52523-1_4

where $u_n(\mathbf{r}, t + T) = u_n(\mathbf{r}, t)$ are eigenstates of the so-called Floquet Hamiltonian $H_F = H(t) - i\partial_t$,

$$H_F u_n(\mathbf{r}, t) = \left(H(t) - i\frac{\partial}{\partial t}\right) u_n(\mathbf{r}, t) = E_n u_n(\mathbf{r}, t), \tag{4.2}$$

where we have assumed $\hbar = 1$ as often in this monograph. The eigenvalues E_n are called quasi-energies of a periodically driven system and time-periodic wavefunctions $u_n(\mathbf{r}, t)$ are dubbed Floquet states. The eigenvalue Eq. (4.2) is obtained when we substitute $\psi(\mathbf{r}, t) = e^{-iE_n t} u_n(\mathbf{r}, t)$ in the time-dependent Schrödinger equation. The time-periodicity of $H(t)$ guarantees that Floquet states fulfill periodic boundary conditions in time, i.e. $u_n(\mathbf{r}, t+T) = u_n(\mathbf{r}, t)$. It is important to stress that the quasi-energy spectrum is periodic with the period of the frequency of the driving ω. That is, having E_n and the corresponding u_n we can obtain a family of Floquet states

$$e^{-iE_n t} u_n(\mathbf{r}, t) = e^{-i(E_n + m\omega)t} e^{im\omega t} u_n(\mathbf{r}, t) = e^{-i\tilde{E}_n t} \tilde{u}_n(\mathbf{r}, t), \tag{4.3}$$

where $\tilde{u}_n(\mathbf{r}, t+T) = \tilde{u}_n(\mathbf{r}, t) = e^{im\omega t} u_n(\mathbf{r}, t)$ is also a Floquet state corresponding to the quasi-energy $\tilde{E}_n = E_n + m\omega$ where m is integer. Consequently, it is sufficient to consider the quasi-energy spectrum restricted to a single Floquet zone of the width ω because quasi-energy levels of all physically distinct Floquet states are located there. The Floquet states form a complete orthogonal basis at any time.

The Floquet formalism is in full analogy to the Bloch formalism used in the condensed matter physics for the description of eigenstates of a spatially periodic time-independent Hamiltonian $H(x + a) = H(x)$. An electron moving in a spatially periodic potential does not have a well- defined momentum but due to the spatial periodicity of the potential one can define a quasi-momentum k, i.e. eigenstates of an electron are plane waves modulated with the spatial periodicity of the potential $e^{ikx} u_k(x)$, where $u_k(x + a) = u_k(x)$. Note that time evolution of a single Floquet state, $e^{-iE_n t} u_n(\mathbf{r}, t)$, has the analogous form, i.e. evolution of a phase like for an ordinary energy eigenstate but modulated with the time periodicity of the Hamiltonian. Similarly like in the case of the quasi-energy spectrum in the Floquet formalism, in the Bloch approach it is also sufficient to restrict to a single Brillouin zone of the quasi-momenta $k \in [-\pi/a, \pi/a]$.

Interpretation of Floquet States

How to interpret quasi-energies and Floquet states of periodically driven systems? In order to gain some intuition it is instructive to analyze an example from the quantum optics.

Let us consider a two-level atom described by the Hamiltonian of the form $H_A = \omega_0(|e\rangle\langle e| - |g\rangle\langle g|)/2$ where ω_0 is the energy difference between the

(continued)

excited and ground state levels (remember that $\hbar = 1$) and $|g\rangle$ and $|e\rangle$ are the ground and excited eigenstates of the atom, respectively. The atom is at rest in the electromagnetic cavity and we focus on interaction of the atom with a single electromagnetic mode of the cavity corresponding to the photon energy $\omega \approx \omega_0$. The Hamiltonian of the entire system, i.e. the atom and the quantum electromagnetic field, in the rotating wave approximation, reads

$$H = \frac{\omega_0}{2}(|e\rangle\langle e| - |g\rangle\langle g|) + \omega\, \hat{a}^\dagger \hat{a} + \lambda \left(|e\rangle\langle g|\, \hat{a} + |g\rangle\langle e|\, \hat{a}^\dagger \right), \tag{4.4}$$

with $\lambda = -\langle e|\mathbf{d} \cdot \mathbf{e}_0|g\rangle$ which is assumed to be real-valued, where $\mathbf{d}$ is the atomic electric dipole moment and the vector $\mathbf{e}_0$ describes the mode of the electromagnetic field, especially its polarization. The second term in the Hamiltonian (4.4) corresponds to the energy of the electromagnetic field ($\hat{a}^\dagger \hat{a}$ is the photon number operator) while the third term describes interaction between the field and the atom where $\hat{a}$ is the photon annihilation operator. The system we consider is the well-known Jaynes–Cummings model [67].

The energy of the entire system is conserved. It is easy to find eigenstates of (4.4) (which in the quantum optics language are called dressed states of an atom) because the Hamiltonian matrix can be written in a block-diagonal form. That is, the Hilbert space can be divided into two-dimensional subspaces $\{|g\rangle|n+1\rangle, |e\rangle|n\rangle\}$, where $|n\rangle$ denotes the state of the electromagnetic field with exactly n photons, which are not coupled to each other by the Hamiltonian. The two-dimensional blocks of the Hamiltonian matrix read

$$H^{(n)} = \begin{pmatrix} \frac{1}{2}\omega_0 + n\omega & \lambda\sqrt{n+1} \\ \lambda\sqrt{n+1} & -\frac{1}{2}\omega_0 + (n+1)\omega \end{pmatrix}, \tag{4.5}$$

and their diagonalization leads to the dressed atom spectrum

$$E_\pm^{(n)} = \left(n + \frac{1}{2}\right)\omega \pm \frac{1}{2}\sqrt{(\omega_0 - \omega)^2 + 4\lambda^2(n+1)}, \tag{4.6}$$

and the corresponding dressed states of an atom which are superposition of $|g\rangle|n+1\rangle$ and $|e\rangle|n\rangle$. For weak interaction between the atom and the field (i.e., $|\lambda|\sqrt{n+1} \ll \omega_0$) and for $|\omega_0 - \omega| \ll \omega$, the spectrum consists of pairs of energy levels which are well separated by the photon energy ω. For $n \gg 1$ the spectrum can be considered as locally periodic with the period ω because $E_+^{(n)} - E_-^{(n)}$ changes little when we change n by $\Delta n \ll n$.

Now assume that we treat the electromagnetic field as a classical wave. Then, its interaction with a two-level atom is described by $-\mathbf{d} \cdot \mathbf{E}_0 \cos \omega t$ where $\mathbf{E}_0$ is the electric field vector. Because the Hamiltonian of the system

(continued)

is time periodic we can define the Floquet Hamiltonian which in the rotating wave approximation reads

$$H_F = \frac{\omega_0}{2}(|e\rangle\langle e| - |g\rangle\langle g|) + \frac{V}{2}\left(|e\rangle\langle g|\, e^{-i\omega t} + |g\rangle\langle e|\, e^{i\omega t}\right) - i\frac{\partial}{\partial t}, \quad (4.7)$$

where $V = -\langle e|\mathbf{d}\cdot\mathbf{E}_0|g\rangle$. Floquet states of the system are time periodic, $|u(t+2\pi/\omega)\rangle = |u(t)\rangle$, and can be expanded in the Fourier series

$$|u(t)\rangle = \sum_k c_k e^{ik\omega t}|u_k\rangle. \quad (4.8)$$

When we define the Fourier basis as $\langle t|k\rangle = e^{ik\omega t}$ we can see that the matrix of the Floquet Hamiltonian in the Hilbert space extended by the time degree of freedom decomposes into two-dimensional uncoupled blocks corresponding to the subspaces $\{|g\rangle|k+1\rangle, |e\rangle|k\rangle\}$, i.e.

$$H_F^{(k)} = \begin{pmatrix} \frac{1}{2}\omega_0 + k\omega & \frac{V}{2} \\ \frac{V}{2} & -\frac{1}{2}\omega_0 + (k+1)\omega \end{pmatrix}. \quad (4.9)$$

Note that the calculations of the matrix elements of H_F involve integration in time over the period of the electromagnetic wave, e.g.

$$\langle k'|\langle g|H_F|g\rangle k\rangle = \frac{\omega}{2\pi}\int_0^{2\pi/\omega} dt\ \langle g|e^{-ik'\omega t}H_F e^{ik\omega t}|g\rangle. \quad (4.10)$$

The corresponding quasi-energy spectrum,

$$E_\pm^{(k)} = \left(k + \frac{1}{2}\right)\omega \pm \frac{1}{2}\sqrt{(\omega_0 - \omega)^2 + V^2}, \quad (4.11)$$

is periodic with the period ω, as it should, and it reproduces the dressed atom spectrum (4.6) if there is a large number of photons ($n \gg 1$), i.e. for $V \approx 2\lambda\sqrt{n}$.

Thus, in the regime of $n \gg 1$ the quasi-energy spectrum of the Floquet formalism, which treats the electromagnetic field as a classical wave, matches the dressed atom spectrum obtained in the full quantum theory of the electromagnetic field. In other words, if the quantum field is strong, its modification due to the interaction with an atom can be neglected and the Floquet approach allows us to obtain good approximation for the eigenstates of the entire atom-field system. Thus, Floquet states can be considered as an approximation of true eigenstates of an entire system when back-action

(continued)

on a subsystem that provides a time-periodic driving (here the quantum electromagnetic field) can be neglected.

Many-Body Floquet Hamiltonian

In the present monograph the Floquet formalism will be often used in description of periodically driven many-body systems, especially when systems consist of bosons. Therefore, it is worth considering how Floquet Hamiltonians look like in the second quantization language.

Let us start with the single-particle case. In order to find T-periodic Floquet states of a driven particle, one has to solve the eigenvalue problem (4.2). To do that, time t is often considered as an additional dimension and it is treated on an equal footing with the space dimensions and consequently the Hilbert space of a system is extended by including *time degree of freedom* [88].

A Floquet state $u(\mathbf{r}, t+T) = u(\mathbf{r}, t)$, where $T = 2\pi/\omega$ is the period of the Hamiltonian $H(t+T) = H(t)$ of a particle, can be expanded in the Fourier series

$$u(\mathbf{r}, t) = \sum_{n,k} c_{n,k}\phi_n(\mathbf{r})e^{ik\omega t}, \tag{4.12}$$

where $\{\phi_n\}$ is a set of states which span the Hilbert space of a particle. We define basis vectors in the extended Hilbert space as

$$\langle\langle \mathbf{r}, t|n, k\rangle\rangle = \phi_n(\mathbf{r})e^{ik\omega t}, \tag{4.13}$$

which are orthonormal according to the scalar product

$$\langle\langle n', k'|n, k\rangle\rangle = \frac{1}{T}\int_0^T dt \int d^3\mathbf{r}\ \phi^*_{n'}(\mathbf{r})e^{-ik'\omega t}\phi_n(\mathbf{r})e^{ik\omega t} = \delta_{nn'}\delta_{kk'}. \tag{4.14}$$

Matrix elements of the Floquet Hamiltonian in this basis read

$$\langle\langle n', k'|H_F|n, k\rangle\rangle = \frac{1}{T}\int_0^T dt \int d^3\mathbf{r}\ \phi^*_{n'}(\mathbf{r})e^{-ik'\omega t}[H(t) - i\partial_t]\phi_n(\mathbf{r})e^{ik\omega t}. \tag{4.15}$$

Diagonalization of the matrix (4.15) yields the quasi-energy spectrum (which is ω-periodic) and the corresponding Floquet states of the system.

Suppose for a moment that we choose the Fourier series not as in Eq. (4.12) but

(continued)

$$u(\mathbf{r},t)=\sum_{n,k}\tilde{c}_{n,k}\phi_n(\mathbf{r})e^{ik\omega t/2}, \tag{4.16}$$

which seems to be also legitimate because if the Hamiltonian is T-periodic, $H(t+T)=H(t)$, it is $(2T)$-periodic too, $H(t+2T)=H(t)$. Then, we get similar expressions for the scalar product and the matrix elements of the Floquet Hamiltonian like in (4.14) and (4.15) but $\frac{1}{T}\int_0^T dt \rightarrow \frac{1}{2T}\int_0^{2T} dt$. Diagonalization of the Floquet Hamiltonian in the new basis yields the same information as the diagonalization of (4.15) but the spectrum is now periodic with the period $\omega/2$ not with the period ω what can be illustrated as follows. Assume that we have a Floquet state expanded in the Fourier series as in (4.12). The same Floquet state in the new basis corresponds to

$$\tilde{c}_{n,2k}=c_{n,k}, \qquad \tilde{c}_{n,2k+1}=0. \tag{4.17}$$

However, if we choose

$$\tilde{c}_{n,2k}=0, \qquad \tilde{c}_{n,2k+1}=c_{n,k}, \tag{4.18}$$

then

$$\sum_{n,k}\tilde{c}_{n,k}\phi_n(\mathbf{r})e^{ik\omega t/2}=e^{i\omega t/2}\sum_{n,k}c_{n,k}\phi_n(\mathbf{r})e^{ik\omega t}, \tag{4.19}$$

which is the same Floquet state as in (4.12) but now the corresponding quasi-energy is shifted by $\omega/2$. Generally if we use the Fourier series with Fourier harmonics $e^{ik\omega t/s}$ where s is an integer number, diagonalization of the resulting matrix of the Floquet Hamiltonian yields quasi-energy spectrum which is periodic with the period ω/s.

Now, let us switch to N-boson systems where particles interact via the two-body Dirac-delta potential $g_0\delta(\mathbf{r})$. The many-body Floquet Hamiltonian takes the form

$$\hat{H}_F=\sum_{i=1}^{N}H(\mathbf{r}_i,\mathbf{p}_i,t)+\frac{g_0}{2}\sum_{i\neq j}^{N}\delta(\mathbf{r}_i-\mathbf{r}_j)-i\partial_t, \tag{4.20}$$

where $H(\mathbf{r},\mathbf{p},t+T)=H(\mathbf{r},\mathbf{p},t)$ is the Hamiltonian for a single particle. Similarly like in the previously considered single-particle case, a complete set of states in the extended Hilbert space can be defined,

$$\langle\langle \mathbf{r}_1,\ldots,\mathbf{r}_N,t|\mathbf{n},k\rangle\rangle=\phi_{\mathbf{n}}(\mathbf{r}_1,\mathbf{r}_2,\ldots,\mathbf{r}_N)e^{ik\omega t}, \tag{4.21}$$

(continued)

where $\phi_{\mathbf{n}}$'s form the basis in the Hilbert space of N particles, and matrix elements of the many-body Floquet Hamiltonian can be calculated. However, for identical bosons it is more convenient to use the second quantization formalism.

We choose a set of T-periodic functions $\varphi_j(\mathbf{r}, t+T) = \varphi_j(\mathbf{r}, t)$, where $j = 1, 2, \ldots$, which form a basis in the extended Hilbert space of the corresponding single particle problem. Next, we define an N-body state in the form of

$$\mathcal{S}\left[\varphi_{j_1}(\mathbf{r}_1, t)\, \varphi_{j_2}(\mathbf{r}_2, t) \ldots \varphi_{j_N}(\mathbf{r}_N, t)\right], \tag{4.22}$$

where the first particle is in a single-particle state $\varphi_{j_1}(\mathbf{r}_1, t)$, the second particle in $\varphi_{j_2}(\mathbf{r}_2, t)$, and so on. The operator $\mathcal{S}$ denotes the symmetrization of a many-boson state under exchange of any $\mathbf{r}_i$ and $\mathbf{r}_j$. The state (4.22) is actually a Fock state,

$$|n_1, n_2, \ldots\rangle, \tag{4.23}$$

where certain numbers n_j of bosons occupy T-periodic modes $\varphi_j(t)$. Matrix elements of the many-body Floquet Hamiltonian (4.20) can be calculated in the basis spanned by many-body wavefunctions (4.22) but it can also be done with the help of the field operator

$$\hat{\psi}(\mathbf{r}, t) = \sum_{j=1}^{\infty} \varphi_j(\mathbf{r}, t)\, \hat{a}_j, \tag{4.24}$$

where $\hat{a}_j$'s are the bosonic annihilation operators which fulfill $[\hat{a}_i, \hat{a}_j^\dagger] = \delta_{ij}$, see the background information in Sect. 2.1.2. Then, the many-body Floquet Hamiltonian (4.20), in the second quantization form, reads

$$\hat{H}_F = \frac{1}{T}\int_0^T dt \int d^3\mathbf{r}\, \hat{\psi}^\dagger(\mathbf{r}, t)\left[H(t) - i\partial_t + \frac{g_0}{2}\hat{\psi}^\dagger(\mathbf{r}, t)\, \hat{\psi}(\mathbf{r}, t)\right]\hat{\psi}(\mathbf{r}, t). \tag{4.25}$$

Diagonalization of the Floquet Hamiltonian (4.25) yields many-body Floquet states and the corresponding quasi-energies which in turn allows one to perform time evolution of any initial many-body state. Indeed, any initial state can be expended in the Floquet state basis and its time evolution is described by an analogues equation like in the single-particle case, cf. Eq. (4.1).

Different sets of T-periodic functions $\varphi_j(\mathbf{r}, t)$ can be chosen as the single-particle basis in (4.22), e.g. $\varphi_j(\mathbf{r}, t)$ can be Floquet states of the corresponding

(continued)

single-particle problem. One can also choose $\varphi_j(\mathbf{r}, t)$ which are not T-periodic functions but (sT)-periodic modes with integer s. Then, a similar Floquet Hamiltonian as in (4.25) is obtained but the range of the integration over time has to be changed, i.e. $\frac{1}{T}\int_0^T dt \to \frac{1}{sT}\int_0^{sT} dt$.

4.1.2 Discrete Time Translation Symmetry of Periodically Driven Systems

We have seen that despite the fact that energy of a system described by a time-periodic Hamiltonian $H(t+T) = H(t)$ is not conserved, it is possible to find eigenstates (Floquet states) of the Floquet Hamiltonian, see Eq. (4.2). A time-periodic Hamiltonian does not possess the continuous time translation symmetry but it commutes with the time translation operator $\mathcal{T}_T$ related to the evolution operator of a system by the period T. Such a discrete time translation symmetry implies that Floquet states are also eigenstates of $\mathcal{T}_T$,

$$\mathcal{T}_T|u_n(t)\rangle = e^{-iE_nT}|u_n(t+T)\rangle = e^{-iE_nT}|u_n(t)\rangle. \tag{4.26}$$

Probability densities of single- or many-body periodically driven systems prepared in a Floquet state and average values of any time-independent operators calculated in a single Floquet state evolve with the period T.

In the present chapter we address a question if in a quantum many-body system the discrete time translation symmetry can be spontaneously broken and if so what kinds of symmetry broken motion can spontaneously emerge? We will see that it is possible to observe spontaneous emergence of periodic motions with integer multiples of the driving period T or with periods which are fractions of T. It is even possible to observe spontaneous emergence of motion which reveals a quasi-crystalline structure in time.

4.2 Discrete Time Crystals in Ultra-Cold Atoms Bouncing on an Oscillating Atom Mirror

The idea of spontaneous breaking of a discrete time translation symmetry into another discrete time translation symmetry in an isolated periodically driven quantum many-body system was proposed in Ref. [173]. It was shown that ultra-cold atoms bouncing on an oscillating atom mirror and prepared in a many-body Floquet state is able to spontaneously reorganize their motion and start moving with the period twice longer than the driving period. This is an example of discrete (or Floquet) time crystals [57]. In the present section we focus on the specific system of ultra-cold atoms bouncing on an oscillating mirror but the approach we use is general and can be applied to any periodically driven system that, in the language of the theory of dynamical systems, reveal nonlinear resonances [19, 27, 82, 87, 123].

4.2.1 A Particle Bouncing Resonantly on an Oscillating Mirror: Classical and Quantum Description

Let us consider a particle bouncing on a mirror that oscillates with the frequency ω and amplitude λ/ω^2 in the presence of the gravitational field in the one-dimensional space. The Hamiltonian of the system reads

$$H = \frac{p^2}{2} + z + F\left(z + \frac{\lambda}{\omega^2}\cos\omega t\right), \tag{4.27}$$

where $F(z)$ describes the mirror, i.e. the profile of its repulsive potential. We use the gravitational units,

$$l_0 = \left(\frac{\hbar^2}{m^2 g}\right)^{1/3}, \quad E_0 = mgl_0, \quad t_0 = \frac{\hbar}{mgl_0}, \tag{4.28}$$

for the length, energy, and time, respectively, where m is the mass of a particle and g the gravitational acceleration. Description of the system is more convenient when we switch from the laboratory frame to the frame oscillating with the mirror because then the position of the mirror is fixed and only the effective gravitational field depends periodically on time. The transformation to the oscillating frame can be done by means of the canonical transformation $z' = z + (\lambda/\omega^2)\cos\omega t$ and $p' = p - (\lambda/\omega)\sin\omega t$ in the classical description or by the corresponding unitary transformation $U = e^{iz(\lambda/\omega)\sin\omega t}e^{ip(\lambda/\omega^2)\cos\omega t}$ in the quantum description. The resulting Hamiltonian reads [27, 62, 89]

$$H = \frac{p^2}{2} + z + \lambda z\cos\omega t, \quad z \geq 0, \tag{4.29}$$

where we have dropped the primes over z and p and omitted unimportant terms in the Hamiltonian which have no influence on the dynamics, and assumed that the mirror is modeled by $F(z)$ which forms an infinite barrier at $z = 0$.

If the mirror does not move ($\lambda = 0$), all trajectories are periodic orbits with the period the longer, the greater energy of a particle. In the presence of the mirror oscillations we are interested in resonant motion, i.e. when the driving frequency ω is an integer multiple of the frequency Ω of an unperturbed motion of a particle. Then, $\omega = s\Omega$ where s is integer and an $s:1$ resonance condition between the driving force and particle motion is fulfilled. Before we switch to analysis of the quantum behavior of a resonantly driven particle, we first consider the classical secular approximation approach [123]. In the following background information we present the formalism of action-angle variables which is a basic ingredient of the classical secular approximation method that will be often used in this monograph.

Action-Angle Variables

Let us assume a particle in the one-dimensional space which performs periodic motion in the presence of a time-independent potential,

$$H_0 = \frac{p^2}{2} + V(z). \tag{4.30}$$

Classical description of the system is very simple when we perform canonical transformation to the so-called action-angle variables, I and $\theta \in [0, 2\pi)$, where the Hamiltonian depends on the new momentum (action I) only, i.e. $H_0 = H_0(I)$. Then, $\dot{I} = 0$ and $\dot{\theta} = \Omega(I)$, where

$$\Omega(I) = \frac{\partial H_0(I)}{\partial I}, \tag{4.31}$$

is the frequency of a periodic motion of a particle. The action I is a constant of motion and parametrizes energy of a particle while the conjugate variable (angle θ) evolves linearly with time, $\theta(t) = \Omega(I)t + \theta_0$, and describes position of a particle on a periodic trajectory. Description of a general action-angle formalism can be found in Ref. [123] and here we present rather a practical way how to obtain a canonical transformation to action-angle variables.

It is easy to find the action I and express the Hamiltonian in terms of it. That is, the action is an area, divided by 2π, inside of a closed trajectory drawn by a particle in the phase space,

$$I = \frac{1}{2\pi} \oint p dz = I(E), \tag{4.32}$$

where $p = \pm\sqrt{2[E - V(z)]}$ and E is energy of a particle. Inverting Eq. (4.32) allows us to obtain $H_0 = E(I)$. It is also easy to obtain the canonical transformation between the Cartesian variables $\{z, p\}$ and the new variables $\{\theta, I\}$ if we know periodic orbits,

$$z = z(E, t - t_0) = z\left(E(I), \frac{\theta - \theta_0}{\Omega(I)}\right),$$

(continued)

$$p = p(E, t - t_0) = p\left(E(I), \frac{\theta - \theta_0}{\Omega(I)}\right), \tag{4.33}$$

where the choice of values of energy E and a parameter t_0 allows one to determine initial conditions of a particle.

The first simple example we consider is a harmonic oscillator potential $V(z) = \Omega^2 z^2/2$. Equation (4.32) yields $H_0(I) = \Omega I$ while Eqs. (4.33) define the canonical transformation,

$$z = \frac{\sqrt{2I}}{\Omega}\cos(\theta - \theta_0), \quad p = -\sqrt{2\Omega I}\sin(\theta - \theta_0). \tag{4.34}$$

The harmonic oscillator problem is exceptional because the period of its all orbits is the same and when we turn on a resonant driving it is able to pump unlimited amount of energy to a particle. In order to build time crystals we employ different systems where period of unperturbed motion of a particle depends on energy and a resonant driving is not able to pump continuously energy to a particle because when a particle absorbs energy, the period of its motion changes, a particle gets out of the resonance and the resonant transfer of energy stops. In the language of the theory of dynamical systems, we are interested in systems that reveal nonlinear classical resonances [123].

A particle bouncing on an oscillating mirror is an example of a system with nonlinear resonances. Let us first consider a particle bouncing on a static mirror located at $z = 0$ which is described by the Hamiltonian

$$H_0 = \frac{p^2}{2} + z, \quad z \geq 0, \tag{4.35}$$

and find a canonical transformation to the action-angle variables. Equation (4.32) leads to

$$H_0(I) = \frac{(3\pi I)^{2/3}}{2}, \tag{4.36}$$

and the frequency of a periodic motion (4.31) is

$$\Omega(I) = \left(\frac{\pi^2}{3I}\right)^{1/3}. \tag{4.37}$$

In order to get the canonical transformation to the action-angle variables we have to find a trajectory of a particle which, between two successive bounces that happen at $t = t_0 \mp \pi/\Omega$, is described by

(continued)

$$z = -\frac{1}{2}(t - t_0)^2 + E = \frac{(3\pi I)^{2/3}}{2}\left[1 - \frac{1}{\pi^2}(\theta - \theta_0)^2\right]$$

$$p = t_0 - t = \left(\frac{3I}{\pi^2}\right)^{1/3}(\theta_0 - \theta), \tag{4.38}$$

where $\theta \in [\theta_0 - \pi, \theta_0 + \pi)$. Because the evolution described by Eqs. (4.38) is repeated periodically, we can express it as Fourier series,

$$z(I, \theta) = \sum_{n=-\infty}^{\infty} h_n(I)\, e^{in\theta}, \tag{4.39}$$

where $h_0 = (\pi^2 I^2/3)^{1/3}$ and

$$h_n(I) = \left(\frac{9I^2}{\pi^4}\right)^{1/3} \frac{(-1)^{n+1}}{n^2} e^{-in\theta_0}, \quad \text{for} \quad n \neq 0. \tag{4.40}$$

In this monograph we choose $\theta_0 = 0$.

Having introduced the action-angle variables for a particle bouncing on a static mirror, Eqs. (4.38) and (4.39), we can now perform analysis of fundamental resonances of a particle perturbed by an oscillating mirror within the secular approximation approach [123] which can be done starting with the Hamiltonian (4.29). However, in the quantum case we will be interested in the Floquet Hamiltonian that corresponds to the classical Hamiltonian of the system but in the extended phase space $H_F = H + p_t$, i.e. extended by the time degree of freedom t and the momentum $p_t = -H$ conjugate to t [88, 123],

$$H_F = \frac{p^2}{2} + z + \lambda z \cos \omega t + p_t, \quad z \geq 0. \tag{4.41}$$

The Hamiltonian H_F is the classical counterpart of the Floquet Hamiltonian and it is a constant of motion as a function of the fictitious time ξ. The fictitious time is related to the time t via the Hamilton equation $dt/d\xi = \partial H_F/\partial p_t = 1$ which implies that the fictitious time is essentially the time t, i.e. $t = \xi + t_0$.

The Hamiltonian (4.41) in the action-angle variables reads

$$H_F = \frac{(3\pi I)^{2/3}}{2} + \lambda \cos \omega t \sum_{n=-\infty}^{\infty} h_n(I)\, e^{in\theta} + p_t, \tag{4.42}$$

where $h_n(I)$ is given in (4.40). We are interested in the resonant driving when the frequency of the mirror oscillations is an integer multiple of a frequency of unperturbed particle motion, i.e. $\omega = s\Omega(I_s)$ where s is integer, Ω is given in (4.37), and I_s is the resonant value of the action. Let us perform an additional canonical transformation to the frame moving with the resonant frequency of a particle $\Omega(I_s) = \omega/s$,

$$I' = I, \tag{4.43}$$

$$\Theta = \theta - \frac{\omega}{s}t, \tag{4.44}$$

$$p_t' = p_t + \frac{\omega}{s}I, \tag{4.45}$$

$$t' = t, \tag{4.46}$$

which leads to the Hamiltonian of the form

$$H_F = \frac{(3\pi I)^{2/3}}{2} - \frac{\omega}{s}I + \frac{\lambda}{2}\sum_{n=-\infty}^{\infty} h_n(I)\, e^{in(\Theta+\omega t/s)}\left(e^{i\omega t} + e^{-i\omega t}\right) + p_t, \tag{4.47}$$

where we have dropped the primes.

If we choose initially the action close to the resonant value, $I \approx I_s$, then p_t, I, and Θ are slowly varying variables provided the time-dependent perturbation is weak. That is, $dp_t/dt = O(\lambda)$, $dI/dt = O(\lambda)$ and

$$\frac{d\Theta}{dt} = \Omega(I) - \frac{\omega}{s} + O(\lambda)\ \Big|_{I=I_s} = O(\lambda), \tag{4.48}$$

and the only fast variable is time t because $dt/d\xi \approx 1$. When we average the original Hamiltonian (4.47) over t while keeping the slowly varying variables fixed, we obtain the effective secular Hamiltonian

$$H_F \approx \frac{P^2}{2m_{\text{eff}}} + V_0\cos(s\Theta) + p_t, \tag{4.49}$$

where

$$m_{\text{eff}} = -\frac{\pi^2 s^4}{\omega^4}, \quad V_0 = \frac{\lambda}{\omega^2}(-1)^{s+1}. \tag{4.50}$$

A constant term in the Hamiltonian (4.49) has been omitted. The first two terms in (4.47) have been expanded in the Taylor series around $P = I - I_s \approx 0$ up to the second order resulting in the effective mass which is negative.

The effective Hamiltonian (4.49) indicates that close to a resonant orbit and in the frame moving with the resonant frequency, cf. (4.44), a driven particle behaves like a particle moving in a potential that is periodic function of Θ and consists of s identical wells. If $s \gg 1$, we deal with a solid state-like system, i.e. a particle behaves like an electron moving in a potential periodic in space. In Chap 5 we will explore this analogy and investigate condensed matter phenomena in the time dimension.

We have obtained the classical effective Hamiltonian (4.49) for a resonantly driven particle but our aim is to describe the quantum resonant behavior. Before proceeding further, let us analyze the scaling symmetry of the classical equations of motion corresponding to the Hamiltonian (4.41). This symmetry tells us that if we rescale appropriately all quantities, the equations of motion do not change. It means that analyzing the system we can eliminate one parameter. For example, we can set the resonant value of the action $I_s = 1$, which implies that the resonant value of the driving frequency $\omega = s\Omega(I_s) = s(\pi^2/3)^{1/3}$. Next we can perform analysis of the behavior of the system for different λ and the results can be used for any other values of ω. Indeed, any trajectory we get for $I_s = 1$ can be used to obtain a similar trajectory for $I_s \neq 1$ when we rescale

$$\omega' = I_s^{-1/3}\omega, \quad \lambda' = \lambda, \quad p' = I_s^{1/3}p, \quad z' = I_s^{2/3}z, \quad t' = I_s^{1/3}t. \tag{4.51}$$

The classical scaling symmetry is broken in the quantum description because the Planck constant sets a scale in the phase space. It can be easily seen when we try to apply the classical scaling transformation in the commutation relation between the position and momentum operators,

$$[z', p'] = i \quad \Rightarrow \quad [z, p] = \frac{i}{I_s}. \tag{4.52}$$

It also shows that the resonant value of the action determines an effective Planck constant $1/I_s$. Thus, for $I_s \gg 1$, the effective Planck constant is small and we may expect that when we quantize the classical effective Hamiltonian (4.49) by substitution $P = -i\partial/\partial\Theta$ and $p_t = -i\partial/\partial t$, we should get an accurate description of the quantum resonant dynamics. Another way to calculate the quantum effective Hamiltonian is to apply the quantum version of the secular approximation method at the very beginning [19]. The eigenvalue problem, $H_0|n\rangle = E_n|n\rangle$, for the unperturbed Hamiltonian (4.35) can be easily solved [27] ($2^{1/3}E_n$ are given by zeros of the Airy function). In order to obtain the quantum secular Floquet Hamiltonian we first switch to the moving frame by means of the unitary transformation $U = e^{i\hat{n}\omega t/s}$, where $\hat{n}|n\rangle = n|n\rangle$, which is an analogue of the classical canonical transformation (4.44) and then we perform averaging over time which results in the desired effective Floquet Hamiltonian

$$\langle n'|H_F|n\rangle \approx \left(E_n - n\frac{\omega}{s}\right)\delta_{n,n'} + \frac{\lambda}{2}\langle n'|z|n\rangle\left(\delta_{n',n+s} + \delta_{n',n-s}\right), \tag{4.53}$$

that describes the quantum resonant dynamics of a particle.

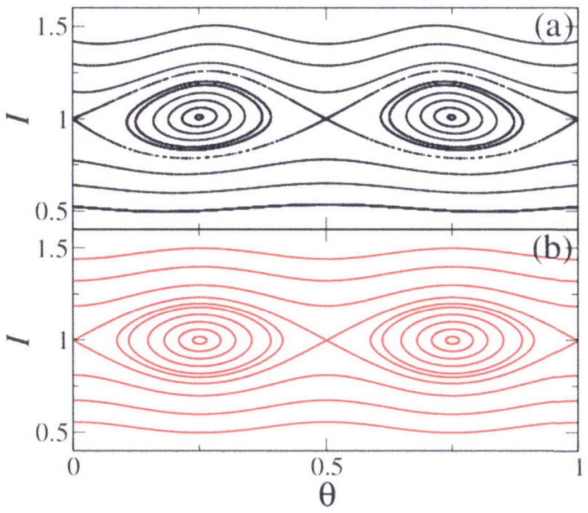

Fig. 4.1 Classical phase space of a particle bouncing on an oscillating mirror. Top panel shows the exact stroboscopic map around the 2 : 1 resonance obtained by plotting points in the $\{\theta, I\}$ space every period of the mirror oscillations. Bottom panel presents a similar phase space portrait but obtained with the help of the effective secular Hamiltonian (4.49) for $s = 2$. Two elliptical resonance islands are visible in the plots. At the centers of the islands, the resonant orbit is located, i.e. it corresponds to jumps between the centers after each period of the mirror oscillations. The amplitude and frequency of the mirror motion are $\lambda = 0.06$ and $\omega = 2\Omega(I_2)$, respectively, where $I_2 = 1$ is chosen as the resonant value of the action, cf. (4.37). Thanks to the classical scaling symmetry, the presented results can be used to obtain the 2 : 1 resonant motion for any value of ω (or equivalently any value of I_2) by means of the scaling transformation (4.51)

We have collected all necessary tools to describe resonant dynamics of a particle bouncing on an oscillating mirror. We have got the classical secular Hamiltonian (4.49) that allows us to get intuition how the resonant dynamics of a particle looks like. Results obtained by means of this effective Hamiltonian can be compared with the exact classical dynamics that allows us to check if a chosen value of λ is sufficiently small to get an accurate quantitative description of the system by means of the secular approximation method, an example is shown in Fig. 4.1. Having determined the suitable parameters of the system we can switch to the quantum description. If the exact solution of the Schrödinger equation is difficult to obtain numerically, we can use the quantized version of the classical secular Hamiltonian or perform the calculations using the quantum version of the secular Hamiltonian (4.53).

Let us complete the analysis of the single-particle problem by presenting an example of the quantum resonant behavior. In the 2 : 1 resonance case it is easy to calculate numerically resonant Floquet states without referring to the secular approximation method [27, 62, 89]. For $s = 2$, the elliptical resonant islands, visible in Fig. 4.1, are sufficiently large to capture quantum states (in the semiclassical sense) even for not very small the effective Planck constant $1/I_2$. Indeed, if we choose $\lambda = 0.06$ and $\omega = 1.1$ (that is related to $I_2 \approx 20$), we get two semiclassical states localized at the bottom of each of the two resonant islands shown in Fig. 4.1. These states form localized wavepackets moving along the resonant orbit with the period of the classical motion of a particle $2\pi/\Omega(I_2)$ that is twice longer than the driving period $T = 2\pi/\omega$. They move on the same orbit but with the delay of T with respect to each other. The wavepackets are localized in the different islands but they are coupled by a quantum tunneling process. Consequently, in the full quantum description there are two quantum resonant Floquet states which are two orthogonal superpositions of the wavepackets. The fact that the wavepackets exchange their role after every period of the mirror oscillations makes the Floquet states periodic with the driving period T, see Fig. 4.2.

Quasi-energies corresponding to the 2 : 1 resonant dynamics repeat themselves on the energy axis with the period $\omega/2$. Why it is so can be explained when we perform the WKB semiclassical quantization of the classical secular Hamiltonian (4.49) [27]. Within the WKB method we have to simultaneously fulfill the following conditions

$$\frac{1}{2\pi}\int_0^{2T} p_t\, dt = k, \tag{4.54}$$

$$\frac{1}{2\pi}\oint P\, d\Theta = n + \frac{1}{2}, \tag{4.55}$$

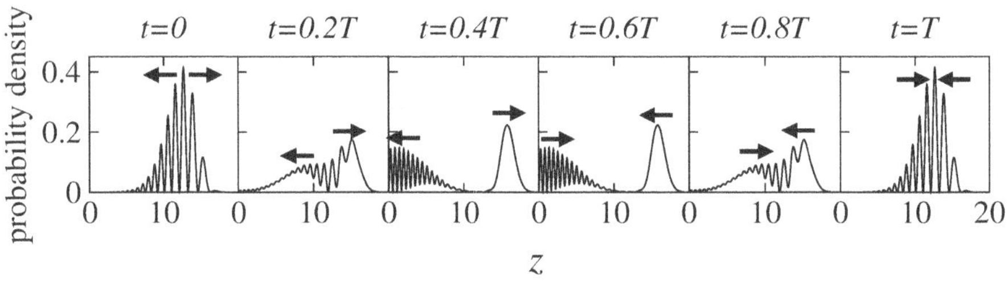

Fig. 4.2 Quantum description of a particle bouncing resonantly on an oscillating mirror in the case of the 2 : 1 resonance. Time evolution of a single Floquet state, $u_1(z, t)$, is presented. In the semiclassical picture, the Floquet state $u_1(z, t)$ corresponds to states localized at the bottom of the elliptical resonant islands shown in Fig. 4.1. The Floquet state is superposition of two localized wavepackets which are moving along the classical 2 : 1 resonant orbit with the period $2T$ where $T = 2\pi/\omega$ is the mirror oscillation period. Despite the fact that the wavepackets evolve with the period $2T$, the Floquet state is periodic with the driving period T because after each period T the wavepackets exchange their role. At $t = 0$ the wavepackets overlap but because they are moving in the opposite directions we can see interference fringes. In the course of time evolution one wavepacket moves towards the mirror (which is located at $z = 0$), bounces off the mirror, and returns. The other wavepacket travels towards the classical turning point (located at $z \approx 16$) and also returns and at $t = T$ we end up with the initial probability density. There is another Floquet state, $u_2(z, t)$, which evolves in a similar way as the state $u_1(z, t)$ presented in the figure but it corresponds to the quasi-energy nearly exactly shifted by $\omega/2$. Pairs of such Floquet states are sometimes called π-states [102]. The parameters of the system are the same as in Fig. 4.1 but $\omega = 1.1$. Arrows indicate directions of motion of the wavepackets

where k is integer and n is non-negative integer. In (4.54) the integration is carried out over the interval $2T$ because a particle returns to an initial point in the phase space after the period of the classical motion. Due to the fact that the secular Hamiltonian does not depend on time, the momentum p_t is a constant of motion and consequently Eq. (4.54) yields

$$p_t = k\frac{\omega}{2}. \tag{4.56}$$

The condition (4.55) allows us to obtain resonant quasi-energies E_n of the system while Eq. (4.56) implies that the 2 : 1 resonant quasi-energies are repeated along the energy axis with the period $\omega/2$, i.e. eigenvalues of the secular Hamiltonian are $E_n + k\omega/2$.[1]

The results of the semiclassical analysis are confirmed in the diagonalization of the exact Floquet Hamiltonian $H_F = H - i\partial_t$ where H is given in (4.29). In Fig. 4.2 we show time evolution of the probability density of the Floquet state $u_1(z,t)$ which in the semiclassical picture is localized at the bottom of the resonant islands. There is another Floquet state $u_2(z,t)$ which evolves in a similar way as the state presented in Fig. 4.2. The quasi-energies $E_{1,2}$ of these two Floquet states are shifted by $\omega/2+J$ where $J \ll \omega$ is a tiny correction responsible for a tunneling process. That is, if at $t = 0$ we prepare superposition $\psi(z,0) = [u_1(z,0) \pm u_2(z,0)]/\sqrt{2}$, we can extract one of the localized wavepackets. Such a state evolves like

$$\psi(z,t) = \frac{e^{-iE_1 t}}{\sqrt{2}}\left[u_1(z,t) \pm e^{-i(\omega/2+J)t}u_2(z,t)\right], \tag{4.57}$$

and initially it follows the 2 : 1 resonant orbit with the period $2T$. However, when t approaches $\pi/J \gg T$, a particle tunnels to the other wavepacket which is initially unoccupied [89]. Next, at $t = 2\pi/J$, a particle returns to the initial wavepacket and so on.

We have analyzed Floquet states which in the semiclassical picture are localized at the bottom of the elliptical resonant islands. However, if the islands are big enough, more Floquet states can be trapped inside the islands. They correspond to excited localized wavepackets which are moving along a resonant classical orbit [134]. These states are located in the same region of the classical phase space as the Floquet states localized at the bottom of the islands. Therefore, when we switch to many particles bouncing resonantly on an oscillating mirror, the particle interactions can couple these states with the Floquet states localized at the bottom of the islands. In order to avoid it we should keep the interaction energy per particle much smaller than the energy gap between the quasi-energy eigenvalues of the Floquet states at

---

[1] Floquet states with the quasi-energy difference $\omega/2$ are sometimes called π-states because the difference between the corresponding eigenphases of the Floquet evolution operator, that evolves the system over the period T, is $(\omega/2)T = \pi$ [102]. For a general s : 1 resonance, the resonant quasi-energy levels repeat themselves with the period ω/s.

the bottom of the islands and Floquet states corresponding to excited states in the islands. In Fig. 4.9 such an energy gap is illustrated for the case of 40 : 1 resonance where the states localized at the bottom of the resonant islands form an energy band. In the present case of the 2 : 1 resonance and for the parameters as in Fig. 4.2, the energy gap is very large, i.e. $\Delta E \approx 1030J$.

4.2.2 Formation of Discrete Time Crystals

In the present section we will see that when we switch from a single particle to many particles bouncing resonantly on an oscillating mirror, the discrete time translation symmetry of the Hamiltonian can be spontaneously broken into another discrete time translation symmetry if the interaction between particles is sufficiently strong. For example, ultra-cold atoms bouncing on an oscillating atom mirror and fulfilling the 2 : 1 resonance condition are able to spontaneously self-reorganize their periodic motion and start moving with the period twice longer than T [173]. It means that a new periodic structure in the time domain can form spontaneously. This kind of spontaneous breaking of the time translation symmetry is dubbed discrete (or Floquet) time crystals [57].

Let us consider N ultra-cold identical bosonic atoms bouncing on an oscillating mirror in the one-dimensional model. Interaction between atoms is described by the contact Dirac-delta potential $g_0\delta(z)$. The parameter g_0 determines the strength of the interaction which in the gravitational units (4.28) reads

$$g_0 = 2\omega_\perp a_s, \tag{4.58}$$

where a_s is the s-wave scattering length of atoms and $\omega_\perp$ is the frequency of the harmonic trap along the transverse direction.[2] We assume attractive interactions, i.e. $g_0 < 0$. For large N, the many-body Hilbert space is huge and in order to carry out analysis of resonant dynamics we have to choose a suitable basis and restrict to a relevant Hilbert subspace. From Sect. 4.2.1 we know that the single-particle 2 : 1 resonant motion is described by two Floquet states $u_{1,2}(z,t)$ corresponding to the quasi-energies $E_{1,2}$ which differ by $\omega/2 + J$, where $J \ll \omega$ is the tunneling amplitude. In Fig. 4.2 time evolution of one of these Floquet states is presented. Superposition of the Floquet states allows us to define two basis vectors which are localized wavepackets,

$$w_{1,2}(z,t) = \frac{1}{\sqrt{2}}\left(u_1(z,t) \pm e^{-i\omega t/2} u_2(z,t)\right), \tag{4.59}$$

[2] In order to deal effectively with the one-dimensional system, strong transverse confinement of ultra-cold atoms has to be performed so that excitation energy along the transverse directions $\omega_\perp$ is much greater than the interaction energy per particle [155].

and which are periodic functions of t with the period $2T$.[3] Let us restrict to the resonant many-body Hilbert subspace, i.e. the subspace spanned by $w_{1,2}$. In the second quantization formalism (see the background information in Sect. 2.1.2 and Eq. (4.24)) it means that the bosonic field operator $\hat{\psi}$ is the sum of two operators $\hat{a}_{1,2}$ only,

$$\hat{\psi}(z,t) \approx w_1(z,t)\,\hat{a}_1 + w_2(z,t)\,\hat{a}_2. \tag{4.60}$$

The many-body Floquet Hamiltonian in this resonant Hilbert subspace can be written as follows: [40, 41, 173], cf. (4.25):

$$\begin{aligned}\hat{H}_F &= \frac{1}{2T}\int_0^{2T} dt \int_0^{\infty} dz\, \hat{\psi}^\dagger \left(H - i\partial_t + \frac{g_0}{2}\hat{\psi}^\dagger\hat{\psi}\right)\hat{\psi} \\ &\approx -\frac{J}{2}\left(\hat{a}_1^\dagger\hat{a}_2 + \hat{a}_2^\dagger\hat{a}_1\right) + \frac{U-2U_{12}}{2}\left(\hat{a}_1^\dagger\hat{a}_1^\dagger\hat{a}_1\hat{a}_1 + \hat{a}_2^\dagger\hat{a}_2^\dagger\hat{a}_2\hat{a}_2\right),\end{aligned} \tag{4.61}$$

with a constant term omitted. In Eq. (4.61), H is the single-particle Hamiltonian (4.29),

$$J = -\frac{1}{T}\int_0^{2T} dt \int_0^{\infty} dz\, w_2^*(z,t)\,(H - i\partial_t)\, w_1(z,t) = E_2 - E_1 - \frac{\omega}{2}, \tag{4.62}$$

and

$$\begin{aligned}U &= \frac{g_0}{2T}\int_0^{2T} dt \int_0^{\infty} dz\, |w_{1,2}(z,t)|^4, \\ U_{12} &= \frac{g_0}{2T}\int_0^{2T} dt \int_0^{\infty} dz\, |w_1(z,t)|^2\, |w_2(z,t)|^2.\end{aligned} \tag{4.63}$$

In the Floquet Hamiltonian (4.61) only the interaction terms which are proportional to overlaps of the probability densities, cf. (4.63), are kept because other terms turn out to be negligible [173]. The Floquet Hamiltonian (4.61) fulfills the U(1) symmetry, i.e. the symmetry related to the unitary transformation $e^{iq\hat{N}}$ where $\hat{N}$ is the operator of the total number of particles and q is an arbitrary real parameter.

In the absence of the interactions ($g_0 = 0$ and thus $U = 0$ and $U_{12} = 0$), the Floquet Hamiltonian in its final form (4.61) constitutes the exact description of non-interacting particles. When the interactions are on, there appears a coupling between the resonant Hilbert subspace and the complementary space that is neglected

[3]Note that in the definition of the basis vectors (4.59) we have not included the additional energy shift J responsible for the tunneling of a particle. Equation (4.59) defines the time periodic basis only—it does not reflect the tunneling process, cf. (4.57). The tunneling phenomenon is fully taken into account but in the Hamiltonian (4.61) and not in the basis vectors.

in (4.61). Coupling to single-particle Floquet states which in the semiclassical picture are localized in the 2 : 1 resonant islands is most dangerous because they are located in the region of the phase space where the states $u_{1,2}$ are. However, there is a gap ΔE between quasi-energies of these Floquet states and the eigenvalues $E_{1,2}$ corresponding to the $u_{1,2}$ states which, in a typical situation, is much bigger than the tunneling amplitude J (in Fig. 4.9 such a gap is illustrated in the case of the 40 : 1 resonant driving). Thus, even if the interaction energy per particle E_I is much greater than J but $E_I \ll \Delta E$, the Floquet Hamiltonian (4.61) yields very good approximation of the exact many-body resonant dynamics on a time scale much longer than the lifetime of a Bose–Einstein condensate in ultra-cold atoms experiments [68, 173]. This approximation is in the same spirit as the two-mode approximation for a Bose–Einstein condensate in a double-well potential [163, 212]. There, one may restrict to a two-mode Hamiltonian if interaction energy per particle is much smaller than the energy gap between the two lowest nearly degenerate single-particle energy levels and other excited levels.

Let us consider the N-body system for the same parameters of the mirror oscillations as those used in Fig. 4.2. We assume that ultra-cold atoms are bouncing on the atom mirror and fulfill the 2 : 1 resonant condition. It means that we restrict to the many-body resonant subspace spanned by Fock states $|n_1, n_2\rangle$ where $n_{1,2}$ are numbers of atoms that occupy the modes $w_{1,2}$, cf. (4.59). If the interactions between atoms are very weak ($g_0 \to 0$) and if the system is in the lowest many-body eigenstate of (4.61), we recover basically the single-particle behavior which is described in Sect. 4.2.1. Indeed, each atom occupies the single-particle Floquet state u_1 and thus the system forms a Bose–Einstein condensate. No breaking of the discrete time translation symmetry of the original Hamiltonian takes place because the system evolves with the driving period T and this evolution is stable. That is, even if there are atomic losses or destructive measurement of positions of particles, the remaining particles still evolve with the period T. The system forms a Bose–Einstein condensate, therefore it can be accurately described by the mean-field approach where all N particles occupy the same single-particle state ϕ,

$$\psi(z_1, z_2, \ldots, z_N, t) = \phi(z_1, t)\, \phi(z_2, t) \ldots \phi(z_N, t), \tag{4.64}$$

where $\phi(z, t)$ is a solution of Gross–Pitaevskii equation [30] (see the background information in Sect 3.1.1),

$$i\frac{\partial \phi}{\partial t} = H\phi + g_0 N |\phi|^2 \phi. \tag{4.65}$$

In the present case when $g_0 \to 0$, the lowest energy state of (4.61) corresponds to $\phi(z, t) = e^{-iE_1 t} u_1(z, t)$.

The situation changes if the attractive particle interactions are sufficiently strong. Both U and U_{12} are negative but $|U| \gg |U_{12}|$ and when the parameter

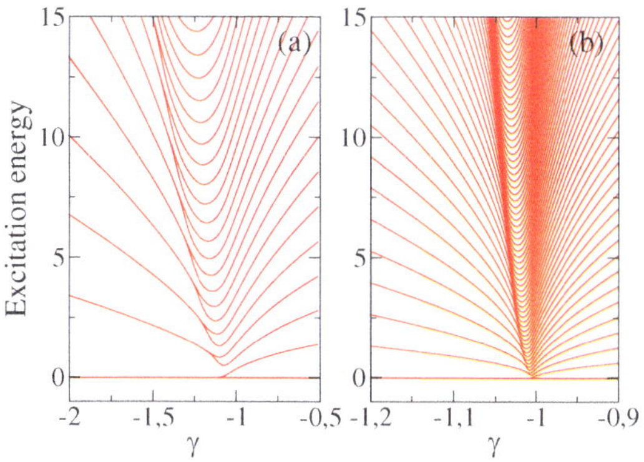

Fig. 4.3 Eigenvalues E of the Hamiltonian (4.61) with respect to the lowest energy level E_0, i.e. $2(E - E_0)/J$. Results for $N = 200$ are presented in (**a**) and for $N = 10^4$ in (**b**). For $\gamma < -1$, pairs of nearly degenerate levels dominate the spectrum for E below the broken symmetry edge E_{edge}, see the text. The figure reproduced from [212]

$$\gamma = \frac{N(U - 2U_{12})}{J}, \tag{4.66}$$

becomes smaller than -1, the state $e^{-iE_1 t}u_1(z,t)$ is no longer a stable solution of the Gross–Pitaevskii equation. At $\gamma = -1$ a quantum phase transition takes place [212]. For $\gamma < -1$ and in the limit when $N \to \infty$ and $g_0 \to 0$ but $g_0 N =$ constant (and thus $\gamma =$ constant), pairs of degenerate eigenvalues of the effective Floquet Hamiltonian (4.61) form which dominate the low energy spectrum, see Fig. 4.3.[4]

For sufficiently strong attractive interactions, in order to get the lowest energy state of (4.61), it is energetically favorable to group all N bosons in one of the localized wavepackets: w_1 or w_2 (4.59). However, such a many-body state cannot be a Floquet state because it would evolve with the period twice longer than the driving period. In order to reconcile the symmetry and energy requirements, the many-body Floquet states corresponding to the lowest energy eigenstates of (4.61) become superpositions of two Bose–Einstein condensates,

$$|\psi_\pm\rangle \approx \frac{1}{\sqrt{2}}\left(|N, 0\rangle \pm |0, N\rangle\right), \tag{4.67}$$

where N atoms occupy the localized wavepacket $w_1(z,t)$ and also N particles occupy the other wavepacket $w_2(z,t)$. The difference between the corresponding quasi-energies tends to zero nearly exponentially fast when $N \to \infty$ and $g_0 \to 0$ but $g_0 N =$ constant, i.e.

[4]In order to diagonalize a Floquet Hamiltonian its matrix representation is prepared usually in the Fourier basis $e^{ik\omega t}$ where k is integer. However, when we use the Fourier basis with harmonics corresponding to integer multiplicity of $\omega/2$, i.e. $e^{ik\omega t/2}$, then the results are still fine but we actually duplicate the Floquet Hamiltonian matrix. That is, the resulting spectrum consists of two identical Floquet spectra shifted with respect to each other by $\omega/2$. Using the $(2T)$-periodic basis (4.59), which is very convenient in the many-body problem considered here, we actually duplicate the many-body Floquet spectrum and quasi-energies of the members of the pairs of the Floquet eigenstates with the Schrödinger cat structure, like in (4.67), do not differ by $\omega/2$ but they are nearly degenerate, see Fig. 4.3 and the background information on *Many-body Floquet Hamiltonian* in Sect. 4.1.1.

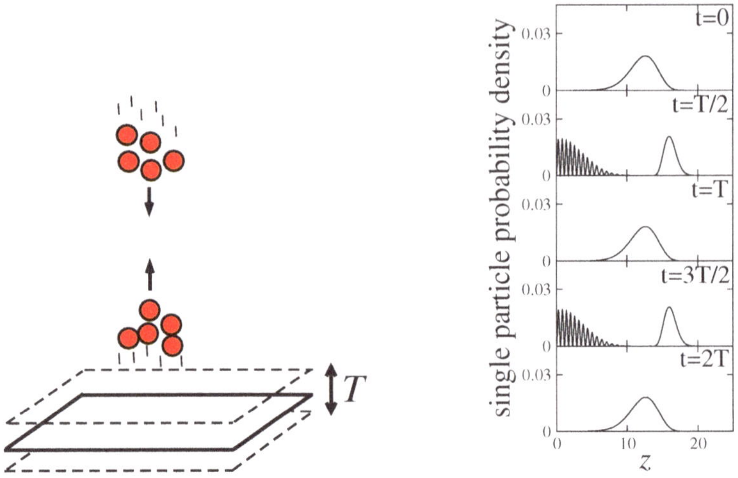

Fig. 4.4 The 2 : 1 resonant dynamics of ultra-cold atoms bouncing on an oscillating atom mirror. Left panel shows a schematic plot of atomic density corresponding to the Schrödinger cat-like state (4.67). There are two clouds of atoms which propagate along the classical 2 : 1 resonant orbit and exchange their role after every period of the mirror oscillations T. Right panel presents the time evolution of the single-particle density (4.69) corresponding to $|\psi_+\rangle$ (4.67). The parameters are the same as in Fig. 4.2, $N = 10^4$ and the particle interaction strength corresponds to $\gamma < -1$. The figure reproduced from [174]. © IOP Publishing. Reproduced with permission. All rights reserved. The figure was adapted from [173]

$$E_- - E_+ \propto N e^{-\alpha N}, \tag{4.68}$$

where α is a positive constant [173]. The many-body Floquet states (4.67) are periodic functions of time with the period T because the wavepackets exchange their role every period T, i.e. $w_2(z, t) = w_1(z, t + T)$. In Fig. 4.4 time evolution of the corresponding single-particle probability density,

$$\rho(z, t) \propto \langle\psi_\pm|\hat{\psi}^\dagger(z, t)\, \hat{\psi}(z, t)|\psi_\pm\rangle \approx \frac{N}{2}\left(|w_1(z, t)|^2 + |w_2(z, t)|^2\right), \tag{4.69}$$

is presented.

However, the Floquet states (4.67) have a form of a Schrödinger cat-like state and destructive measurement of the position of even only one atom can lead to a collapse of the many-body wavefunctions (4.67) to one of the states that form the superpositions, i.e. $|N - 1, 0\rangle$ or $|0, N - 1\rangle$ depending on the result of the measurement. The state $|N - 1, 0\rangle$ (or $|0, N - 1\rangle$) evolves with the period $2T$ and thus breaks the discrete time translation symmetry of the Hamiltonian. Its lifetime τ goes nearly exponentially quickly to infinity when $N \to \infty$ and $g_0 \to 0$ but $g_0 N$ =constant,

$$\tau \propto \frac{1}{E_- - E_+} \propto \frac{e^{\alpha N}}{N}, \tag{4.70}$$

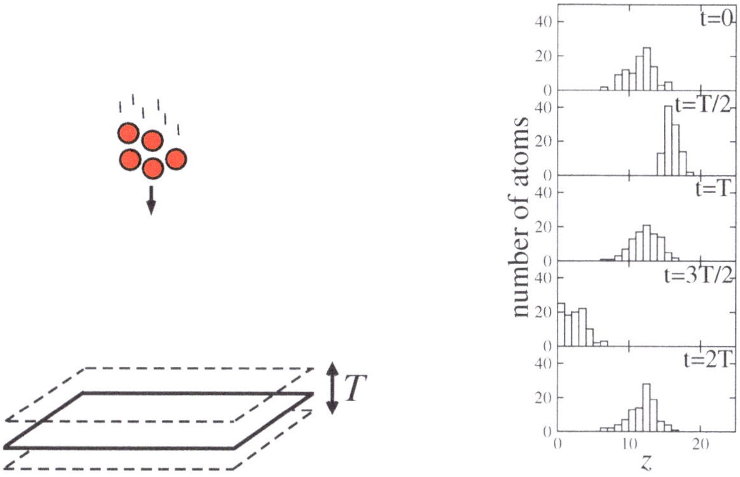

Fig. 4.5 Left panel: a similar plot as in Fig. 4.4 but the evolution of the symmetry broken state is presented. Right panel: time evolution of the atomic density. The initial state is the Floquet state $|\psi_+\rangle$ (4.67), however, at $t = 0$ and after: $T/2$, T, $3T/2$ and $2T$, measurements of atomic densities are performed. That is, at $t = 0$ positions of 100 atoms are measured and the results are presented as a histogram. Then, the remaining $(N - 100)$ atoms are evolved according to the many-body Hamiltonian (4.61) and at $t = T/2$ again 100 atoms are measured and again a histogram is prepared and so on. The parameters are the same as in Fig. 4.4. The figure reproduced from [174]. The figure was adapted from [173]

where $\alpha > 0$. In Fig. 4.5 we present the results of the measurement of positions of atoms at different moments of time when the system is initially prepared in the Floquet state $|\psi_+\rangle$ (4.67). The initial state preserves the time translation symmetry but if we want to get information about its time evolution and we measure positions of atoms, the symmetry is immediately broken into another discrete time translation symmetry corresponding to the period $2T$ and a symmetry broken state lives forever in the limit of $N \to \infty$.

The spontaneous symmetry breaking can be also illustrated with the help of the two-time correlation function which, thanks to the fact that $E_- - E_+ \to 0$, is a periodic function of t and t' with the period $2T$, i.e.

$$
\begin{aligned}
G_2(t', t) &= \langle \psi_\pm | \hat{\psi}^\dagger(z, t)\ \hat{\psi}^\dagger(z, t')\ \hat{\psi}(z, t')\ \hat{\psi}(z, t) | \psi_\pm \rangle \\
&\approx \frac{N(N-1)}{2} \left(|w_1(z, t')|^2\ |w_1(z, t)|^2 + |w_2(z, t')|^2\ |w_2(z, t)|^2 \right).
\end{aligned}
\tag{4.71}
$$

Spontaneous breaking of the discrete time translation symmetry is not restricted to the Floquet states corresponding to the ground state manifold of the Hamiltonian (4.61). Indeed, all many-body Floquet states related to the eigenstates of (4.61) with the eigenenergies below the so-called broken symmetry edge E_{edge} appear in degenerate doublets (see Fig. 4.3) and can reveal spontaneous breaking of the discrete time translation symmetry in the limit when $N \to \infty$ but $g_0 N$ =constant. It can be shown by rewriting the Hamiltonian H_F (4.61) into the form of the Lipkin–Meshkov–Glick model [165],

$$\hat{H}_F = J\left(-\hat{S}_x + \frac{\gamma}{N}\hat{S}_z^2\right), \tag{4.72}$$

with a constant term omitted, where the spin operators read

$$\hat{S}_x = \frac{\hat{a}_1^\dagger \hat{a}_2 + \hat{a}_2^\dagger \hat{a}_1}{2}, \tag{4.73}$$

$$\hat{S}_z = \frac{\hat{a}_2^\dagger \hat{a}_2 - \hat{a}_1^\dagger \hat{a}_1}{2}. \tag{4.74}$$

The total spin $\hat{\mathbf{S}}^2 = \hat{S}_x^2 + \hat{S}_y^2 + \hat{S}_z^2$ commutes with H_F and the ground state of (4.72) belongs to the total spin sector $S = N/2$. When we represent the total spin operators as sums of N spins-1/2, $\hat{S}_{x,z} = \sum_{j=1}^N s_j^{x,z}$ with $\hat{s}_j^{x,z} = \sigma_j^{x,z}/2$ where $\sigma_j^{x,z}$ are the Pauli matrices, we obtain the Lipkin–Meshkov–Glick model in the form of the chain of spins-1/2 with infinite-range interactions

$$H_F = J\left(-\sum_{j=1}^N s_j^x + \frac{\gamma}{N}\sum_{i=1}^N\sum_{j=1}^N s_i^z s_j^z\right). \tag{4.75}$$

The model possesses the cyclic $\mathbb{Z}_2$ symmetry, i.e. the Hamiltonian H_F commutes with the spin flip operator $e^{i\pi\hat{S}_x}$ which fulfills $(e^{i\pi\hat{S}_x})^2 = 1$ and thus H_F and the spin flip operator have common eigenstates. The mean field limit of the model, i.e. $N \to \infty$ but $\gamma =$ constant, coincides with the mean field limit of the original Bose system, i.e. $N \to \infty$ but $g_0 N =$ constant. In the mean field limit, the total spin operator can be substituted by the classical angular momentum $\hat{S}_{x,y,z} \to S_{x,y,z}$ with $|\mathbf{S}| = N/2$. Then, for $\gamma > -1$, one obtains that the classical ground state energy is $-JN/2$ and corresponds to $S_x = N/2$ and $S_z = 0$. For $\gamma < -1$ the classical approach predicts the degenerate ground state energy corresponding to $S_x = -N/(2\gamma)$ and $S_z = \pm(N/2)\sqrt{1 - 1/\gamma^2}$. The latter solutions break the $\mathbb{Z}_2$ symmetry of the Hamiltonian H_F. That is, H_F should have common eigenvectors with the spin flip operator $e^{i\pi\hat{S}_x}$ while the mean field results choose the spin magnetization either with $S_z > 0$ or $S_z < 0$. For $\gamma < -1$ the exact quantum many-body approach shows that the two lowest energy levels are nearly degenerate (the energy splitting goes to zero nearly exponentially quickly with increasing N, cf. (4.68)) and correspond to symmetric and antisymmetric superposition of the states with positive and negative z component of the spin magnetization. These superposition states easily decohere and the $\mathbb{Z}_2$ symmetry is spontaneously broken and the mean field properties of the system emerge. Actually, for large N, all quantum energy levels below the broken symmetry edge $E_{\rm edge} = -JN/2$ form doublets with the corresponding eigenstates that are not immune to decoherence and reveal spontaneous breaking of the $\mathbb{Z}_2$ symmetry. Note that the z component of the magnetization is equaled to the population imbalance in the Bose system (4.74). Thus, the spontaneous breaking of the $\mathbb{Z}_2$ symmetry in the Lipkin–Meshkov–Glick model corresponds to the spontaneous breaking of the discrete time translation symmetry of the Floquet Hamiltonian which in the time-periodic basis is represented by the Hamiltonian H_F (4.61).

It should be also stressed that the mean field approximation is valid but not in a close vicinity of the critical point at $\gamma = -1$. Indeed, the symmetry preserving phase and the discrete time crystal region are connected by the quantum phase transition where the mean field approximation breaks down because the second order terms of an effective potential disappear, see [149, 212].

4.2.3 Phase Diagram

We have considered the perfect 2 : 1 resonant driving of ultra-cold atoms by an atom mirror that oscillates with the frequency ω. However, period of bouncing of atoms on the mirror can be also chosen slightly off the resonance. Can imperfections in the resonant driving and imperfections in state preparation spoil the time crystal evolution? Following the approach of Ref. [204] we examine this issue in the present section [111]. We also analyze quantum many-body fluctuations within the approach that goes beyond the resonant Hilbert subspace, i.e. beyond the two-mode approximation, cf. (4.61).

The symmetry broken states $|N, 0\rangle$ and $|0, N\rangle$ in (4.67) are actually Bose–Einstein condensates where all N bosons occupy the same single-particle state $\phi = w_1$ or $\phi = w_2$, see (4.64). It means that when the time translation symmetry is broken the system can be described by the mean-field approach where the macroscopically occupied single-particle mode ϕ evolves according to the Gross–Pitaevskii equation (4.65) [30, 155] (see the background information in Sect 3.1.1). Thus, in order to investigate how the detuning from the perfect resonance and imperfections in the initial state preparation affect time crystal behavior we may use the mean-field approach.

Assume that at $t = 0$ all N atoms are prepared in the same single-particle state $\phi(z, 0)$ which is a Gaussian wavefunction

$$\phi(z, 0) = \left(\frac{\Omega_0}{\pi}\right)^{1/4} e^{-\Omega_0(z-h)^2/2+ipz}, \tag{4.76}$$

where h and p denote initial average position of atoms and their average momentum, respectively. Such a state can be realized in the experiment if ultra-cold atoms condense in a harmonic trap of the frequency Ω_0 which is located above the mirror

[68]. For $p = 0$, $h = h_0$, where

$$h_0 = \frac{2\pi^2}{\omega^2}, \tag{4.77}$$

and suitably chosen Ω_0, the Gaussian wavefunction reproduces very well the localized wavepacket w_1 (4.59) at the position h_0 of the classical turning point. If the trapping potential is turned off at the moment when the mirror is in its downwards position, the released cloud of atoms starts moving along the 2 : 1 resonant orbit [68].

In Fig. 4.6 we present how the squared overlap between $\phi(z, t)$ and the initial state $\phi(z, 0)$,

$$F(t) = |\langle\phi(t)|\phi(0)\rangle|^2, \tag{4.78}$$

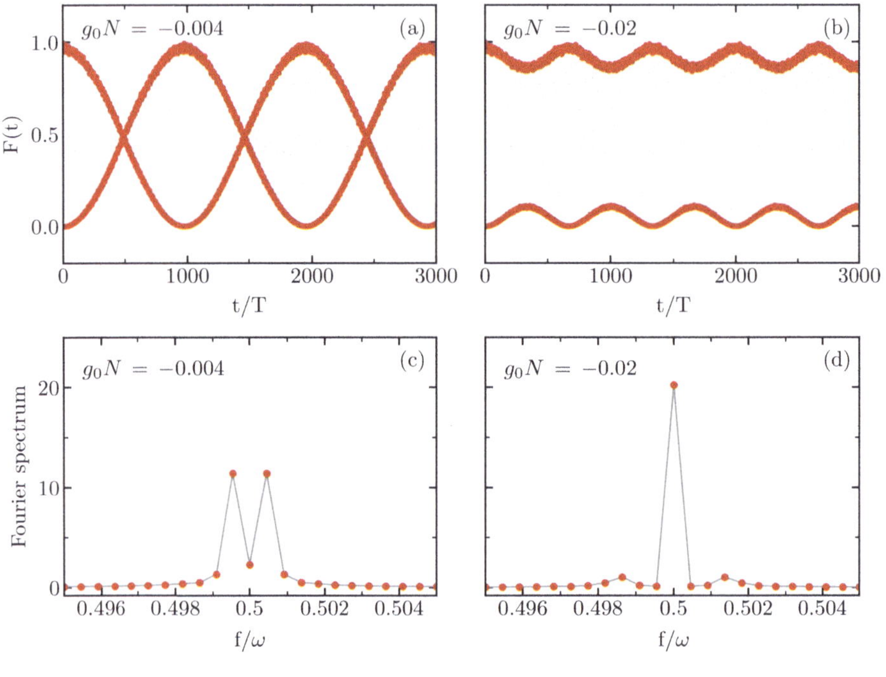

Fig. 4.6 Top panels: time evolution of the return probability (4.78) to the initial state $\phi(z, 0)$ at the stroboscopic times $t = 0, T, 2T, \ldots$ where $T = 2\pi/\omega$ and $\phi(z, 0)$ is the initial Gaussian wavefunction (4.76) and $\phi(z, t)$ is evolved according to the Gross–Pitaevskii equation (4.65). Bottom panels show Fourier transforms of $F(t)$ presented in the corresponding top panels. Left part of the figure is related to too weak interactions for a time crystal to form ($|g_0N| < |g_{cr}N| \approx 0.006$), i.e. atoms initially perform motion with the period $2T$ but they slowly tunnel between two localized wavepackets that propagate along the 2 : 1 resonant orbit—for non-interacting particles the tunneling time is $\pi/J \approx 1000T$. The right part of the figure corresponds to the time crystal regime where the tunneling process is suppressed. The parameters of the Hamiltonian (4.29) are $\lambda = 0.12$, $\omega = 1.4$, and the parameters of the initial Gaussian state (4.76): $h = h_0 = 9.82$, $\Omega_0 = 0.68$ and $p = 0$. The figure reproduced from [111]

changes in time for the perfect 2 : 1 resonant driving (i.e., for $h = h_0$, $p = 0$, and optimally chosen Ω_0). If the attractive interactions are too weak (i.e., $|g_0 N| < |g_{cr} N| \approx 0.006$), then atoms initially evolve with the period $2T$ but at $t = \pi/J \gg T$ all of them tunnel to the other wavepacket w_2 which was initially unoccupied and such a transfer of atoms between w_1 and w_2 occurs every period π/J. It results in the beating phenomenon in the behavior of the return probability $F(t)$ as presented in Fig. 4.6a. The Fourier transform of $F(t)$ shows two peaks located symmetrically around $\omega/2$. The situation changes if the interactions are strong enough. Then, the tunneling stops and there is uninterrupted periodic evolution with the period twice longer than the driving period that is reflected by the presence of the strong peak at $\omega/2$ in the Fourier spectrum of $F(t)$, see Fig. 4.6.[5]

We can introduce detuning ϵ from the 2 : 1 resonance by releasing the atomic cloud at a slightly higher or lower position with respect to the optimal turning point h_0,

$$h = h_0 + \epsilon, \tag{4.79}$$

and investigate if the particle interactions are able to stabilize time crystal evolution even if the driving is not perfectly resonant. Moreover we introduce imperfections in the initial state by choosing randomly the initial momentum p and also introduce fluctuations of the width $1/\sqrt{\Omega_0}$ of the Gaussian wavepacket (4.76). The parameters p and Ω_0 are chosen randomly from the uniform distributions in the intervals $p/p_{\max} \in [-0.02, 0.02]$ and $\Omega_0/\Omega_{0,\,\mathrm{opt}} \in [0.98, 1.02]$. We have chosen $p_{\max} = 4.4$ which corresponds to the maximal momentum of atoms at the moment of a reflection from the mirror and $\Omega_{0,\,\mathrm{opt}} = 0.68$ which is related to the best approximation of the wavepacket w_1 at the classical turning point by the Gaussian wavefunction (4.76).

If $\epsilon \neq 0$, it is still possible to observe time crystal evolution provided the interactions are sufficiently strong. However, for a given $g_0 N$ one can find a certain ϵ for which the system has a difficulty to decide if it is a time crystal or not. This difficulty is reflected by the maximal fluctuations of the height of the peak at $\omega/2$ in the Fourier spectra of $F(t)$ for different random choices of p and Ω_0 in the initial Gaussian wavefunction [204]. The set of the values of $g_0 N$ and ϵ corresponding to the maximal variance of the height of the peak at $\omega/2$ determines the boundary between the time crystal and symmetry unbroken regimes, see Fig. 4.7 - such phase diagram was observed in the experiments on discrete time crystals in spin systems [209] which we discuss in Sect. 4.6. Note that we present the phase diagram for $|g_0 N| > 1.3$ only. For $|g_{cr} N| < |g_0 N| < 1.3$, symmetry broken solutions are superposition of the two wavepackets w_1 and w_2 with different weights and the detuning cannot be described by the single parameter ϵ only. Indeed, one should

[5]Weak modulation of the periodic evolution of $F(t)$ is visible in Fig. 4.6b because the initial Gaussian wavefunction (4.76) does not perfectly reproduce the exact (2T)-periodic solution of the Gross-Pitaevskii equation.

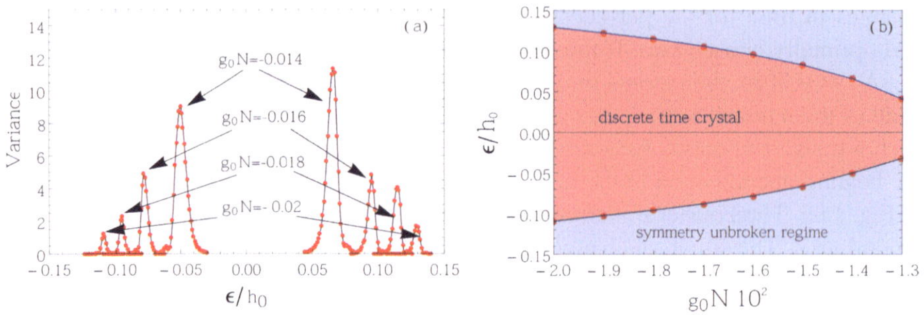

Fig. 4.7 Phase diagram for the discrete time crystal in ultra-cold atoms bouncing on an oscillating atom mirror. For given interaction strength g_0N and the detuning ϵ, see (4.79), time evolution of $F(t)$, Eq. (4.78), is obtained and the amplitude of the peak at $\omega/2$ in the Fourier spectrum of $F(t)$ is determined. Such a procedure is performed many times for different initial Gaussian wavepackets (4.76), with the parameters p and Ω_0 chosen randomly which allows one to calculate the variance of the amplitude of the peak at $\omega/2$. The variance is plotted in the left panel as a function of the detuning ϵ for different interaction strength g_0N. Values of ϵ corresponding to the maximal variance determine the phase boundary between the time crystal and symmetry unbroken regimes. That is, at the boundary between the regimes, the system has a difficulty to decide in which phase it is and therefore the fluctuations of the subharmonic peak at $\omega/2$ are the largest. The phase diagram is presented in the right panel. The parameters p and Ω_0 of the initial Gaussian wavefunction (4.76) are chosen randomly from the uniform distribution in the intervals $p/p_{\max} \in [-0.02, 0.02]$ and $\Omega_0/\Omega_{0,\,\mathrm{opt}} \in [0.98, 1.02]$ where $p_{\max} = 4.4$ and $\Omega_{0,\,\mathrm{opt}} = 0.68$. All other parameters are the same as in Fig. 4.6. The figure reproduced from [111]

also consider the relative displacement of the wavepackets and deviation from the optimal relative phase between them.

It should be stressed that if the interactions are too weak to form the discrete time crystal, the beating in the plot of $F(t)$ and the double peak structure around $\omega/2$ in the Fourier transform of $F(t)$ (see Fig. 4.6) are present even if there is no detuning from the perfect resonant driving, i.e. for $\epsilon = 0$. It is in contrast to discrete time crystals in spin systems, which we discuss in Sect 4.3, where for $\epsilon = 0$, time evolution of the non-interacting systems is still periodic with the period $2T$. In other words, in the spin systems in order to demonstrate formation of a discrete time crystal the detuning is necessary while in the case of ultra-cold atoms bouncing on an oscillating atom mirror we can see differences between interacting and non-interacting cases even for $\epsilon = 0$.

So far the robustness of the discrete time crystal was demonstrated within the mean-field approach where the time evolution of the system is described by the Gross–Pitaevskii equation. If we restrict to the resonant many-body Hilbert subspace, we know that the time crystal formation and its properties are identical to the phenomenon of the spontaneous breaking of the $\mathbb{Z}_2$ symmetry in the Lipkin–Meshkov–Glick model (4.72). However, we have not analyzed yet many-body effects that go beyond the resonant subspace. In the following, such an analysis is performed within the Bogoliubov approach [29, 30, 53]. However, first we would like to make a short introduction to the Bogoliubov formalism.

Bogoliubov Formalism for Bosons

Here, we make a short description of the U(1) symmetry-conserving version of the Bogoliubov formalism for Bose systems, i.e. the version that deals with a well-defined number of bosons N [29, 30, 53].

Ideally a Bose–Einstein condensate is a state of N bosons where all of them occupy the same single-particle wavefunction ϕ, i.e.

$$\psi(\mathbf{r}_1, \mathbf{r}_2, \ldots, \mathbf{r}_N, t) = \phi(\mathbf{r}_1, t)\, \phi(\mathbf{r}_2, t) \ldots \phi(\mathbf{r}_N, t). \tag{4.80}$$

For interacting bosons such a many-body state is an approximation only because even if we prepare initially N particles in a state like (4.80), time evolution results in corrections to a product state (4.80) which are caused by interactions between particles. For weak interactions, the expected corrections are small and we can simplify description of a Bose system by deriving an effective Hamiltonian that describes small quantum many-body fluctuations around a product state (4.80).

In the U(1) symmetry-conserving Bogoliubov approach, the bosonic field operator is decomposed into [29, 30]

$$\hat{\psi}(\mathbf{r}, t) = \phi(\mathbf{r}, t)\, \hat{a}_0 + \delta\hat{\psi}(\mathbf{r}, t), \tag{4.81}$$

where $\phi(\mathbf{r}, t)$ is the condensate mode which evolves according to the Gross–Pitaevskii equation (4.65) and which is the eigenstate of the reduced single-particle density matrix corresponding to the largest eigenvalue. The operator $\delta\hat{\psi}(\mathbf{r}, t)$ describes all other modes which are orthogonal to the condensate wavefunction, i.e. $\langle\phi(t)|\delta\hat{\psi}(t)\rangle = 0$. The reduced single-particle density matrix of a Bose system corresponding to a many-body state $|\psi\rangle$, in the second quantization formalism, reads

$$\begin{aligned}\rho_1(\mathbf{r}', \mathbf{r}; t) &\propto \langle\psi|\hat{\psi}^\dagger(\mathbf{r}', t)\, \hat{\psi}(\mathbf{r}, t)|\psi\rangle \\ &= \phi^*(\mathbf{r}', t)\phi(\mathbf{r}, t)\, \langle\hat{a}_0^\dagger\hat{a}_0\rangle + \langle\delta\hat{\psi}(\mathbf{r}', t)\, \delta\hat{\psi}(\mathbf{r}, t)\rangle,\end{aligned} \tag{4.82}$$

where we have assumed that ϕ is an eigenstate of ρ_1.[6] If the corresponding eigenvalue $\langle\hat{a}_0^\dagger\hat{a}_0\rangle \approx N$, the depletion of a condensate (i.e., the average number of bosons which occupy other modes orthogonal to ϕ) is negligible,

(continued)

[6] Actually, not ϕ but ϕ^* is an eigenstate of ρ_1 given in (4.82).

$$dN = \mathrm{Tr}[\langle \delta\hat{\psi}(\mathbf{r}', t)\delta\hat{\psi}(\mathbf{r}, t)\rangle] = N - \langle \hat{a}_0^\dagger \hat{a}_0\rangle \approx 0, \tag{4.83}$$

and $\delta\hat{\psi}$ can be considered as a *small parameter* [29, 30].

Let us start with the time-independent case where the many-body Hamiltonian $\hat{H}$ of a Bose system does not depend on time. If $\delta\hat{\psi}$ can be treated as a *small parameter*, one may expand the many-body Hamiltonian, written in the second quantization formalism (see the background information in Sect. 2.1.2), up the second order in $\delta\hat{\psi}$ and neglect higher order terms. For ultra-cold atoms, which interact via the Dirac-delta contact potential $g_0\delta(\mathbf{r})$, it results in $\hat{H} \approx \hat{H}_B +$ constant where the Bogoliubov Hamiltonian [29, 30]

$$\hat{H}_B = \frac{1}{2}\int d^3\mathbf{r}\, \left[\hat{\Lambda}^\dagger, -\hat{\Lambda}\right] \mathcal{L} \begin{bmatrix}\hat{\Lambda} \\ \hat{\Lambda}^\dagger\end{bmatrix}, \tag{4.84}$$

with $\hat{\Lambda} = \hat{a}_0^\dagger \delta\hat{\psi}/\sqrt{N}$ and

$$\mathcal{L} = \begin{bmatrix} \hat{Q}\left(H + 2g_0 N|\phi|^2\right)\hat{Q} & g_0 N \hat{Q}\phi^2 \hat{Q}^* \\ -g_0 N \hat{Q}^* \phi^{*2} \hat{Q} & -\hat{Q}^*\left(H + 2g_0 N|\phi|^2\right)\hat{Q}^* \end{bmatrix}, \tag{4.85}$$

where H in (4.85) is a single-particle Hamiltonian and $\hat{Q} = 1 - |\phi\rangle\langle\phi|$ is the projection operator on the subspace orthogonal to the condensate mode ϕ. The latter fulfills the stationary Gross–Pitaevskii equation. The Bogoliubov Hamiltonian $\hat{H}_B$ describes excitations of a system and many-body fluctuations of a condensate. It is easy to diagonalize $\hat{H}_B$ by means of the Bogoliubov transformation

$$\begin{bmatrix}\hat{\Lambda} \\ \hat{\Lambda}^\dagger\end{bmatrix} = \sum_n \left(\begin{bmatrix} u_n \\ v_n \end{bmatrix} \hat{b}_n + \begin{bmatrix} v_n^* \\ u_n^* \end{bmatrix} \hat{b}_n^\dagger \right), \tag{4.86}$$

where $[u_n, v_n]$ and $[v_n^*, u_n^*]$ are right eigenstates of the non-Hermitian operator $\mathcal{L}$ corresponding to eigenvalues E_n and $-E_n$, respectively, which fulfill $\langle u_n|u_m\rangle - \langle v_n|v_m\rangle = \delta_{nm}$—these eigenstates are often called Bogoliubov modes[7]. The operators $\hat{b}_n$ are bosonic annihilation operators, i.e. $[\hat{b}_n, \hat{b}_m^\dagger] = \delta_{nm}$. The Bogoliubov Hamiltonian in the diagonal form reads

$$\hat{H}_B = \sum_n E_n\, \hat{b}_n^\dagger \hat{b}_n, \tag{4.87}$$

(continued)

[7]We have assumed here that all E_n are real. It can be not the case if a solution ϕ of the stationary Gross–Pitaevskii equation is dynamically unstable [29, 30].

where a constant term is omitted. The many-body eigenstate of $\hat{H}_B$ with no excitation (often called the Bogoliubov vacuum sate), i.e. $\hat{b}_n|0_B\rangle = 0$ for all n, is the eigenstate with the smallest quantum depletion,

$$dN = \mathrm{Tr}[\langle 0_B|\delta\hat{\psi}(\mathbf{r}')\delta\hat{\psi}(\mathbf{r})|0_B\rangle] \approx \int d^3\mathbf{r}\, \langle 0_B|\hat{\Lambda}(\mathbf{r})\hat{\Lambda}(\mathbf{r})|0_B\rangle = \sum_n \langle v_n|v_n\rangle. \tag{4.88}$$

Now let us switch to description of dynamics of a Bose system within the Bogoliubov formalism [30, 53]. Suppose that initially a system is prepared as a perfect Bose–Einstein condensate with no depletion,, i.e. the initial state is a product state,

$$\psi(\mathbf{r}_1, \mathbf{r}_2, \ldots, \mathbf{r}_N, 0) = \phi(\mathbf{r}_1, 0)\, \phi(\mathbf{r}_2, 0) \ldots \phi(\mathbf{r}_N, 0), \tag{4.89}$$

and we ask the question if interactions between bosons induce depletion of a condensate in the course of time evolution. This is the question we would like to ask also in the case of the time crystal in ultra-cold atoms bouncing on an oscillating atom mirror. In order to describe the time evolution, the initial perfect Bose–Einstein condensate state (4.89) has to be defined within the Bogoliubov approach what turns out to be quite easy. Indeed, no depletion means that we have the Bogoliubov vacuum state and all v_n components of the Bogoliubov modes vanish, i.e.

$$\begin{bmatrix} u_n(\mathbf{r}, 0) \\ v_n(\mathbf{r}, 0) \end{bmatrix} = \begin{bmatrix} \chi_n(\mathbf{r}) \\ 0 \end{bmatrix}, \tag{4.90}$$

where $\langle \phi(0)|\chi_n\rangle = 0$ and $\langle \chi_n|\chi_m\rangle = \delta_{nm}$ [53]. The initial condensate mode $\phi(\mathbf{r}, 0)$ has to be evolved in time according to the time-dependent Gross–Pitaevskii equation (4.65) while all initial Bogoliubov modes (4.90) must be evolved according to the Bogoliubov–de Gennes equations

$$i\partial_t \begin{bmatrix} u_n(\mathbf{r}, t) \\ v_n(\mathbf{r}, t) \end{bmatrix} = \mathcal{L} \begin{bmatrix} u_n(\mathbf{r}, t) \\ v_n(\mathbf{r}, t) \end{bmatrix}. \tag{4.91}$$

The set of the functions $\chi_n(\mathbf{r})$ is arbitrary but if they are chosen suitably to the problem under consideration, a small number of the Bogoliubov modes have to be evolved in time in order to get converged results for the reduced single-particle density matrix and for the depletion dN. Having $\phi(\mathbf{r}, t)$ and the Bogoliubov modes at a certain time t, the single-particle density matrix can be calculated and diagonalized [53]

(continued)

$$\rho_1(\mathbf{r}', \mathbf{r}; t) \propto \langle\psi|\hat{\psi}^\dagger(\mathbf{r}', t)\,\hat{\psi}(\mathbf{r}, t)|\psi\rangle$$
$$\approx N\,\phi^*(\mathbf{r}', t)\phi(\mathbf{r}, t) + \sum_n v_n(\mathbf{r}', t)v_n^*(\mathbf{r}, t)$$
$$= N\,\phi^*(\mathbf{r}', t)\phi(\mathbf{r}, t) + \sum_j dN_j\,\varphi_j^*(\mathbf{r}', t)\varphi_j(\mathbf{r}, t), \quad (4.92)$$

where the eigenvalues dN_j are average numbers of bosons which occupy eigenmodes $\varphi_j(\mathbf{r}, t)$ orthogonal to the condensate mode, i.e. $\langle\phi(t)|\varphi_j(t)\rangle = 0$. The sum of the eigenvalues dN_j tells us what is the total average number of atoms depleted from a Bose–Einstein condensate,

$$dN = \sum_j dN_j. \quad (4.93)$$

The Bogoliubov approach is valid provided $dN \ll N$. Note that the Gross–Pitaevskii equation (4.65) and the Bogoliubov operator (4.85) do not depend on the strength of the interaction g_0 and on the total number of bosons N separately but only on the product g_0N. Thus, in the limit when $N \to \infty$ and $g_0 \to 0$ but $g_0N =$ constant we always obtain $dN/N \ll 1$ if the total depletion $dN < \infty$. However, in the experiment with ultra-cold atoms one deals with a large but finite N and if a solution ϕ of the Gross–Pitaevskii equation is unstable [29, 30], the depletion increases exponentially in time and very quickly a Bose–Einstein condensate is destroyed because dN becomes comparable with N.

We return now to the discrete time crystal in ultra-cold atoms bouncing on an oscillating atom mirror. To describe experimental realization of the discrete time crystal and also to obtain the phase diagram we have assumed that a Bose–Einstein condensate is initially prepared in a harmonic trapping potential and next the trap is turned off and atoms start bouncing on an oscillating mirror. The time evolution of the condensate has been described within the mean-field approximation where a condensate wavefunction evolves according to the Gross–Pitaevskii equation (4.65). However, in such an approach we have not tested if interactions between atoms do not destroy the condensate. In other words, if depletion of bosons from the condensate is negligible or it becomes substantial and comparable to the total particle number N? With the help of the Bogoliubov formalism, which we have just described in the background information, it is straightforward to answer this question.

Starting with the condensate mode $\phi(z, 0)$ which is given by the Gaussian wavefunction (4.76) with $p = 0$, $h = h_0$, cf. (4.77), and the width $1/\sqrt{\Omega_0}$ that best fits the width of the wavepacket w_1 at the classical turning point, and with the interaction strength in the time crystal regime (i.e., $g_0N = -0.02$ like in Fig. 4.6b), time evolution of $\phi(z, t)$ is obtained by integration of the Gross–Pitaevskii equation.

Integration of the Bogoliubov–de Gennes equations (4.91) allows us to get the Bogoliubov modes $[u_n(z,t), v_n(z,t)]$ where initially they correspond to (4.90). At any time t, the reduced single-particle density matrix (4.92) can be calculated and diagonalized. It turns out that during the entire time evolution (i.e., up to $t_f = 1999T$ which corresponds to two tunneling periods of non-interacting atoms between the wavepackets $w_{1,2}$) there is only one eigenmode $\varphi_1(z,t)$ of the single-particle density matrix which dominates the depletion of the condensate, i.e. $dN(t) \approx dN_1(t)$. At $t = t_f$ the corresponding eigenvalue is $dN_1(t_f) \approx 1.6$, so less than two atoms are depleted from the condensate and occupy the mode φ_1. All other modes are negligibly populated, i.e. $dN_{j>1}(t_f) \leq 6 \times 10^{-3}$. The total depletion $dN(t_f) \approx 1.6$ means that for N of the order of 10^4 (which is a typical number of atoms in the experiment) there is no noticeable correction to the prediction obtained within the mean-field approach and the Gross–Pitaevskii equation is sufficient to describe the experiment.

In Fig. 4.8 time evolution of the total depletion $dN(t)$ is shown. The figure presents also probability densities of the condensate mode $|\phi(z,t_f)|^2$ and the

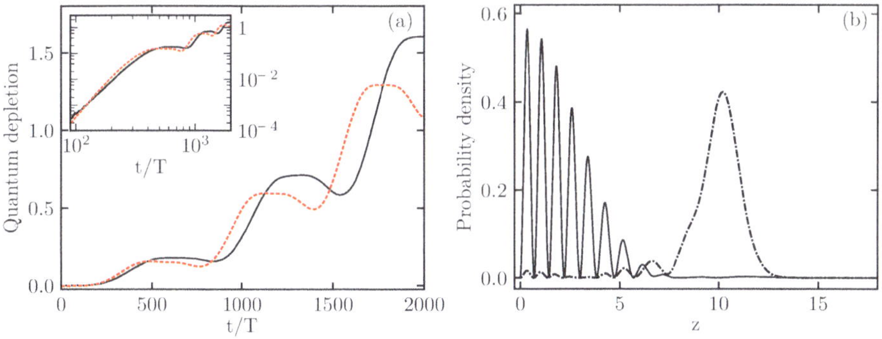

Fig. 4.8 Left panel: solid black line shows the total number of atoms, $dN(t)$, depleted from the condensate mode. The results are obtained within the Bogoliubov approach. The depletion is dominated by one mode $\varphi_1(z,t)$ only, $dN(t) \approx dN_1(t)$. Initially, the many-body state is a perfect Bose–Einstein condensate, i.e. $dN(0) = 0$, and in the course of time evolution only about two atoms become depleted from a condensate, i.e. $dN(t_f) \approx dN_1(t_f) \approx 1.6$ where $t_f = 1999T$. Thus, for a typical experimental value of N, which is of the order of 10^4, the depletion effect is totally negligible. Red dashed line presents the depletion obtained within the many-body two-mode approach, cf. (4.61), where the initial state was chosen as a perfect Bose–Einstein condensate $|N, 0\rangle$ with $N = 600$. Inset shows the same as the main figure but in the log-log scale indicating the initial algebraic increase of the depletion. Right panel: probability densities of the condensate mode $|\phi(z,t_f)|^2$ (solid line) and the dominant mode $|\varphi_1(z,t_f)|^2$ (dotted-dashed line) at $t = t_f$ obtained within the Bogoliubov approach. The overlap $|\langle\phi(t_f \pm T)|\varphi_1(t_f)\rangle|^2 \approx 0.87 \pm 0.02$ implies that the mode $\varphi_1(z,t)$ is essentially the same wavepacket as the condensate mode but it travels along the 2 : 1 resonant orbit with the delay T with respect to the condensate wavepacket. The parameters of the mirror oscillations are $\lambda = 0.12$ and $\omega = 1.4$, cf. (4.29). The parameters of the two-mode Hamiltonian (4.61) are $J = 7.26 \times 10^{-4}$, $U/g_0 = 0.23$ and $U_{12}/g_0 = 0.05$, and the interaction strength is chosen so that $g_0 N = -0.02$. The two-mode results are obtained for $N = 600$ but they remain the same for $N > 600$ provided $g_0 N = -0.02$. The figure reproduced from [111]

dominant mode $|\varphi_1(z, t_f)|^2$. It turns out that $|\langle\phi(t_f \pm T)|\varphi_1(t_f)\rangle|^2 \approx 0.87 \pm 0.02$ what implies that atoms depleted from the condensate occupy the wavepacket that travels along the $2 : 1$ resonant orbit but is delayed (or advance depending on a point of view) with respect to the condensate mode by the period T of the mirror oscillations. Thus, the many-body evolution of the system is restricted to two modes and these modes are similar to the modes w_1 and w_2 used to define the resonant many-body Hilbert subspace in Sect. 4.2.2, cf. Eq. (4.60).

Let us compare the quantum many-body effects of the time crystal evolution obtained within the Bogoliubov approach and with the help of the two-mode Hamiltonian (4.61). In the two-mode case, a perfect Bose–Einstein condensate, $|N, 0\rangle$, is chosen as the initial state where all atoms occupy the mode w_1. At $t = 0$ the quantum depletion of the condensate is zero but because the initial state is not an eigenstate of the Hamiltonian, the depletion increases in time which is shown in the left panel of Fig. 4.8. Despite the fact that in the Bogoliubov description the initial condensate wavefunction $\phi(z, 0)$ is not exactly the mode w_1 (i.e., $\phi(z, 0)$ is the Gaussian approximation of the mode w_1 only), the results for the quantum depletion obtained with the help of the both methods follow each other quite well. The depletion obtained within the two-mode approach corresponds to $N = 600$ but it is the same for any $N > 600$ provided $g_0 N = -0.02$.

The two-mode description allows us also to investigate what happens in the extremely long time scale. It turns out that at $t \approx 500T\sqrt{N}$, the depletion saturates at $dN \approx 0.02N$ and next, for much longer time evolution, the N-body system shows a revival and returns close to the initial perfect Bose–Einstein condensate.

The initial Bose–Einstein condensate states used in the Bogoliubov description and in the two-mode approach are generic uncorrelated states of the resonantly driven many-body system. The presented results of the many-body time evolution of these states show that heating effects are negligible.

4.2.4 Dramatic Breaking of the Discrete Time Translation Symmetry and Formation of n-Tupling Discrete Time Crystals

We have seen that ultra-cold bosonic atoms bouncing on an oscillating atom mirror are able to break spontaneously the discrete time translation symmetry of the Hamiltonian and start moving with the period twice longer than the driving period $T = 2\pi/\omega$. The same system can reveal dramatic breaking of the time translation symmetry by spontaneously switching to motion with a period s times longer than T where s is a large integer number [68]. It corresponds to the formation of the n-tupling discrete time crystal where $n = s$. In order to observe it, the $s : 1$ resonance condition has to be fulfilled where the driving frequency ω is s times greater than the frequency of the bounces of ultra-cold atoms on the mirror.

In Sect 4.2.1 we have derived a classical effective Hamiltonian (4.49) and its quantum version (4.53) which describe a single particle for any $s : 1$ resonant

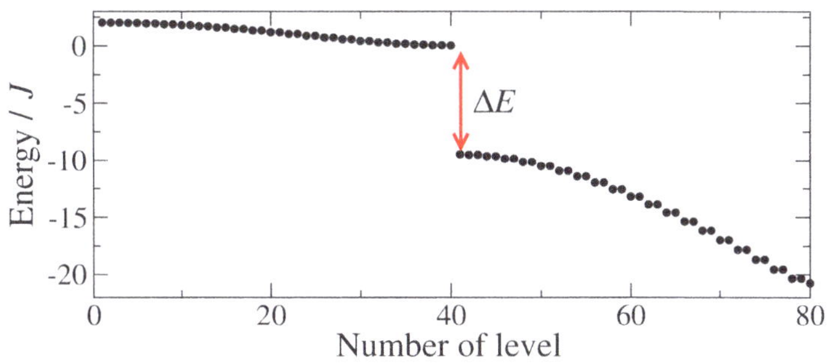

Fig. 4.9 Band structure of quasi-energy levels of a particle bouncing resonantly on an oscillating mirror—energies are given in the units of the tunneling amplitude J, Eq. (4.96). The energy levels are obtained by diagonalization of the quantized version of the classical secular Hamiltonian (4.49). The $40:1$ resonance condition is fulfilled, i.e. the period of an unperturbed particle motion is $s=40$ times longer than the period of the mirror oscillations. The quantum resonant motion can be described by the quantum secular Hamiltonian (4.53) or the quantized version of the classical secular Hamiltonian (4.49). For $s \gg 1$, the quasi-energy spectrum reveals a band structure. Because the effective mass in (4.49) is negative, the first band is the highest on the energy scale. Floquet states $u_n(z,t)$ corresponding to the first energy band form the resonant Hilbert subspace. In the many-body case, one may restrict to the resonant subspace if the interaction energy per particle is much smaller than the gap ΔE between the first and second energy bands that is visible in the plot. The parameters of the system are $\lambda = 0.2$ and $\omega = 4.9$, cf. the Hamiltonian (4.29), which result in the tunneling amplitude $J = 8.6 \times 10^{-4}$

driving. Thus, the generalization of the approach presented in Sect. 4.2.2 is straightforward. For $s \gg 1$, the spectrum of the quantum secular Hamiltonian (4.53), or the quantized version of the classical secular Hamiltonian (4.49), reveals a band structure, see Fig. 4.9. Because the effective mass in (4.49) is negative, the first band is the highest on the energy scale. The Floquet states $u_n(z,t)$ corresponding to the first energy band form the $s:1$ resonant Hilbert subspace which we restrict to in the many-body case. The restriction is valid provided the interaction energy per particle is much smaller than the gap ΔE between the first and second energy bands shown in Fig. 4.9. Such a condition can be easily fulfilled because the interaction energy per particle we need to realize discrete time crystals is of the order of the width of the first energy band, $2J$, which can be much smaller than the gap.

The resonant Floquet states $u_n(z,t)$ are superposition of s localized wavepackets $w_j(z,t)$ evolving along the $s:1$ classical resonant orbit. Proper superposition of the resonant Floquet states allows one to extract the localized wavepackets, $w_j(z,t)$, which are periodic functions of time with the period sT but after each period T they exchange their role, i.e. $w_j(z,t+T) = w_{j+1}(z,t)$. These wavepackets form a time-periodic basis similar to (4.59). Restricting to the many-body resonant Hilbert subspace, i.e. truncating the bosonic field operator

$$\hat{\psi}(z,t) \approx \sum_{j=1}^{s} w_j(z,t)\,\hat{a}_j, \tag{4.94}$$

we obtain the many-body Floquet Hamiltonian in the second quantization formalism, cf. (4.25) and (4.61),

$$\hat{H}_F = \frac{1}{sT}\int_0^{sT} dt \int_0^{\infty} dz\, \hat{\psi}^\dagger \left(H - i\partial_t + \frac{g_0}{2}\hat{\psi}^\dagger\hat{\psi}\right)\hat{\psi},$$
$$\approx -\frac{J}{2}\sum_{i=1}^{s}(\hat{a}_{i+1}^\dagger \hat{a}_i + \text{h.c.}) + \frac{1}{2}\sum_{i,j=1}^{s} U_{ij}\hat{a}_i^\dagger \hat{a}_j^\dagger \hat{a}_j \hat{a}_i, \tag{4.95}$$

where a constant term has been omitted. In (4.95), H is the single-particle Hamiltonian (4.29), the coefficient g_0 is given by (4.58),

$$J = -\frac{2}{sT}\int_0^{sT} dt \int_0^{\infty} dz\, w_{j+1}^*(z,t)\,(H - i\partial_t)\, w_j(z,t), \tag{4.96}$$

and

$$U_{ij} = \frac{2g_0}{sT}\int_0^{sT} dt \int_0^{\infty} dz\, |w_i(z,t)|^2\, |w_j(z,t)|^2, \tag{4.97}$$

for $i \neq j$ and similar expression for U_{ii} but twice smaller.

The many-body resonant Hilbert space, spanned by Fock states $|n_1, \ldots, n_s\rangle$ where n_i's are numbers of bosons occupying the modes w_i, is still too big to calculate numerically many-body eigenstates of (4.95) for $s \gg 1$ and for a large number of bosons. However, when $g_0 N \to 0$, the Bose system forms a Bose–Einstein condensate in the ground state of (4.95), i.e. in the lowest energy state of the system within the resonant subspace. When the attractive interactions between particles are sufficiently strong we expect that symmetry broken states that represent time crystal evolution are also Bose–Einstein condensates [173]. Both these situations can be described by the mean field approach (see the background information in Sect. 2.1.2).

The mean field approximation means that we are looking for time-periodic N-body states in the form of a product state $\phi(z_1,t)\phi(z_2,t)\ldots\phi(z_N,t)$. The resonant subspace is spanned by s single-particle Floquet states $u_n(z,t)$ or equivalently by s localized wavepackets $w_j(z,t)$. Thus, the mean field time-periodic solutions within that subspace can be expanded in the time-periodic basis as follows:

$$\phi(z,t) = \sum_{j=1}^{s} a_j w_j(z,t). \tag{4.98}$$

In (4.98) the complex-valued coefficients a_j, which fulfill $\sum_{j=1}^{s}|a_j|^2 = 1$, correspond to extremal values of the quasi-energy of the system per particle

$$E_F \approx -\frac{J}{2}\sum_{i=1}^{s}(a_{i+1}^* a_i + \text{c.c.}) + \frac{N}{2}\sum_{i,j=1}^{s} U_{ij}|a_i|^2|a_j|^2. \tag{4.99}$$

The extremal values of E_F can be found by solving the Gross–Pitaevskii equation that in the present case reduces to the set of the following equations

$$\frac{\partial}{\partial a_j^*}\left(E_F - \mu\sum_{i=1}^{s}|a_i|^2\right) = 0, \tag{4.100}$$

where μ is a Lagrange multiplier.[8]

Let us consider an example where ultra-cold atoms are bouncing on an oscillating mirror and the 40 : 1 resonance condition is fulfilled (i.e., $s = 40$). If, in the Hamiltonian (4.29), we choose the parameters of the mirror oscillations $\lambda = 0.2$ and $\omega = 4.9$, the tunneling amplitude in (4.95) $J = 8.6\times10^{-4}$. When the attractive interactions are sufficiently weak, i.e. $g_0N \gtrsim -1.64\times10^{-3}$, the lowest energy state of the Hamiltonian (4.95) is well approximated by the mean field solution, i.e. by a product state $\psi(z_1,\ldots,z_N,t) = \prod_{i=1}^{N}\phi(z_i,t)$ where

$$\phi(z,t) = \frac{1}{\sqrt{s}}\sum_{j=1}^{s} w_j(z,t). \tag{4.101}$$

Despite the fact that $w_j(z,t)$'s are periodic with the period sT, the wavefunction $\phi(z,t)$ is periodic with the period T because after each period T the wavepackets exchange sequentially their role, i.e. $w_{j+1}(z,t) = w_j(z,t+T)$. Actually the uniform superposition of w_j in (4.101) is the Floquet state $u_1(z,t)$ of the single-particle system. The mean field time-periodic solution $\psi(z_1,\ldots,z_N,t)$ preserves the discrete time translation symmetry of the many-body Hamiltonian.

If the attractive interactions are sufficiently strong, i.e. $g_0N \lesssim -1.64\times10^{-3}$, the Gross–Pitaevskii solution (4.101) becomes dynamically unstable and new s stable periodic solutions are born which, however, are non-uniform superposition of w_j and therefore they are evolving with the period $s = 40$ times longer than T and consequently break the discrete time translation symmetry of the many-body Hamiltonian. For $g_0N < -0.1$ the stable periodic solutions reduce practically to the single localized wavepackets $\phi(z,t) \approx w_j(z,t)$ where $1 \le j \le s$. In order to recover the discrete time translation symmetry of the many-body system we can prepare the superposition of the mean field product states

$$\psi(z_1,\ldots,z_N) \approx \frac{1}{\sqrt{s}}\sum_{j=1}^{s}\prod_{i=1}^{N} w_j(z_i,t), \tag{4.102}$$

[8]Comparing (4.95) and (4.99) we can see that the mean field quasi-energy per particle E_F can be obtained from the many-body Floquet Hamiltonian $\hat{H}_F$ by substituting the operators by complex numbers, $\hat{a}_n \to \sqrt{N}a_n$. The latter operation is a shortcut (in the spirit of the U(1) symmetry breaking Bogoliubov approach [155]) of a more formal way of obtaining the average energy (or quasi-energy) of a Bose system in the subspace of N-body product states [29].

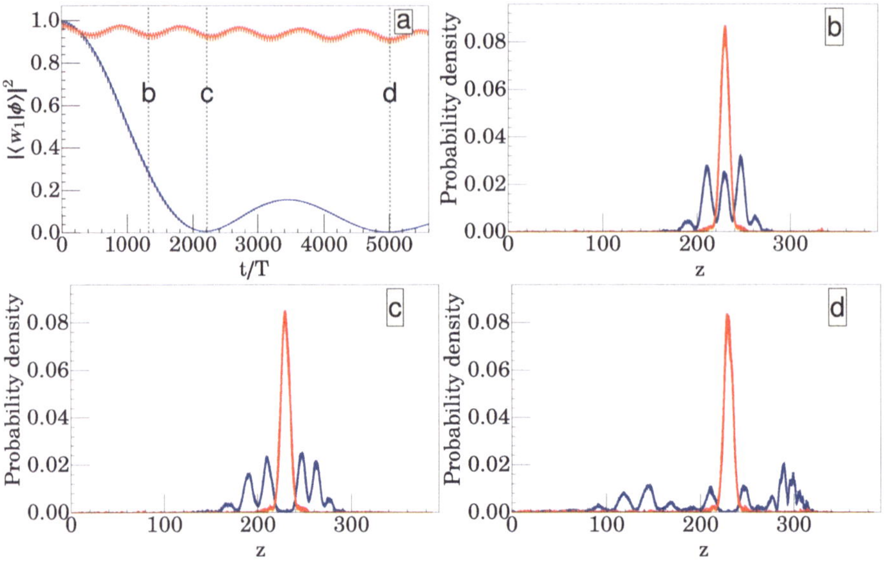

Fig. 4.10 Proposal for the experimental demonstration of a *big* time crystal where there is a dramatic breaking of the discrete time translation symmetry. Initially a Bose–Einstein condensate of ultra-cold atoms are prepared in a harmonic trap at the classical turning point above the oscillating atom mirror. The trap is turned off at a moment when the mirror is in its downwards position. The cloud bounces off the mirror but if the attractive interactions between particles are too weak, atoms tunnel from the initial wavepacket to the neighboring wavepackets moving along the 40 : 1 resonant orbit. It is indicated by the drop of the squared overlap of the evolving solution of the mean field Gross–Pitaevskii equation $\phi(z,t)$ with the wavepacket $w_1(z,t)$ which was initially highly occupied, see blue curve in (**a**). In (**b**)–(**d**) blue lines show the probability density $|\phi(z,t)|^2$ at time moments indicated with dashed vertical lines in (**a**). At $t = 2210T$ and $t = 5010T$ no atoms occupy the periodically evolving wavepacket w_1. The situation changes when the attractive interactions are sufficiently strong ($g_0 N = -0.12$). Then, the wavepacket prepared initially in the harmonic trap well reproduces a symmetry broken state that evolves with the period 40 times longer than the period of the mirror oscillations and no decay is observed even for time evolution as long as 5000 of T, see red curves in the panels. The figure reproduced from [68]

that in the second quantization formalism reads

$$|\psi\rangle \approx \frac{1}{\sqrt{s}}\left(|N,0,\ldots,0\rangle + |0,N,0,\ldots,0\rangle + \cdots + |0,\ldots,0,N\rangle\right). \tag{4.103}$$

The state (4.102) is a good approximation of the many-body Floquet state corresponding to the lowest energy state of the Hamiltonian (4.95). It is a superposition of $s = 40$ Bose–Einstein condensates and because $\langle w_j|w_{j'}\rangle = \delta_{jj'}$, it is also a Schrödinger cat-like state. It evolves with the driving period T but after, e.g., measurement of the position of a particle, the Schrödinger cat state collapses, the discrete time translation symmetry is broken and the system starts evolving with the period $s = 40$ times longer than T, cf. similar discussion in Sect. 4.2.2.

In the experiment it will not be possible to prepare the Schrödinger cat state (4.102). However, it should be quite easy to prepare a symmetry broken state $\psi(z_1, \ldots, z_N, 0) = \prod_{i=1}^{N} \phi(z_i, 0)$ with $\phi(z, 0) \approx w_j(z, 0)$, if a Bose–Einstein condensation of ultra-cold atoms is achieved in a harmonic trap located at the classical turning point above the oscillating mirror and the trap is turned off at the moment when the mirror is in its downwards position [68]. Then, the Bose–Einstein condensate starts evolving along the classical 40 : 1 resonant orbit. If the attractive interactions are too weak, then atoms start tunneling to the neighboring wavepackets w_{j-1} and w_{j+1} and at $t \approx 2.4/J \approx 2210T$ they totally leave the initial wavepacket w_j, see blue curves in Fig. 4.10. However, when the attractive interactions are sufficiently strong, the initial wavefunction $\phi(z, 0) \approx w_j(z, 0)$ well reproduces the symmetry broken state and no tunneling of atoms to the neighboring wavepackets is observed even for time evolution as long as $5000T$, see red curves in Fig. 4.10 which corresponds to $g_0 N = -0.12$.

It should be mentioned that for $g_0 N = -0.12$ the interaction energy per particle $|U_{ii}|N/2 \approx 3.8J$ which is smaller than the energy gap $\Delta E \approx 10J$ and consequently, the tight-binding approximation, i.e. the restriction to the resonant Hilbert subspace corresponding to the first energy band shown in Fig. 4.9, is valid.

In the laboratory it could be difficult to realize the hard-wall mirror that we have assumed in all theoretical analyses. However, if a realistic Gaussian shape mirror (that can be produced by a repulsive light-sheet) is used, the same time crystal phenomena as in the hard-wall case can be realized [70]. The experiment demonstrating discrete time crystals in ultra-cold atoms bouncing on a mirror is on the way [68, 70].

Discrete time crystals can reveal a dynamical quantum phase transition when the strength of the interactions between particles is quickly switched from the time crystal regime to the symmetry unbroken regime [68, 108]. In time-independent systems signature of the dynamical quantum phase transition is observed in the time domain as non-analytical behavior of the return probability to the initial ground state [85]. In the time crystal case, it corresponds to non-analytical behavior of the return probability to an initial Floquet state [68, 108, 109].

4.2.5 *Stability of Discrete Time Crystals in Ultra-Cold Atoms Bouncing on an Oscillating Mirror*

There is an important question how long the discrete time crystals, realized with the help of bosons bouncing on an oscillating mirror, can live? There are a few aspects of this issue which we are going to discuss in the present section. For simplicity we will concentrate on the case when the spontaneous breaking of the discrete time translation symmetry of the system corresponds to motion with the period twice longer than the driving period, i.e. we will focus on the 2 : 1 resonant dynamics.

1. *Mean field description.* If a many-body Bose system forms a Bose–Einstein condensate, its description simplifies dramatically because a many-body wave-

function is very well approximated by a product state $\psi(z_1, \ldots, z_N, t) \approx \prod_{i=1}^{N} \phi(z_i, t)$ where $\phi(z, t)$ is a solution of the Gross–Pitaevskii equation (see the background information in Sect. 2.1.2). In order to find such a mean field approximation for many-body Floquet states, we are looking for time-periodic solutions of the Gross–Pitaevskii equation

$$\left(H(t) + g_0 N|\phi(z, t)|^2 - i\partial_t\right)\phi(z, t) = \mu\phi(z, t), \tag{4.104}$$

where $H(t + T) = H(t)$, cf. (4.65). If we assume that $\phi(z, t + T) = \phi(z, t)$, i.e. we are looking for a solution in the subspace of T-periodic functions, then the operator on the left-hand side of (4.104) can be considered as a Floquet Hamiltonian and the Floquet theorem guarantees time-period solutions with the period T. However, when switch from the subspace of T-periodic functions to the full space, it may happen that a T-periodic solution we have found is not dynamically stable. It is precisely the case when a discrete time crystal forms [173]. Indeed, the 2 : 1 resonant dynamics is described by stable periodic solutions of (4.104) but with the period twice longer than T. Note that if $|\phi(z, t + T)| \neq |\phi(z, t)|$ but $|\phi(z, t + 2T)| = |\phi(z, t)|$, the operator on the left-hand side of (4.104) can still be considered as a time-periodic Floquet Hamiltonian but with the period $2T$ and $(2T)$-periodic solutions are expected. The stable $(2T)$-periodic solutions of (4.104) break the discrete time translation symmetry of the original many-body Hamiltonian and describe time crystal evolution. Their stability guarantees that within the mean field approach, the time crystal lives forever. It should be stressed that the stability of the $(2T)$-periodic solutions of the Gross–Pitaevskii equation (4.104) in the time crystal regime has been *proven* numerically only [68, 111, 173]

2. *Many-body description within the resonant Hilbert subspace.* Within the mean field approximation we observe the stability of the discrete time crystal. However, we have to keep in mind that the mean field approach assumes that the many-body Hilbert subspace is spanned by the product states $\psi(z_1, \ldots, z_N, t) = \prod_{i=1}^{N} \phi(z_i, t)$ only. The interaction terms in the many-body Hamiltonian couple the product state subspace to the complementary Hilbert subspace [155]. Therefore, the stability of the time crystals with respect to the exact many-body dynamics is not guaranteed. Many-body effects beyond the mean field approximation can be quite easily analyzed when we restrict to the resonant many-body Hilbert subspace. It leads to the two-mode many-body Hamiltonian (4.61) which corresponds to the many-body Floquet Hamiltonian written in the time-periodic basis (4.59). In the time crystal regime, the mean field solutions related to the two-mode many-body Hamiltonian (4.61) can be found analytically and they perfectly agree with the results of the full mean field approach that we have descried in the previous paragraph [173]. The advantage of the two-mode approximation is that it allows also for the full many-body analysis within the resonant Hilbert subspace. If the attractive interactions are sufficiently strong, the lowest energy levels of (4.61) form degenerate pairs (see Fig. 4.3)

and the corresponding many-body eigenstates become Scrödinger cat-like states. For example, the lowest energy eigenstates are of the form of the noon states $|N, 0\rangle \pm |0, N\rangle$ (4.67) and breaking of the discrete time translation symmetry means that the system chooses spontaneously one of the states: $|N, 0\rangle$ or $|0, N\rangle$. The symmetry broken state $|N, 0\rangle$ (or $|0, N\rangle$) evolves periodically with the period $2T$ and in the large N limit it does not decay. More precisely, its lifetime goes to infinity nearly exponentially fast when $N \to \infty$ but $g_0 N =$ constant [173], see (4.70), which is the limit when the mean field approximation is expected to be exact [124].

It should be mentioned that the spontaneous breaking of the discrete time translation symmetry described in the time-periodic basis by means of the two-mode Hamiltonian (4.61) corresponds to the spontaneous breaking of the $\mathbb{Z}_2$ symmetry in the Lipkin–Meshkov–Glick model [165]. Indeed, the latter model can be transformed to the Bose Hamiltonian (4.61). It is also known that the spontaneous breaking of the $\mathbb{Z}_2$ symmetry in the Lipkin–Meshkov–Glick model (and thus the spontaneous breaking of the discrete time translation symmetry in the ultra-cold atoms bouncing on an oscillating mirror) is revealed by all eigenstates of the model up to the so-called broken symmetry edge, see the discussion at the end of Sect 4.2.2.

3. *Many-body problem in the full Hilbert space.* The many-body results in the resonant Hilbert subspace show that the symmetry broken states that describe time crystal evolution live forever when $N \to \infty$. However, this is not the end of the story because the interaction terms in the many-body Hamiltonian couple the resonant subspace to the complementary space.

   In order to discuss this problem let us start with the classical analysis of a single particle bouncing resonantly on an oscillating mirror. It is known that a resonantly driven harmonic oscillator can absorb unlimited amount of energy. However, it is not the case for a particle bouncing on a mirror because the period of the unperturbed motion of a particle (i.e., when the mirror is fixed and does not oscillate) depends on its energy. When a resonantly bouncing particle absorbs energy, the period of its motion changes, a particle gets out of the resonance and the resonant transfer of energy stops. This phenomenon is called a nonlinear resonance in the theory of dynamical systems and it is reflected by the presence of stable elliptical resonance islands in the classical phase space [123], see Fig. 4.1. Switching from a single particle to N particles, we enlarge the dimension of the classical phase space from 2 to $2N$. When the interactions between particles are turned on, we expect that most of the Kolmogorov–Arnold–Moser (KAM) tori survive if the interaction terms in the Hamiltonian constitute a weak perturbation of the system [123]. However, for a periodically driven system with $N > 1$, KAM tori do not divide the phase space into disconnected regions and primary and higher order resonances form the Arnold web [123]. A system can diffuse along such a network of the resonances and explore the phase space but such an Arnold diffusion may be astronomically long [164]. Switching to the quantum description, the role of the Arnold diffusion in the dynamics of the discrete time crystal is an open and interesting question.

What can be done and has been done is the many-body description of the discrete time crystal within the Bogoliubov approach. This approach takes into account coupling of a Bose–Einstein condensate, that describes evolution of the discrete time crystal, to the entire many-body Hilbert space. It turns out that the long time dynamics (at least as long as thousands of the driving periods) does not reveal noticeable coupling of the system to the Hilbert subspace complementary to the subspace spanned by two modes. In other words, such many-body description shows that there are no heating effects in the long-term time evolution.

4. *Attractive interactions.* Formation of the discrete time crystal in ultra-cold atoms bouncing on an oscillating atom mirror takes place in the presence of attractive interactions between atoms. In the two- and three-dimensional cases, attractive contact interactions may lead to a collapse of a Bose–Einstein condensate. In order to get rid of this problem, the experiment has to be performed with sufficiently strong confinement of an atomic cloud along the transverse directions. Realization of such confinement in the laboratory is a standard technique nowadays.

   In the one-dimensional case, a bright soliton can form if the interactions are attractive. However, the discrete time crystals we consider here are realized in the very weak interaction regime where the formation of bright solitons is not observed. For example, for the parameters used in Fig. 4.8, the interaction strength is $g_0N = -0.02$ and consequently the bright soliton would have the width of $4/|g_0N| = 200$, Eq. (3.8), which is about 40 times larger than the maximal width of the wavepacket that describes the time crystal dynamics. One can expect that if the attractive interactions became very strong, a narrow soliton would form which would bounce on an oscillating mirror like a single massive object. Then, we would deal effectively with a one-body problem described by the center of mass of the soliton while all other degrees of freedom of the many-body system would not be accessible due to very large bounding energy [175, 202] and the system would behave like a pendulum in a clock [103].

5. *Experimental imperfections.* Imperfections in the initial state preparation and effects of a small detuning from the resonance condition have been discussed in Sect 4.2.3 where the phase diagram for the discrete time crystal has been presented. The time crystal evolution is robust and the necessary experimental conditions for its realization are attainable in the laboratory [68, 70, 111]. It is all under the assumption that although the system is driven by an external force, there is no dissipation. However, in ultra-cold atom experiments there are atomic losses due to, for example, three-body collisions. The three-body recombination can be reduced if an atomic cloud is sufficiently dilute but there is another source of dissipation: bounces off a mirror also introduce losses of atoms which can significantly reduce lifetime of the time crystals. The final answer concerning the role of this process will be provided by the experiment. In the case of time crystals in spin systems, that we are going to describe in the next paragraph, the role of dissipation was analyzed in Refs. [61, 114, 116] which show that contact with environment makes lifetime of time crystals finite.

4.3 Discrete Time Crystals in Periodically Driven Lattice Systems

Spontaneous breaking of a discrete time translation symmetry of the Hamiltonian into another discrete time translation symmetry in an isolated periodically driven quantum many-body system was predicted in ultra-cold atoms bouncing on an oscillating atom mirror [173]. The same phenomenon was later proposed in spin systems [57, 102] and afterwards demonstrated in the laboratory [33, 209]. In the present section we describe discrete time crystals in periodically driven spin systems.

4.3.1 Discrete Time Crystals in Spin Systems with Disorder

In general an isolated time-independent many-body system is expected to follow the eigenstate thermalization hypothesis [38, 45, 166, 187]. The hypothesis says that for any eigenstate of the system Hamiltonian corresponding to an eigenvalue E, the behavior of a small subsystem of the entire system can be described by the canonical ensemble whose average energy is equal to E. In other words, expectation values of local observables look like in a thermal state. Integrable systems may not follow the eigenstate thermalization hypothesis but such a resistance to thermalization does not survive if an integrability-breaking perturbation is added. Another example are systems that reveal many-body localization [15, 95, 143, 148, 213]. In the presence of strong disorder in space, an isolated one-dimensional translationally invariant many-body system may be characterized by a complete set of quasi-local integrals of motion and consequently is integrable and also breaks the eigenstate thermalization hypothesis [6, 9, 31, 42, 43, 56, 72, 74, 94, 101, 126, 143, 152, 162, 167, 177, 179–181, 185]. This many-body localization phase is accompanied by vanishing dc transport [4, 10, 13, 14, 18, 110, 188], extremely slow dynamics of various correlation functions [126, 136, 139, 140, 152, 160], and the logarithmic growth of the entanglement entropy [11, 106, 126, 181, 213].

When switching to periodically driven problems, a common wisdom is that interacting periodically driven systems should absorb energy from the driving force and heat up to infinite temperature where all correlations are trivial and do not depend on an initial state. Floquet eigenstates of such systems look like the infinite temperature states when we focus on small subsystems [37, 115, 159]. In order to prevent the heating, a mechanism of the energy localization is needed. The energy localization in the discrete time crystals based on ultra-cold atoms bouncing on an oscillating mirror was related to classical nonlinear resonances, see Sect. 4.2.5.

In Ref. [102] Khemani et al. investigated novel phases of periodically driven translationally invariant one-dimensional systems in the presence of strong disorder. With the help of spin models they found, in particular, a phase which they called a π-spin glass phase and which was actually a discrete time crystal [103]. Else et al. considered a similar system and showed that such a system broke discrete time translation symmetry of the Hamiltonian and actually formed a novel crystalline structure in time [57, 59]. They introduced the name *Floquet time crystals*, which today is used interchangeably with *discrete time crystals* [204].

Let us consider a one-dimensional chain of spins-1/2 which is driven periodically in time in a stroboscopic way. The time-periodic Hamiltonian of the system is defined as follows: [57]

$$H(t) = H(t+T) = \begin{cases} H_1 & \text{for} \quad 0 \le t < t_1 \\ H_2 & \text{for} \quad t_1 \le t < T, \end{cases} \tag{4.105}$$

with

$$H_1 = -\frac{\pi}{2t_1}(1-\epsilon)\sum_j \sigma_j^x, \tag{4.106}$$

$$H_2 = \sum_j \left(J_j \sigma_j^z \sigma_{j+1}^z + h_j^z \sigma_j^z + h_j^x \sigma_j^x \right), \tag{4.107}$$

where σ_j^x and σ_j^z are Pauli matrices associated with a spin located at a j-site of the chain and J_j, h_j^x, and h_j^z are real-valued parameters. The time evolution Floquet operator (i.e., the operator that evolves the system over one period of the time-periodic driving) is the product of two operators

$$U_F = e^{-i(T-t_1)H_2} e^{-it_1 H_1}. \tag{4.108}$$

For $\epsilon = h_j^x = h_j^z = 0$ it is easy to find eigenstates of U_F which are Floquet states of the time-periodic system.[9] Indeed, for $\epsilon = 0$ the operator on the right-hand side of (4.108),

$$e^{-it_1 H_1} = \exp\left(i\frac{\pi}{2}\sum_j \sigma_j^x \right) = \prod_j i\sigma_j^x, \tag{4.109}$$

is the spin flip operator, $\prod_j i\sigma_j^x |m_1, m_2, \ldots, m_L\rangle_z = i^L |-m_1, -m_2, \ldots, -m_L\rangle_z$ where $m_j = \pm 1$ are eigenvalues of the Pauli matrices σ_j^z and $|m_1, m_2, \ldots, m_L\rangle_z$ are their eigenvectors and L is the length of the chain. For $h_j^x = 0$ eigenstates of H_2 are eigenvectors of the individual σ_j^z and if additionally $h_j^z = 0$ we obtain

[9]Actually it is also easy to find the eigenstates when $h_j^z \neq 0$. That is, one obtains the same eigenphases of the Floquet evolution operator as in (4.112) and the slightly modified Floquet eigenstates $|\psi_\pm\rangle = \left[e^{-i(T-t_1)\sum_j h_j^z m_j/2} |m_1, \ldots\rangle_z \pm e^{i(T-t_1)\sum_j h_j^z m_j/2} |-m_1, \ldots\rangle_z \right] / \sqrt{2}$, cf. (4.111). For simplicity, in the main text we begin with the $h_j^z = 0$ case.

$$H_2|m_1, m_2, \ldots\rangle_z = \left(\sum_j J_j m_j m_{j+1}\right) |m_1, m_2, \ldots\rangle_z. \tag{4.110}$$

Thus, if ϵ and all h_j^x and h_j^z vanish, one can check that

$$|\psi_\pm\rangle = \frac{1}{\sqrt{2}} \left(|m_1, m_2, \ldots\rangle_z \pm |-m_1, -m_2, \ldots\rangle_z\right), \tag{4.111}$$

are eigenstates of the Floquet operator U_F with the corresponding eigenvalues

$$e^{i\phi_\pm} = \pm \exp\left(-i(T-t_1)\sum_j J_j m_j m_{j+1}\right), \tag{4.112}$$

where a common constant phase shift has been omitted. We have thus obtained pairs of Floquet eigenstates (4.111) which have the form of Schrödinger cat-like states similarly as in the case of the discrete time crystal in ultra-cold atoms bouncing on an oscillating atom mirror, cf. (4.67). The corresponding eigenphases differ by $\phi_+ - \phi_- = \pi$ and that is why the many-body phase described in Ref. [102] was called π-phase.[10]

The Floquet states (4.111) evolve with the period of the driving T (apart from a global phase) and if we calculate time evolution of the average value of a single spin operator, we obtain trivial dependence on time,

$$\langle\psi_\pm|\,\sigma_j^z(nT)\,|\psi_\pm\rangle = \langle\psi_\pm|\left(U_F^\dagger\right)^n \sigma_j^z(0)\,U_F^n\,|\psi_\pm\rangle = \langle\psi_\pm|\,\sigma_j^z(0)\,|\psi_\pm\rangle = 0. \tag{4.113}$$

However, when we calculate the two-time correlation function which reduces to

$$\langle\psi_\pm|\,\sigma_j^z(nT)\,\sigma_i^z(0)\,|\psi_\pm\rangle = m_j m_i (-1)^n, \tag{4.114}$$

we see that it evolves with the period $2T$. Thus, spontaneous breaking of the discrete time translation symmetry of the system Hamiltonian is expected. Indeed,

[10]In order to obtain quasi-energies (or Floquet eigenphases) one usually diagonalizes a Floquet Hamiltonian (or Floquet evolution operator) in a time-periodic basis, e.g. in the Fourier basis $e^{ik\omega t}$ where $\omega = 2\pi/T$ and k is integer. However, if one uses the basis $e^{ik\omega t/2}$ also with integer k, then the quasi-energy spectrum will consist of the two identical Floquet spectra shifted with respect to each other by $\omega/2$. The eigenphases of the Floquet evolution operator will be also duplicated and in the time crystal case considered here, the pairs of the Schrödinger cat-like states will correspond to the degenerate eigenphases $\phi_+ - \phi_- = 0$. It is the case in Sect. 4.2.2, cf. (4.68), where a $(2T)$-periodic basis is used because such a basis is very convenient for the description of the many-body system. In spins systems one can use the so-called toggling frame of reference where the eigenphases of the pairs of the Floquet states (4.111) are not shfited by π but they are degenerate, see Sect. 4.6.3. We recommend also a reader the background information on *Many-body Floquet Hamiltonian* in Sect. 4.1.1.

the Floquet states are Schrödinger cat-like states and it is sufficient to measure the z-component of one spin only and the system collapses to one of the states that form the superposition, $|\psi_\pm\rangle \to |m_1,\ldots\rangle_z$. After the measurement, the time evolution of a single spin operator reveals that the discrete time translation symmetry is broken because it evolves with the period twice longer than the driving period,

$$_z\langle m_1,\ldots|\,\sigma_j^z(nT)\,|m_1,\ldots\rangle_z = m_j(-1)^n. \tag{4.115}$$

In the spin system considered here, the Floquet states (4.111) reveal long-range correlations, i.e.

$$\langle\psi_\pm|\,\sigma_j^z\,\sigma_i^z\,|\psi_\pm\rangle - \langle\psi_\pm|\,\sigma_j^z\,|\psi_\pm\rangle\langle\psi_\pm|\,\sigma_i^z\,|\psi_\pm\rangle = m_j m_i \neq 0, \quad \text{for } |j-i|\to\infty, \tag{4.116}$$

and such Schrödinger cat-like states cannot be prepared in the laboratory because for a long spin chain they quickly decohere. Short-range correlated symmetry broken states $|m_1,\ldots\rangle_z$ are resistant to decoherence what enables experimental observation of the discrete time crystal evolution.

If we consider the case of the decoupled spins, i.e. $J_j = 0$, then for $\epsilon = h_j^x = h_j^z = 0$, the states (4.111) are still Floquet eigenstates of the system and the symmetry broken states $|m_1,\ldots\rangle_z$ still evolve with the period twice longer than the driving period T. One can thus question the role of the spin–spin coupling and the need for self-organization of the many-body system in the formation of the discrete time crystals? However, the independent spins system reveals the $(2T)$-periodic evolution of the symmetry broken states $|m_1,\ldots\rangle_z$ in the ideal case of $\epsilon = 0$ only. If we change ϵ slightly from zero value, then the z-components of the spins no longer evolve with the period $2T$ but instead reveal beating, i.e.

$$_z\langle m_1,\ldots|\,\sigma_j^z(nT)\,|m_1,\ldots\rangle_z = m_j(-1)^n\cos(n\epsilon\pi), \tag{4.117}$$

similarly like in the case of non-interacting ultra-cold atoms bouncing resonantly on an oscillating mirror, cf. Fig. 4.6. The Fourier transform of (4.117) consists of two peaks at $\pi(1\pm\epsilon)/T$. We will see in a moment that when the spin–spin coupling is turned on, even in the presence of a small detuning $\epsilon \neq 0$, the spins are able to self-organize their motion and evolve with the frequency π/T. In the case of ultra-cold atoms bouncing on an oscillating mirror, differences between the interacting and non-interacting cases are visible even without introducing a detuning from the perfect resonant driving because the tunneling process destroys $(2T)$-periodic time evolution of non-interacting atoms. Here, one has to introduce the detuning in order to demonstrate differences between the coupled spins and independent spins cases and such detuning was used in the experiments that we describe in Sect. 4.6.

The essence of discrete time crystals is their stability with respect to perturbations respecting the time-translation symmetry. Time crystals in the periodically driven spin systems were introduced in the many-body localization regime because such a regime should guarantee resistance to heating and to local perturbations [57, 102]. The stability of the time crystal described by the Hamiltonian (4.105)

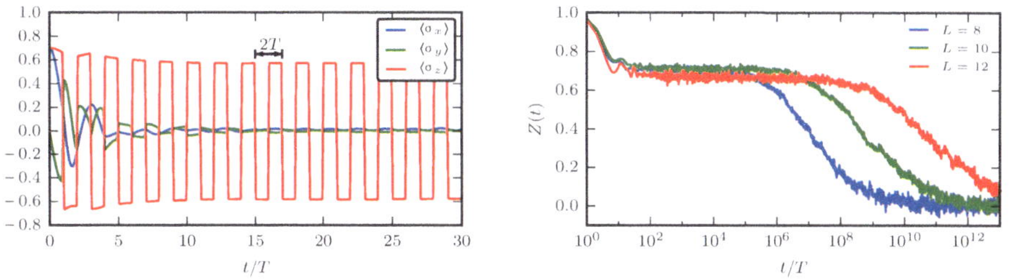

Fig. 4.11 Left panel: time evolution of $\langle\sigma_j^{x,y,z}\rangle$ averaged over 146 realizations of the randomly and uniformly chosen parameters $J_j \in [1/2, 3/2]$, $h_j^z \in [0, 1]$, $h_j^x \in [0, 0.3]$ and over the spins $j \in [50, 150]$ for the length of the spin chain $L = 200$. The detuning $\epsilon = 0$ and the initial short-range correlated tensor product state is chosen where each spin is in the superposition state $\cos(\pi/8)|1\rangle +$ $\sin(\pi/8)|-1\rangle$. After an initial transient evolution, the average value of $\langle\sigma_j^z\rangle$ oscillates with the period twice longer than the driving period T demonstrating the discrete time crystal behavior. Right panel: the magnetization along the z-axis of a single spin, $Z(t = nT) = (-1)^n\langle\sigma_j(nT)\rangle$, averaged over 500 realizations of the random parameters for different lengths of the spin chain $L = 8, 10, 12$. The lifetime of the time crystal increases exponentially with the length of the chain. The spin flips were chosen to be instantaneous, i.e. $t_1 \to 0$ in (4.106). Reprinted figure with permission from Else et al. [57]. Copyright (2016) by the American Physical Society

was demonstrated numerically [57]. The parameters J_j and h_j^z of the Hamiltonian were chosen as random numbers from the uniform distributions, $J_j \in [1/2, 3/2]$ and $h_j^z \in [0, 1]$, so that the many-body localization regime is ensured.[11] Additionally a perturbation was turned on. The perturbation was the disorder transverse field, i.e. h_j^x were chosen randomly from the uniform distribution $h_j^x \in [0, h]$ with $h \lesssim 1$. It turned out that in the presence of such a perturbation, the time crystal dynamics was still observed and its lifetime increased exponentially with the increase of the length of the spin chain L, see Fig. 4.11.

Analysis of the stability of the spin time crystal with respect to the detuning ϵ from the perfect spin flips, cf. (4.106), allowed Yao et al to obtain the phase diagram [204]. In the absence of the spin–spin coupling and for small non-zero ϵ, the two-point correlation function,

$$R(nT) = {}_z\langle m_1, m_2, \ldots |\, \sigma_j^z(nT)\, \sigma_j^z(0)\, |m_1, m_2, \ldots\rangle_z, \tag{4.118}$$

reveals beating, cf. (4.117), and consequently its Fourier transform possesses two symmetric peaks around $\omega/2 = \pi/T$ separated by $2\pi\epsilon/T$. In the presence of the sufficiently strong coupling, despite the presence of the detuning, the spins self-organize their motion and perform $(2T)$-periodic evolution which is reflected by the single peak at $\omega/2$ in the Fourier transform of $R(nT)$. However, when we increase

[11] In order to check if the chosen parameters correspond to the many-body localization regime, one can analyze level statistics (here statistics of eigenphases of the Floquet evolution operator U_F). Entering the localization regime, the distribution of nearest level spacings switches from the circular orthogonal ensemble prediction to the Poisson distribution [102].

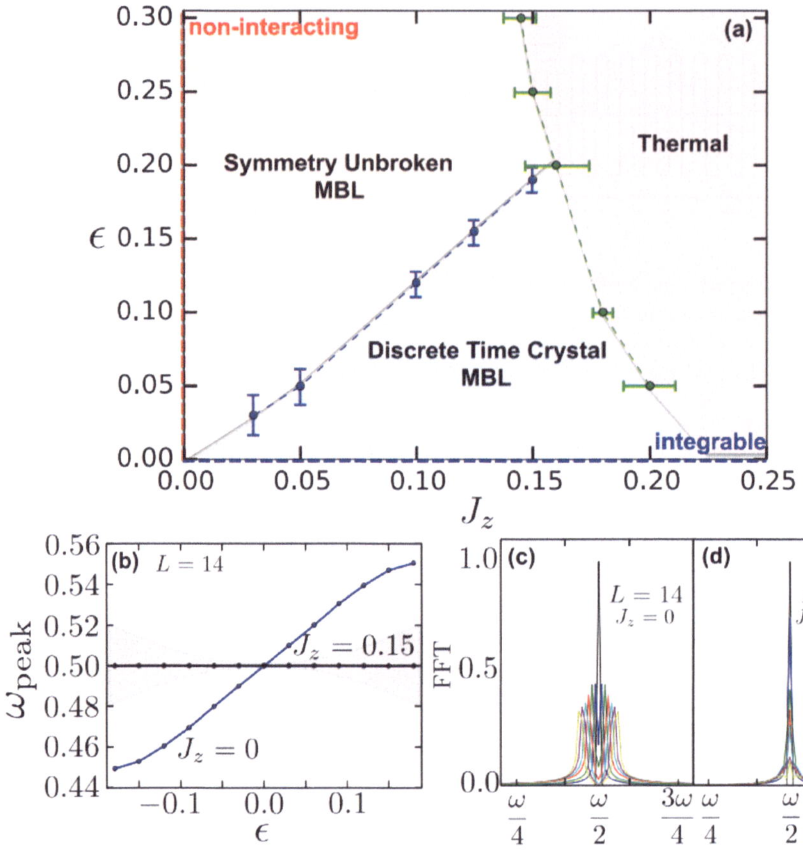

Fig. 4.12 Panel (**a**): phase diagram of the periodically driven spin system (4.105) (for $J_j \in [0.8J_z, 1.2J_z]$, $h_j^x = 0$, $h_j^z \in [0, 2\pi]$ and $t_1 = T/2 = 1$) as a function of the average spin–spin coupling J_z and the detuning ϵ from the perfect spin flips. Panel (**b**): location of one of the two subharmonic peaks of the Fourier transform of (4.118) as a function of ϵ for the non-interacting system ($J_z = 0$) and location of the peak in the presence of the spin–spin coupling ($J_z = 0.15$)—the shaded region indicates the FWHM of the peak at $\omega/2$. Panels (**c**) and (**d**) show examples of the Fourier spectrum of (4.118) for ϵ corresponding to the results presented in (**b**). The length of the spin chain $L = 14$. The results in (**a**) and (**b**) are obtained by averaging over 100 realizations of the disorder. Reprinted figure with permission from Yao et al. [204]. Copyright (2017) by the American Physical Society

ϵ, while keeping the strength of the coupling fixed, at some point the detuning becomes too large for the system to establish the synchronization and the beating in time evolution of $R(nT)$ appears and the time crystal phenomenon is lost. At the border between the time crystal regime and the symmetry-unbroken regime, the system has difficulty to decide on which side it is [60]. Then, one observes the strongest fluctuations of the height of the Fourier peak at $\omega/2$ in different realizations of the disorder parameters of the Hamiltonian (4.107). The maximal variance of the peak height corresponds to the border between the two regimes. In Fig. 4.12 the phase diagram is presented together with examples of the Fourier spectra related to the two regimes [204]. The results are obtained for an initial short-range correlated state $|m_1, m_2, \ldots\rangle_z$ with random $m_j = \pm 1$. The transverse field is not present ($h_j^x = 0$) and the disorder longitudinal field is chosen uniformly from the interval $h_j^z \in [0, 2\pi]$ what, for $t_1 = T/2 = 1$, corresponds to the maximal disorder for the Floquet unitary operator. The spin–spin couplings are also random $J_j \in [0.8J_z, 1.2J_z]$ where J_z is the average coupling strength. Apart from the time-crystal and symmetry unbroken regimes, Yao et al. identified a thermal phase which corresponded to so strong spin–spin coupling that the disorder related to the transverse field was too weak for the system to be in the many-body localization phase. The latter appears when the level statistics of the Floquet evolution operator changes from the Poisson distribution to the circular orthogonal ensemble [204].

4.3.2 Clean Discrete Time Crystals and Prethermal Character of Discrete Time Crystals in Lattice Systems

There is a debate in the literature if it is necessary to introduce spatial disorder in a periodically driven translationally invariant system in order to realize discrete time crystals [174]. Sufficiently strong disorder in a static system leads to the many-body localization and the resulting integrability prevents a system from thermalization. In a periodically driven system with strong spatial disorder, the integrability and the consequent non-ergodicity are inherited from the many-body localization of a static Hamiltonian [57, 102]. The experiments demonstrating discrete time crystals, which we discuss in Sect. 4.6, show that lifetime of time crystals, in terms of the driving periods, is similar both in the presence or absence of the strong disorder [33, 153, 169, 170, 209]. However, in the laboratory there are many different imperfections which do not allow for a very long observation of time crystal evolution and discrimination if an isolated time crystal could live forever or a periodically driven system is actually in a prethermal state, that is, a state that does not reveal thermalization for a very long time but finally it does [1–3, 58, 112, 130, 138, 141, 193]. In the present section we discuss a few examples of discrete time crystals in lattice systems in the absence of spatial disorder and also the approaches for the description of prethermal discrete time crystal phases.

Clean Discrete Time Crystals

Huang et al. proposed a discrete time crystal in a periodically driven translationally invariant system without disorder and argued that it did not thermalize because it exhibited emergent integrability of the Floquet evolution operator [90]. They called it a *clean* discrete time crystal because no disorder in the system was present. Let us consider the following time-periodic Hamiltonian

$$H(t) = H(t+T) = \begin{cases} H_1 & \text{for} \quad 0 \le t < t_1 \\ H_2 & \text{for} \quad t_1 \le t < T, \end{cases} \tag{4.119}$$

with

$$H_1 = -\frac{\pi}{2t_1}(1+\epsilon)\sum_j \left(\hat{a}_j^\dagger \hat{b}_j + \hat{b}_j^\dagger \hat{a}_j\right), \tag{4.120}$$

$$H_2 = \frac{1}{T-t_1}\left[-J\sum_j \left(\hat{a}_{j+1}^\dagger \hat{a}_j + \hat{b}_{j+1}^\dagger \hat{b}_j + \text{h.c.}\right) + U\sum_j \left(\hat{n}_j^A \hat{n}_{j+1}^A + \hat{n}_j^B \hat{n}_{j+1}^B\right) \right.$$
$$\left. + \Delta \sum_j \left(\hat{n}_j^A - \hat{n}_j^B\right)\right], \tag{4.121}$$

where $\hat{a}_j$, $\hat{b}_j$ and $\hat{a}_j^\dagger$, $\hat{b}_j^\dagger$ are bosonic annihilation and creation operators, $\hat{n}_j^A = \hat{a}_j^\dagger \hat{a}_j$ and similarly $\hat{n}_j^B$. The corresponding Floquet evolution operator reads

$$U_F = e^{-i(T-t_1)H_2} e^{-it_1 H_1}. \tag{4.122}$$

The system consists of two chains (or legs): A and B, see Fig. 4.13. The operator $U_1 = e^{-it_1 H_1}$ in (4.122) transfers bosons from the chain A to B and vice versa and for $\epsilon = 0$ the transfers are perfect[12]—compare the spin flip operator in (4.109). During the second part of the period, i.e. for $t_1 \le t < T$, the chains evolve independently and bosons in the each chain can tunnel between sites and experience nearest neighbor interactions, see (4.121). There are also energy shifts of the chains proportional to $\pm\Delta$.

In order to illustrate time crystal evolution, the polarization $P(t)$ is defined,

$$P(t) = \frac{1}{L}\sum_{j=1}^{L} \langle \psi(t)| \left(\hat{n}_j^A - \hat{n}_j^B\right) |\psi(t)\rangle, \tag{4.123}$$

which is actually the population imbalance of the chains, L is the length of each of the chains. For $\epsilon = 0$, if one starts with a state where only one chain is occupied, the polarization evolves with the period $2T$ independently of the parameters of the Hamiltonian H_2. However, for $J = U = 0$, the $(2T)$-periodic oscillations are not stable. That is when we set small $\epsilon \neq 0$, the polarization $P(t)$ is no longer a $(2T)$-periodic function, see Fig. 4.13. Sufficiently strong interactions (large U) lead to self-organization of the many-body system which is able to evolve with the period

[12]For an arbitrary ϵ, one can show that $U_1^\dagger \hat{a}_j^\dagger U_1 = i\hat{b}_j^\dagger \cos\epsilon - \hat{a}_j^\dagger \sin\epsilon$ and $U_1^\dagger \hat{b}_j^\dagger U_1 = i\hat{a}_j^\dagger \cos\epsilon - \hat{b}_j^\dagger \sin\epsilon$.

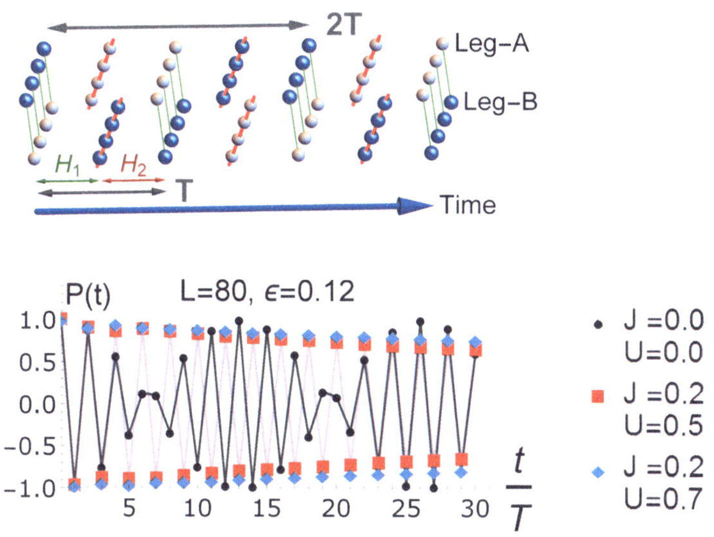

Fig. 4.13 Top panel: schematic plot of the action of the Floquet evolution operator. There are two chains (legs), A and B, which consist of $L = 4$ sites. Initially the chain A is occupied (blue dots) and the chain B is empty (grey dots) but after the action of the operator $e^{-it_1 H_1}$, the population is transferred (green lines) to the chain B. The action of the operator $e^{-i(T-t_1)H_2}$ introduces interactions between bosons in each of the chains independently (red lines) that completes a single period of the driving. For the chosen initial state, the evolution of the system is periodic with the period $2T$. Bottom panel: time evolution of the polarization (4.123) obtained with the help of the density matrix renormalization group for the initial state where the chain A is fully occupied, for $L = 80$, $\Delta = 0.1$ and different J and U as indicated in the panel. Open boundary conditions and the hard core bosons are assumed. Reprinted figure with permission from Huang et al. [90]. Copyright (2018) by the American Physical Society

twice longer than the driving period despite the fact the transfers of bosons between the chains is not perfect, i.e. $\epsilon \neq 0$. Non-zero tunneling amplitude J of bosons between sites of each of the chains improves the oscillations with the subharmonic frequency π/T but it should be kept smaller than U in order not to enhance other frequencies [90]. The parameter Δ is not essential - it breaks the integrability of the system for the special case when $J = 0$. The Hilbert space is restricted to $n_j^A + n_j^B = 1$ [90].

Figure 4.13 illustrates the geometry of the system and evolution of the polarization $P(t)$ in the non-interacting and interacting regimes for the initial short-range correlated state where one of the chains is fully occupied $|\psi(0)\rangle = \prod_j \hat{a}_j^\dagger|0\rangle$. Huang et al. show that although there is no disorder in the system, the lifetime of the discrete time crystal increases exponentially with the increase of the lengths of the chains. As the reason for the lack of thermalization they point out emergent integrability of the Floquet dynamics that is supported by an analysis of the level statistics of the Floquet evolution operator. That is, for $J = 0.1$, $U = 0.2$ and $\Delta = 0.1$, going from the thermalizing regime ($\epsilon = -1.35$) deeply to the time crystal one ($\epsilon = 0.05$), the nearest neighbor level spacing distribution goes from the orthogonal ensemble to the Poisson distribution [90]. The former is typical for chaotic systems while the latter for integrable ones [75].

Another discrete time crystal with short-range interactions and no disorder was proposed by Yu et al. [206] who considered the Ising model described by the Hamiltonian

$$H_0 = -J\sum_{j=1}^{L}\sigma_j^z\sigma_{j+1}^z - h\sum_{j=1}^{L}\sigma_j^x, \tag{4.124}$$

where $J > 0$, $\sigma_j^{x,z}$ are the Pauli matrices and L is the length of the chain. Every period T, the rotation of the spins about the x axis is performed, i.e. the Floquet evolution operator of the system reads

$$U_F = e^{i\phi\sum_j \sigma_j^x} e^{-iH_0 T}, \tag{4.125}$$

where $\phi = \pi(1/2 - \epsilon)$ and the operator $e^{i\phi\sum_j\sigma_j^x}$ is responsible for the rotation of the spins.

The static Ising Hamiltonian H_0 exhibits the quantum phase transition when the length L of the spin chain goes to infinity. For $h > J$, the ground state of the system describes a quantum paramagnetic phase with the spins directed along the x axis. For $h < J$, the interactions between neighboring spins favor magnetization of the system parallel or antiparallel to the z axis—in this case the cyclic $\mathbb{Z}_2$ symmetry is spontaneously broken.[13]

For $h = 0$, the ground state level of the static Hamiltonian H_0 is doubly degenerate and the corresponding degenerate Hilbert subspace is spanned by the long-range correlated states $|\psi_\pm\rangle = (|+1, +1, \ldots\rangle_z \pm |-1, -1, \ldots\rangle_z)/\sqrt{2}$. The states $|\psi_\pm\rangle$ are the $\mathbb{Z}_2$ symmetric but for a large L they quickly decohere and in the experiment one ends up with short-range correlated $\mathbb{Z}_2$ symmetry-broken states $|\pm 1, \pm 1, \ldots\rangle_z$. In the periodically driven case when $h = 0$ and additionally we set $J = 0$, the time evolution of the z component of the magnetization is easy to obtain

$$M_z(nT) = \frac{1}{L}\langle\psi(nT)|\sum_{j=1}^{L}\sigma_j^z|\psi(nT)\rangle = (-1)^n\left[2\cos^2(\epsilon\pi n) - 1\right], \tag{4.126}$$

where as the initial state $|\psi(0)\rangle = |+1, +1, \ldots\rangle_z$ has been chosen. Thus, for $H_0 = 0$, the time evolution of the magnetization is $(2T)$-periodic for $\epsilon = 0$ only

[13] The $\mathbb{Z}_2$ symmetry of the Ising Hamiltonian (4.124) is related to the fact that H_0 commutes with the spin flip operator $e^{i\pi/2\sum_j\sigma_j^x}$ which is a $\mathbb{Z}_2$ symmetry operator because $\left(e^{i\pi/2\sum_j\sigma_j^x}\right)^2 = 1$. Thus, all eigenstates of H_0 can be chosen to be also eigenstates of $e^{i\pi/2\sum_j\sigma_j^x}$.

otherwise there is beating and the Fourier spectrum of $M_z(nT)$ consists of two peaks at frequencies $\pi/T \pm 2\epsilon\pi/T$.

The situation changes if there are couplings between the spins. For $J \neq 0$ and $h = 0$, time evolution of the magnetization can be analyzed perturbatively and the Lth-order terms result in merging the main two peaks located symmetrically around π/T into a single one in the thermodynamic limit. That is, for ϵ smaller than some critical value ϵ^*, the frequency splitting of the peaks goes to zero with $L \to \infty$ like $(\epsilon/\epsilon^*)^{\alpha L}$ where the positive constant α is of the order of unity [206]. Thus, the $(2T)$-periodic time evolution of the symmetry broken states are stable in the thermodynamic limit despite the fact there is no disorder in the system. When both J and h are non-zero, the range of the parameters values where the time crystal behavior is observed can be determined numerically [206].

Discrete time crystals with no disorder can be also realized in spin systems with long-range interactions. The extremal version of the long-range interactions is the Lipkin–Meshkov–Glick model where all spins interact with all with the same strength [165],

$$H_{\rm LMG} = -\frac{J}{2N}\sum_{i=1}^{N}\sum_{j=1}^{N}\sigma_i^z\sigma_j^z - h\sum_{j=1}^{N}\sigma_j^x. \tag{4.127}$$

The Hamiltonian can be rewritten in terms of the x and z components of the total spin operator,

$$H_{\rm LMG} = -\frac{2J}{N}\hat{S}_z^2 - 2h\hat{S}_x, \tag{4.128}$$

where $\hat{S}_{x,y,z} = (1/2)\sum_{j=1}^{N}\sigma_j^{x,y,z}$. The total spin operator commutes with the Hamiltonian $H_{\rm LMG}$ and the spin sector $S = N/2$ contains the ground state of the system. It is easy to analyze the system within the mean field approach which is valid for large N and where the total spin operator is substituted by the classical angular momentum, see the end of Sect. 4.2.2. For $h < J$ the mean field approach predicts the degenerate ground state energy with the solutions $S_x = Nh/(2J)$ and $S_z = \pm(N/2)\sqrt{1-h^2/J^2}$ which break the $\mathbb{Z}_2$ symmetry of the system. That is, $H_{\rm LMG}$ should have common eigenvectors with the spin flip operator $e^{i\pi\hat{S}_x}$ while in the mean field approach we get $S_z > 0$ or $S_z < 0$. For large N, all quantum energy levels below $-hN$ form doublets and the corresponding symmetry-preserving many-body eigenstates are not resistant to decoherence and spontaneous breaking of the $\mathbb{Z}_2$ symmetry takes place, see Fig. 4.3 and the discussion at the end of Sect. 4.2.2.

It is known that the Lipkin–Meshkov–Glick model can be mapped to the system of N bosons in the two-mode approximation, cf. (4.73), (4.74) and (4.61), which for attractive contact interactions reveal a self-trapping phenomenon—a bosonic version of the spontaneous breaking of the $\mathbb{Z}_2$ symmetry considered here. The Lipkin–Meshkov–Glick model describes the discrete time crystal for ultra-cold atoms bouncing resonantly on an oscillating atom mirror, see Sect. 4.2.2. However, we should remember that there, the Lipkin–Meshkov–Glick Hamiltonian is the description of the system in the time-periodic basis and the $\mathbb{Z}_2$ symmetry breaking is related to spontaneous breaking of the discrete time translation symmetry of the system. Here, so far there is no driving and we will observe a discrete time crystal behavior only when we turn on a periodic drive.

Russomanno et al. considered the Lipkin–Meshkov–Glick model and turned on T-periodic driving of the system that corresponded to the spin rotation around x axis by an angle ϕ [171]. That is, the Floquet evolution operator of the driven system was the following

$$U_F = e^{i\phi\hat{S}_x} e^{-iTH_{\text{LMG}}}. \tag{4.129}$$

In the regime where $h < J$, starting with a $\mathbb{Z}_2$ symmetry broken state, discrete time crystal phenomenon can be observed where the z component of the system magnetization evolves with the period twice longer than the driving period and such a time evolution is stable if ϕ is close to π ($\phi = \pi$ corresponds to the perfect spin flip). That is, for a small $|\phi - \pi|$, the splitting of the main Fourier peaks located around π/T goes to zero with an increasing N exponentially quickly. The classical phase space portrait presented in Ref. [171] suggests that even if the angle ϕ is too far from π for the stable $(2T)$-periodic oscillations of the magnetization for the chosen initial state, there should exist another manifold of the Floquet eigenstates that can reveal time crystal properties.

In Ref. [192] Surace et al. consider discrete time crystal phenomena in the so-called clock models where sites of a one-dimensional chain are related to n-dimensional Hilbert spaces—the case of $n = 2$ reduces to a chain of periodically driven spins-1/2. Short-range and infinite-range clock models are analyzed and it is shown that period-n-tupling discrete time crystals can be realized similarly like in the case of higher primary resonances of ultra-cold atoms bouncing on an oscillating atom mirror, see Sect. 4.2.4. Moreover, the infinite-range clock models can reveal transition between the time-crystal phase with period-n-tupling to the $n/2$-tupling phase. It turns out that n-tupling discrete time crystals can be also realized in the Lipkin–Meshokov–Glick model of spins-1/2 when the collective nature of the spins is explored [157], see Sect 4.4.2.

Prethermal Character of Discrete Time Crystals in Lattice Systems

A generic time-independent many-body system obeys the eigenstate thermalization hypothesis which says that a system prepared in an eigenstate corresponding to eigenenergy E looks locally like in the thermal state described by the Gibbs state $\rho = e^{-\beta H}/Z$, where H is the system Hamiltonian and $Z = \text{Tr}[e^{-\beta H}]$ is the partition function corresponding to the canonical ensemble [38, 45, 166, 187]. The inverse temperature β is determined by eigenenergy, i.e. $E = \text{Tr}[He^{-\beta H}]/Z$. In other words, a small subsystem experiences the rest of a system as a heat bath and thermalizes, i.e. time evolution of a system erases memory about an initial state if we investigate local properties of a system.

Switching to periodically driven isolated many-body systems we get also a kind of the eigenstate thermalization hypothesis which says that any Floquet eigenstate of a generic time-periodic many-body system (i.e., excluded integrable systems or systems whose heating can be protected by many-body localization phenomena which in essence are also integrable) looks locally like the infinite temperature state [37, 115, 159]. In other words, a periodically driven system gains energy from the drive and heats to the infinite temperature. However, it turns out that a lattice system which is periodically driven with a frequency ω much larger than the frequency $J_{\rm local}$ associated with local interaction energy in a system can reveal a prethermalization phenomenon [1–3, 112, 141]. It is due to the fact that there are two different time scales related to slow internal dynamics of a system and fast external driving. Consequently, we first observe that a system reaches quickly a prethermal steady state and next, after a long thermalization time τ^*, a system is eventually heated to the infinite temperature. The lifetime τ^* of a prethermal state increases exponentially with $\omega/J_{\rm local}$, i.e. $\tau^* \propto e^{\omega/J_{\rm local}}$, because a system needs $\omega/J_{\rm local}$ local rearrangements in order to absorb a single quantum portion of energy from time-periodic driving. In the present section we introduce a formalism which allows one to analyze prethermal nature of discrete time crystals in periodically driven lattice systems.

Description of a periodically driven system during the prethermal regime leads to an approximation of the Floquet unitary evolution operator [1–3, 112, 141]

$$U_F = \mathcal{T} e^{-i\int_0^T dt H(t)} \approx e^{-iD^*T}, \tag{4.130}$$

where $\mathcal{T}$ is the time-ordering operator and $H(t+T) = H(t)$ is the system Hamiltonian with $T = 2\pi/\omega$. The operator D^* is called a prethermal effective Hamiltonian which approximately generates evolution of a periodically driven system during the prethermal phase, i.e. up to the thermalization time τ^*. In the case of discrete time crystals we are not only interested in how long does it take to prethermalize and eventually thermalize a system but most importantly if during the prethermal phase, periodic evolution that breaks the discrete time translation symmetry of the Hamiltonian does not decay and if so under what conditions.

Analysis of prethermal properties of discrete time crystals requires a specific approach which can be performed for any d-dimensional system [58, 130, 138, 193]. Let us assume a spin system described by the Hamiltonian $H(t)$ where the spin–spin interaction vanishes with the relative distance l between lattice sites like $1/l^\alpha$ where $\alpha > d$. In the case of discrete time crystals in spin-1/2 systems, there is a part of the Floquet evolution operator that is responsible for periodic flips of spins which we denote by X and which fulfills $X^2 = 1$.[14] Thus, the Floquet evolution operator for such systems can be written in the following general form

$$U_F = X\,\mathcal{T}e^{-i\int_0^T dt[D+E+V(t)]}, \tag{4.131}$$

where D and E are time-independent operators but the first commutes with the spin flip operator, i.e. $[D, X] = 0$, while the other does not. $V(t)$ is an additional time-periodic contribution which in general can be present. The idea is to perform a series of time-independent locality-preserving unitary operations (see [3, 58, 130] for details), which we denote by U, that transforms the Floquet evolution operator to a similar form as (4.131),

$$UU_FU^\dagger = X\,\mathcal{T}e^{-i\int_0^T dt[D^*+E^*+V^*(t)]}, \tag{4.132}$$

but the magnitude of E^* and $V^*(t)$ is reduced while the magnitude of D^*, which commutes with X, increases [130]. If, after such unitary transformations, the magnitudes of E^* and $V^*(t)$ are small, we may neglect them and approximate the Floquet evolution operator by

$$U_F \approx U_F^{\text{app}} = U^\dagger X e^{-iTD^*} U. \tag{4.133}$$

For high frequency driving, U is close to unity, i.e. $U \approx 1 + O(\omega^{-1})$, and E^* and $V^*(t)$ are exponentially suppressed with an exponent proportional to ω/J_{local} where J_{local} is an upper bound for the magnitudes of D, E and $V(t)$, see [130] for the precise definition of J_{local}. The exponential suppression implies that the energy density $\langle D^*\rangle/\Lambda$, where Λ is the volume of a system, remains exponentially well conserved during the time evolution,

$$\frac{1}{\Lambda}\left|\left\langle(\tilde{U}_F^\dagger)^n D^* \tilde{U}_F^n\right\rangle - \langle D^*\rangle\right| = O\left(nTe^{-\omega/J_{\text{local}}}\right), \tag{4.134}$$

where $\tilde{U}_F = UU_FU^\dagger$. That is, only after $nT \approx \tau^* \propto e^{\omega/J_{\text{local}}}$ the energy density based on the static effective Hamiltonian D^* signals thermalization effects. The relation (4.134) is valid if the exponent α of the algebraic decay of the long-range spin–spin interactions is greater than the dimension d of a system. Similar exponential bound can be derived for the differences between local properties of a system obtained with the help of the exact U_F and the approximate Floquet evolution operator U_F^{app} provided a stronger condition is fulfilled, i.e. $\alpha > 2d$. For $d < \alpha < 2d$ the dynamics of local observables is well approximated by the unitary Floguet operator U_F^{app} up to time $t \approx \tau^* \propto e^{\omega/J_{\text{local}}}$ but only if we deal with a finite system. In the case of the thermodynamic limit, it has been possible to prove

[14]The consideration can be carried out for a general cyclic rotation operator X, where $X^n = 1$, [58, 130, 138, 193] but for simplicity we restrict ourselves to the case of $n = 2$.

only that the evolution of local observables is governed by U_F^{app} but for a short time period that scales linearly with ω—see also the discussion in Ref. [130].

Equation (4.133) implies that according to U_F^{app}, in the frame transformed by the time-independent unitary operator $U \approx 1 + O(\omega^{-1})$, a system gets rotated by the spin flip operator X every period of the drive. However, when we consider time evolution every second period of the drive,

$$U_F^2 \approx \left(U_F^{\text{app}}\right)^2 = U^\dagger\, e^{-i2TD^*}\, U, \tag{4.135}$$

we can see that it is approximately generated only by the static effective Hamiltonian D^* and properties of this effective Hamiltonian determine properties of the dynamics of a system in the prethermal phase. Thus, the analysis of a system within the time period shorter than τ^* reduces to the analysis of D^*. We know that $[D^*, X] = 0$, thus, the effective Hamiltonian D^* possesses the $\mathbb{Z}_2$ symmetry related to the spin flip transformation X. It is the emergent symmetry because the original Hamiltonian of a system does not necessarily commute with X, i.e. $[H(t), X] \neq 0$. It is said that this emergent symmetry is protected by the discrete time translation symmetry of the original system Hamiltonian [58, 130, 138, 193]. The Hamiltonian D^* preserves the $\mathbb{Z}_2$ symmetry, however, it may happen that this symmetry is spontaneously broken, then, a discrete time crystal can form.

Let us try to sum up what we have obtained so far. We know that for $t < \tau^* \propto e^{\omega/J_{\text{local}}}$, properties of a system driven with a high-frequency are determined by a static effective Hamiltonian D^* that respects the $\mathbb{Z}_2$ symmetry corresponding to the spin flip operator X. If D^* is a generic time-independent Hamiltonian, it obeys the eigenstate thermalization hypothesis which says that starting with an initial state corresponding to energy E, after a certain time period τ_{pre} (provided $\tau_{\text{pre}} \ll \tau^*$), a system thermalizes. Then, its local properties can be described by the Gibbs state $\rho = e^{-\beta D^*}/Z$, where the inverse temperature β is determined by E, i.e. $E = \text{Tr}[D^* e^{-\beta D^*}]/Z$ provided long-range spin–spin interactions decay algebraically with the exponent $\alpha > 2d$ (for $d < \alpha < 2d$ it is at least true for a finite system [130]). In general, time evolution of the Gibbs state after every period of the drive is trivial because at the equilibrium, the $\mathbb{Z}_2$ symmetry is respected [130],

$$\rho(t+T) = Xe^{-iTD^*}\rho(t)e^{iTD^*}X^\dagger = X\rho(t)X^\dagger = \rho(t). \tag{4.136}$$

However, while D^* preserves the $\mathbb{Z}_2$ symmetry, it can be a kind of the Hamiltonian for which such a symmetry is spontaneously broken in the thermodynamic limit. Then, prethermalization of a system that happens after a time period τ_{pre} leads to the Gibbs state that breaks spontaneously the $\mathbb{Z}_2$ symmetry,, i.e. $X\rho X^\dagger = \rho' \neq \rho$. Then, after every second period of the drive, a system returns to its initial equilibrium state,

$$\rho(t+2T) = \left(Xe^{-iTD^*}\right)^2 \rho(t) \left(e^{iTD^*}X^\dagger\right)^2 = (X)^2\, \rho(t) \left(X^\dagger\right)^2 = \rho(t), \tag{4.137}$$

but after every single period it switches between the two symmetry-breaking sectors,

$$\rho(t+T) = Xe^{-iTD^*}\rho(t)e^{iTD^*}X^\dagger = X\rho(t)X^\dagger = \rho'(t) \neq \rho(t). \tag{4.138}$$

Such evolution lasts up to the thermalization time τ^*. Thus, if spontaneous breaking of the $\mathbb{Z}_2$ symmetry of the effective Hamiltonian D^* is possible and $\tau_{\rm pre} \ll \tau^*$, there is a long prethermalization period, $\tau_{\rm pre} < t < \tau^*$, when we can observe discrete time crystal evolution.

Hence, we have reduced the question if a discrete time crystal can be observed in the prethermal phase to the question if spontaneous breaking of the $\mathbb{Z}_2$ symmetry of the effective Hamiltonian D^* can happen and under which conditions. Let us discuss first one-dimensional systems with a time-periodic Hamiltonian $H(t)$ and with short-range interactions only. The corresponding effective Hamiltonian D^* possesses also short-range interactions only because the unitary transformation U in (4.132) is locality-preserving. If D^* obeys the eigenstate thermalization hypothesis, the spontaneous breaking of the $\mathbb{Z}_2$ symmetry in the thermodynamic limit can be observed for the zero-temperature initial state only, i.e. for $E = \mathrm{Tr}[D^* e^{-\beta D^*}]/Z$ where $1/\beta \to 0$ (which is the ground state of D^*) because there are only short-range interactions in D^* [130]. In such a case time crystallization, i.e. formation of a discrete time crystal, takes place at zero temperature only. It implies that the clean discrete time crystal considered by Yu et al. [206], see Eq. (4.124), is an example of the zero-temperature time crystallization. However, if there are long-range interactions that in one-dimensional space vanish with the relative distance l between lattice sites like $1/l^\alpha$ where $1 < \alpha < 2$, spontaneous breaking of the $\mathbb{Z}_2$ symmetry can be observed also for non-zero temperature $1/\beta$ below a certain critical temperature needed for the symmetry-breaking to occur.

In Ref. [130] Machado et al. present examples that illustrate behavior of discrete time crystals in prethermal evolution of a finite periodically driven one-dimensional systems with short- and long-range interactions. In the long-range interaction case, they consider a system that is described by the following Floquet evolution operator

$$U_F = e^{-i(\pi/2)\sum_j \sigma_j^x} e^{-iTH}, \tag{4.139}$$

with

$$H = J\sum_{i<j}^{L} \frac{\sigma_i^z\sigma_j^z}{|i-j|^\alpha} + \sum_{j=1}^{L}\left(h_x\sigma_j^x + h_y\sigma_j^y + h_z\sigma_j^z\right) + J_x\sum_{j=1}^{L}\sigma_j^x\sigma_{j+1}^x, \tag{4.140}$$

where L is the length of the one-dimensional spin chain. They focus on the case where $\alpha = 1.13$. This Floquet operator can be written in the form

$$U_F = Xe^{-iT(D+E)}, \tag{4.141}$$

where $X = e^{-i(\pi/2)\sum_j \sigma_j^x} = \prod_j i\sigma_j^x$ is the spin flip operator, the operator

$$D = J\sum_{i<j}^{L}\frac{\sigma_i^z\sigma_j^z}{|i-j|^\alpha} + h_x\sum_{j=1}^{L}\sigma_j^x + J_x\sum_{j=1}^{L}\sigma_j^x\sigma_{j+1}^x \tag{4.142}$$

commutes with X while $E = \sum_j \left(h_y\sigma_j^y + h_z\sigma_j^z\right)$ does not commute with the spin flip transformation X. For $\alpha > 1$ and $\omega = 2\pi/T$ much greater than J, $h_{x,y,z}$ and J_x, the dynamics can be approximated by

$$U_F \approx U^\dagger X e^{-iTD^*}U, \tag{4.143}$$

where the unitary U operator is close to unity and the effective static Hamiltonian D^* commutes with X. The approximation (4.143) is valid for $t < \tau^*$ where τ^* increases exponentially with ω. The effective static Hamiltonian D^* can be obtained by means of the iterative procedure described in Refs. [3, 58, 130] and in the limit of $\omega \to \infty$ one obtains $D^* \approx D$.

In Ref. [130] in order to analyze dynamics of the system, the expectation value of the operator D and the half-chain entanglement entropy,

$$S_{L/2} = -\mathrm{Tr}\left[\rho_{L/2}\log\rho_{L/2}\right], \tag{4.144}$$

are analyzed versus time, see Fig. 4.14 which corresponds to the finite system of $L = 22$. In (4.144), $\rho_{L/2} = \mathrm{Tr}_{1<j\le L/2}|\psi\rangle\langle\psi|$ which is the partial trace of the density matrix of the system prepared in a pure state $|\psi\rangle$ over a half of the spin chain. The expectation value $\langle D\rangle$ reduces to the average value of the effective prethermal Hamiltonian D^* if $\omega \to \infty$. Therefore, $\langle D\rangle$ is plotted instead of $\langle D^*\rangle$ because the latter is much more difficult to get in numerical simulations. In Fig. 4.14 there are results for *cold* and *hot* initial states of the static Hamiltonian D. The *cold* state corresponds to the ferromagnetic phase while in the *hot* state case the z component of the spin magnetization is close to zero.[15] In Fig. 4.14 there are also results for the case where the Hamiltonian (4.140) is exchanged by its short-range interaction version.

As visible in the top panels of Fig. 4.14, the average energy $\langle D(t)\rangle$ is conserved until the short-range or long-range interacting systems thermalize to infinite temperature at $t = \tau^*$ where τ^* increases exponentially with ω. The initial states $|\psi\rangle$ are pure states and consequently the corresponding von Neumann entropy is zero, $S = -\mathrm{Tr}[|\psi\rangle\langle\psi|\log|\psi\rangle\langle\psi|] = 0$. However, when we focus on a subsystem it shows thermalization to infinite temperature at $t > \tau^*$. Indeed, properties of a subsystem can be described by the Gibbs state corresponding to infinite temperature

[15] It is worth noting that due to the fact that the authors of Ref. [130] chose $J > 0$ in (4.140), the ferromagnetic phase is related to the top of the spectrum of $D^* \approx D$. Consequently starting initially from the ferromagnetic phase, prethermalization and final thermalization imply that the average value of the effective Hamiltonian D^* decreases, not increases, with time, cf. top panels of Fig. 4.14.

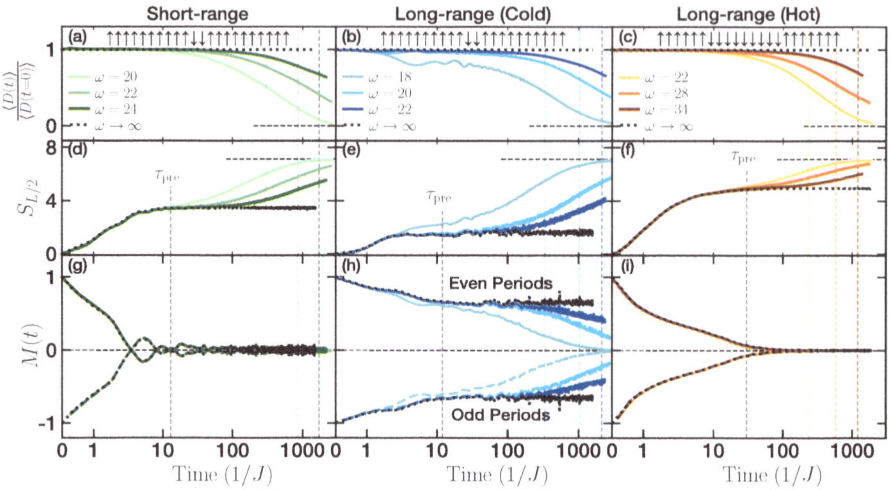

Fig. 4.14 Numerical simulations of dynamics of periodically driven $L = 22$ chain of spins-1/2 with short-range interactions (left column) and long-range interactions (middle and right columns). The initial state is a *cold* state of the static Hamiltonian (4.142) (or its short-range interaction version) in the left and middle columns and a *hot* state in the right column as indicated schematically in the top of the panels. Evolution of the energy $\langle D(t)\rangle$ (top row) and the half-chain entanglement entropy $S_{L/2}$ (middle row) indicate that prethermalization and final thermalization processes occur similarly in the all cases. While the prethermalization times $\tau_{\rm pre}$ do not depend on the frequency ω of the periodic driving, the lifetime of the prethermalization phase τ^*, and thus the time needed for the final thermalization, increases exponentially with ω. The energy and the entropy are similar in the all cases, however, discrete time crystal evolution during the prethermal regime is observed only for the long-range interactions and a *cold* initial state as indicated in the plot of $M(nT)$, Eq. (4.145), (bottom row)—even and odd periods nT are plotted separately. The parameters in (4.140) are the following: $J = \hbar = 1$, $J_x = 0.75$, $\{h_x, h_y, h_z\} = \{0.21, 0.17, 0.13\}$ and $\alpha = 1.13$ in the long-range interaction case. Reprinted from [130]

what implies that the probability of any eigenstate of D is the same and the entropy $S_{L/2}$ of the subsystem corresponding to a half of the chain (i.e., with the site index j that fulfills $1 < j \leq L/2$) is expected to be maximal and equal to $(L \log 2 - 1)/2$. The entanglement entropy $S_{L/2}$ reaches this limit for $t > \tau^*$ as visible in Fig. 4.14. However, for $t < \tau^*$, we observe the prethermalization where an initial value of $\langle D\rangle$ is conserved and the entropy $S_{L/2}$ saturates at a smaller value than the value related to the infinite temperature limit.

While the prethermalization properties (i.e., behavior of the energy density and entanglement entropy) of the systems with short- and long-range interactions are the same for any short-range correlated initial state, the discrete time crystal evolution depends on initial energy density. In Fig. 4.14 time crystal dynamics is analyzed with the help of the following correlation function

$$M(t) = \frac{1}{L} \sum_{j=1}^{L} \langle \sigma_j^z(0) \rangle \langle \sigma_j^z(t) \rangle. \tag{4.145}$$

In the short-range interaction case, only the ground state of D reveals spontaneous symmetry breaking and the resulting discrete time crystal evolution. For the long-range interactions, $M(t)$ reveals time crystal periodic evolution with the period twice longer than the driving period T for initial energy density $\langle D \rangle / L$ corresponding to temperatures $1/\beta$ below the critical value for the ferromagnetic-paramagnetic symmetry breaking phase transition—compare middle and left column of Fig. 4.14. It means that the discrete time crystallization is a finite temperature effects in the case of long-range interactions.

It is worth commenting the role of the terms proportional to h_x and J_x in the Hamiltonian D, Eq. (4.142), which well approximates the effective static Hamiltonian D^* if $\omega \to \infty$. They ensure that $D^* \approx D$ is not integrable and it obeys the eigenstate thermalization hypothesis and h_x and J_x control the time scale τ_{pre} needed for the prethermalization.

In the present section discrete time crystals in lattice systems with long-range interactions are discussed. In Sect. 4.3.2 we have described the periodically driven Lipkin–Meshokov–Glick model and the resulting spontaneous breaking of the discrete time translation symmetry [171]. The driven Lipkin–Meshkov–Glick model is also an example of a system with long-range interactions (actually it is a system with all to all infinite-range interactions) that reveals the discrete time crystal dynamics but which cannot be described by means of the prethermal approach that we present in the current section due to the infinite-range of the interactions.

4.4 Fractional Time Crystals

We have seen that isolated periodically driven quantum many-body systems can spontaneously break discrete time translation symmetry of the time-periodic Hamiltonian and start moving with a period which is an integer multiple of the driving period. One may ask if an isolated many-body system can also spontaneously start moving with a period that is a rational multiple of the period of the external drive? The answer is yes and such a fractional time crystal can be realized if a many-body system is loaded to a resonant Hilbert subspace corresponding to higher order classical resonant dynamics [134]. This kind of time crystals can be realized in any system that in the classical description reveals higher order nonlinear resonances [123]. In the following we will illustrate the phenomenon with atoms bouncing on an oscillating atom mirror in the presence of the gravitational field and also in periodically driven spin systems.

4.4.1 *Fractional Time Crystals in Ultra-Cold Atoms Bouncing on a Mirror*

Let us begin with the classical description of a single particle bouncing on an oscillating mirror that, in the frame oscillating with the mirror, is described by the Hamiltonian (4.29) where $\omega = 2\pi/T$ and λ determines the amplitude of the mirror oscillations.[16] If the period of a particle motion is an integer multiple of the period T, we deal with a principal resonance and such a periodic orbit is stable and it is surrounded by elliptical resonant islands in the classical phase space, see Fig. 4.1 for the 2 : 1 resonant case and Sect. 4.2.4 for the analysis of the 40 : 1 resonance. The description of the resonant dynamics is best performed when we switch from the Cartesian position and momentum z and p to the angle-action variables θ and I, see the Hamiltonian (4.42). Then, going to the frame moving along the classical resonant orbit and averaging over the fast time *degree* of freedom we obtain the effective Hamiltonian that constitutes the first order approximation in λ, see (4.43)–(4.46) and (4.49).

In the present section we are interested in a resonant trajectory that evolves with a period that is a rational multiple of the mirror oscillations period, e.g. $3T/2$ as shown in Fig. 4.15. If we switch, similarly as in (4.43)–(4.46), to the frame moving along such a resonant orbit and try to average the resulting Hamiltonian over the time, we obtain a null contribution from the external drive. It means that in order to get an effective Hamiltonian that describes dynamics in the vicinity of such a resonant orbit we have to refer the higher-order secular approach [123]. However, it turns out that for a particle bouncing on an oscillating mirror an effective Hamiltonian can be obtained much more easily without applying the machinery of the systematic perturbative expansion [134].

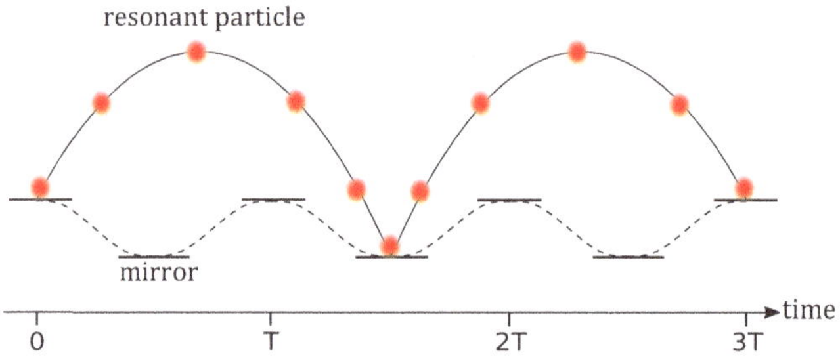

Fig. 4.15 Schematic plot of a particle bouncing resonantly on a mirror that oscillates with the period T. The period of the particle motion is equaled to $3T/2$ which means that the resonant orbit corresponds to consecutive bounces from the uppermost and lowermost positions of the mirror

[16]More precisely the amplitude of the mirror oscillations in the laboratory frame is given by λ/ω^2.

Let us focus on the 3 : 2 resonance but the results can be generalized to any $(2s + 1) : 2$ resonance with integer s. The period of the resonant bounces of a particle and the corresponding resonant value of the action are

$$T_r = \frac{3}{2}T, \qquad I_r = \frac{1}{3\pi\omega^3}\left[\frac{3\pi}{2} + \frac{2\lambda}{3\pi}\right]^3, \tag{4.146}$$

respectively, where it is taken into account that in subsequent bounces the mirror is in the uppermost and lowermost positions, see Fig. 4.15. Due to the fact that between bounces, the dynamics reduces to a very simple motion of a particle in the gravitational field, it is not difficult to calculate the angle $\theta(2T_r)$ and the action $I(2T_r)$ after two consecutive bounces if initially $\theta(0)$ and $I(0)$ correspond to small deviation with respect to the resonant orbit. It allows one to get approximate equations of motion

$$\frac{d\theta}{dt} \approx \frac{\theta(2T_r) - \theta(0)}{2T_r} = \frac{\partial H_{\text{eff}}}{\partial I}, \tag{4.147}$$

$$\frac{dI}{dt} \approx \frac{I(2T_r) - I(0)}{2T_r} = -\frac{\partial H_{\text{eff}}}{\partial \theta}, \tag{4.148}$$

which turn out to be the Hamilton equations generated by the effective Hamiltonian,

$$H_{\text{eff}} = \frac{[P - A(\theta)]^2}{2m_{\text{eff}}} + \frac{\lambda^2}{4\omega^2}\cos(3\theta), \tag{4.149}$$

where $P = I - I_r$ and

$$m_{\text{eff}} = -\frac{81\pi^2}{16\omega^4}, \qquad A(\theta) = \frac{9\pi\lambda}{4\omega^3}\sin(3\theta/2). \tag{4.150}$$

The effective Hamiltonian (4.149) is the second order approximation in λ of the exact Hamiltonian and it reproduces accurately the exact stroboscopic dynamics of a particle. The latter is obtained by integrating the exact equations of motion and collecting $\theta(t)$ and $I(t)$ at time moments $t = n2T_r$ where n's are integer numbers, see Fig. 4.16.

The phase space in the vicinity of the 3 : 2 resonant trajectory consists of three elliptical islands. Thus, from the semiclassical reasoning (see Sect. 4.2.1), in the quantum description we expect three Floquet states that are superposition of three localized wavepackets $w_j(z, t)$ propagating along the resonant trajectory. Strictly speaking, the wavepackets $w_j(z, t)$ evolve with the period $2T_r$ but they are bouncing on the mirror with the period T_r similarly as a classical particle illustrated in Fig. 4.15. Due to the fact that the wavepackets bounce once off the mirror in the uppermost position and next off the mirror in the lowermost position and so on, there is a tiny difference in the propagation of the wavepackets after each period T_r. The difference disappears when the amplitude of the mirror oscillations divided by

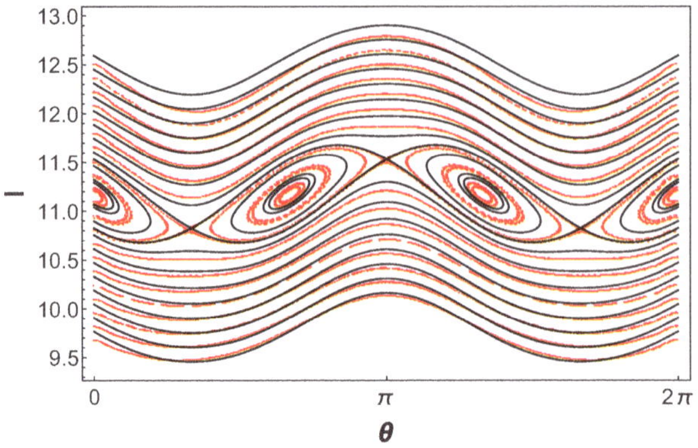

Fig. 4.16 Phase space portrait of a particle bouncing on an oscillating mirror in the vicinity of the 3 : 2 resonance. Black curves correspond to the effective Hamiltonian (4.149) and red dots to the stroboscopic picture obtained by numerical integration of the exact equations of motion and collecting values of the action $I(t)$ and the angle $\theta(t)$ at $t = n2T_r = n6T/2$ where n is integer. The frequency and amplitude of the mirror oscillations are $\omega = 1$ and $\lambda/\omega^2 = 0.05$, respectively. The figure reproduced from [134]

the length of the resonant orbit tends to zero, that is when $\lambda/9 \to 0$. However, in order not to let the width of the resonant elliptical islands shrink to zero, we have to compensate $\lambda \to 0$ by decreasing the frequency ω of the mirror oscillations so that, speaking in the semiclassical language, the elliptical islands are always large enough to capture quantum states. The number of quantum states trapped in the resonant islands can be estimated as [134]

$$n_{\text{trapped}} \approx \frac{3\pi}{4\sqrt{2}} \frac{\lambda}{\omega^3}, \tag{4.151}$$

which shows that if $\lambda \to 0$, the frequency ω must also tend to 0 so that the ratio λ/ω^3 =constant and consequently $n_{\text{trapped}} > 1$ can be always ensured.

To sum up the single-particle quantum resonant dynamics: for suitable choice of the system parameters we observe three Floquet resonant states that evolve with the driving period T but they are superposition of three localized wavepackets $w_j(z,t)$ which are bouncing off the mirror with the period $T_r = 3T/2$.

Having the single-particle problem described we switch to N particles which are bouncing on the oscillating mirror. We focus on ultra-cold bosonic atoms that perform the 3 : 2 resonant motion. The many-body resonant Hilbert subspace is spanned by the Fock states $|n_1, n_3, n_3\rangle$ where n_j's are numbers of bosons that occupy the localized wavepackets $w_j(z,t)$ propagating along the 3 : 2 resonant classical trajectory. When we restrict to this subspace, i.e. when we truncate the bosonic field operator $\hat{\psi}(z,t) \approx \sum_{j=1}^{3} w_j(z,t)\hat{a}_j$ (see Sect. 4.2.2), we obtain the many-body Floquet Hamiltonian

$$
\begin{aligned}
\hat{H}_F &= \frac{1}{3T}\int_0^{3T} dt \int_0^{\infty} dz\, \hat{\psi}^{\dagger}\left(H + \frac{g_0}{2}\hat{\psi}^{\dagger}\hat{\psi} - i\partial_t\right)\hat{\psi} \\
&\approx -\frac{1}{2}\sum_{i\neq j} J_{ij}\hat{a}_i^{\dagger}\hat{a}_j + \frac{g_0}{2}\sum_{i,j=1}^{3} U_{ij}\hat{a}_i^{\dagger}\hat{a}_j^{\dagger}\hat{a}_j\hat{a}_i,
\end{aligned} \tag{4.152}
$$

where H is the single-particle Hamiltonian (4.29), $g_0 < 0$ is the strength of the attractive contact interactions between ultra-cold atoms and

$$
J_{ij} = -\frac{2}{3T}\int_0^{3T} dt \int_0^{\infty} dz\, w_i^*(z,t)\,(H - i\partial_t)\, w_j(z,t), \tag{4.153}
$$

$$
U_{ij} = \frac{2}{3T}\int_0^{3T} dt \int_0^{\infty} dz\, |w_i|^2|w_j|^2, \tag{4.154}
$$

for $i \neq j$ and similar U_{ii} but by the factor 2 smaller, cf. (4.61) and (4.95). The restriction to the resonant Hilbert subspace is valid provided the interactions are not too strong and they are indeed weak because we need the interaction energy per particle to be of the order of the tunneling amplitude $|J_{ij}|$ which is a tiny energy as compared to other energy scales of the system (see Sect. 4.2.2).

If there are no interactions between particles ($g_0 = 0$), then the many-body Floquet state Ψ_0 corresponding to the lowest quasi-energy within the resonant subspace is a Bose–Einstein condensate where all N atoms occupy the single-particle Floquet state which is a uniform superposition of the localized wavepackets w_j, i.e.

$$
\Psi_0(z_1,\ldots,z_N,t) = \prod_{i=1}^{N}\phi_1(z_i,t), \quad \text{with} \quad \phi_1(z_i,t) = \frac{1}{\sqrt{3}}\sum_{j=1}^{3} w_j(z_i,t). \tag{4.155}
$$

However, when the attractive interactions are turned on and they are sufficiently strong, the lowest quasi-energy many-body Floquet state within the resonant Hilbert subspace becomes a Schrödinger cat-like state,

$$
\Psi_0(z_1,\ldots,z_N,t) \approx \frac{1}{\sqrt{3}}\sum_{j=1}^{3}\prod_{i=1}^{N} w_j(z_i,t). \tag{4.156}
$$

That is, it is the superposition of three Bose–Einstein condensates where each condensate is formed by N atoms occupying one of the wavepackets w_j. The Schrödinger cat structure implies that in the limit when $N \to \infty$ but g_0N =constant, spontaneous breaking of the time translation symmetry occurs and the state Ψ_0 collapses to one of the Bose–Einstein condensates,

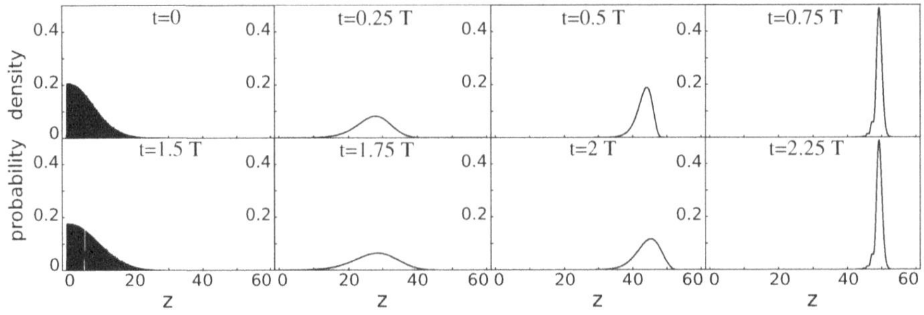

Fig. 4.17 Time evolution of a fractional time crystal. The probability density of ultra-cold atoms corresponding to the symmetry broken state (4.157) is shown in different moments of time as indicated in the plots. The atomic cloud is bouncing on the mirror (which is located at $z = 0$) with the period $T_r = 3T/2$. The motion within two consecutive periods T_r is slightly different because the cloud hits the mirror once in its uppermost position and next in its lowermost position (see Fig. 4.15). The differences disappear in the limit when $\lambda \to 0$ and $\omega \to 0$ but λ/ω^3 =constant. The parameters of the systems are the following: $\lambda = 0.0825$, $\omega = 0.47168$ and $g_0 N U_{ii}/|J_{ij}| = -3$. The figure reproduced from [134]

$$\Psi_0(z_1, \ldots, z_N, t) \to \Psi(z_1, \ldots, z_N, t) \approx \prod_{i=1}^{N} w_j(z_i, t). \tag{4.157}$$

The state (4.157) describes an ultra-cold atomic cloud bouncing on the oscillating atom mirror with the period $T_r = 3T/2$. The lifetime of such a symmetry broken state increases exponentially with N for $g_0 N$ =constant [134, 173]. It is important to note that not only the lowest quasi-energy Floquet state within the resonant Hilbert subspace but a finite range of the spectrum is dominated by symmetry broken states, see Sect. 4.2.2. In Fig. 4.17 time evolution of a symmetry broken state (4.157) is shown. It is obtained within the mean-field approach similarly like in the case of the dramatic breaking of the time translation symmetry described in Sect. 4.2.4. The figure illustrates that the discrete time translation symmetry of the Hamiltonian can be spontaneously broken and time evolution of the system with a period which is a fractional multiple of the driving period emerges [134]. The stability of such a fractional time crystal with respect to a detuning of the driving from the resonant condition and due to experimental imperfections is similar as discussed in Sects. 4.2.3 and 4.2.5.

4.4.2 *Fractional Time Crystal in Spin Systems*

Formation of fractional time crystals can be also observed in periodically driven spin-1/2 systems [157]. The Hilbert space of a single spin is two-dimensional and one could have impression that only discrete time crystals with period doubling can be realized in spin-1/2 systems. However, Pizzi et al. show that when a many-spin system behaves collectively, the dynamics can reveal not only time crystals evolving with a period twice longer than the driving period T but also with many different integer and fractional multiplicities of T [157].

Let us consider a one-dimensional chain of N spin-1/2 which interact via long-range and short-range potentials and which are periodically driven,

$$H = \frac{J}{\mathcal{N}_{N,\alpha}} \sum_{i \neq j=1}^{N} \frac{\sigma_i^z \sigma_j^z}{r_{ij}^{\alpha}} + \lambda \sum_{i=1}^{N} \sigma_i^z \sigma_{i+1}^z - \pi h[1 + \sin(2\pi t)] \sum_{i=1}^{N} \sigma_i^x, \tag{4.158}$$

where $\sigma_i^{x,y,z}$ are the Pauli matrices, $J > 0$ determines the strength of the long-range interactions which decay algebraically like $1/r_{ij}^{\alpha}$—for periodic boundary conditions chosen here, $r_{ij} = \min(|i-j|, N-|i-j|)$. The nearest neighbor interactions are characterized by $\lambda > 0$ while h stands for the average transverse magnetic field that tries to rotate spins if they are not parallel to the x axis. In contrast to the previously analyzed discrete time crystal protocols in spin systems (Sect. 4.3), where an abrupt spin flip driving is often used, here a continuous monochromatic drive is applied with the period $T = 1$. The so-called Kac normalization, $\mathcal{N}_{N,\alpha} = \sum_{i=2}^{N} r_{1i}^{-\alpha}$, is used to ensure the extensivity of the system [22]. That is, there is the double sum in (4.158) and without the Kac normalization the average energy per particle would not be independent of N in the thermodynamic limit.

For $\alpha = 0$ and $\lambda = 0$ there are infinite-range (all-to-all) interactions only and if we allow for $i = j$ in the double sum in (4.158), the Hamiltonian of the system can be written in terms of the total (collective) spin operators, $\hat{S}^{x,y,z} = \sum_{i=1}^{N} \sigma_i^{x,y,z}/2$,

$$H = \frac{4J}{N} \hat{S}^z \hat{S}^z - 2\pi h[1 + \sin(2\pi t)]\hat{S}^x, \tag{4.159}$$

where we assume that the total spin $S = N/2$. The Hamiltonian (4.159) is actually a time-periodic version of the Lipkin–Meshkov–Glick model [165]. If $N \to \infty$, we may apply the mean-field approximation where the components of the collective spin operator become the components of the classical angular momentum, $\hat{S}^{x,y,z} \to S^{x,y,z}$. The mean-field equations can be written in the form of Hamilton equations where the magnetization m along the z axis and the azimuthal angle θ are canonically conjugate variables and they parameterize the angular momentum components,

$$S^x = \frac{N}{2}\sqrt{1-m^2}\cos\theta, \quad S^y = \frac{N}{2}\sqrt{1-m^2}\sin\theta, \quad S^z = \frac{N}{2}m. \tag{4.160}$$

Hamilton equations of motion for the canonically conjugate magnetization m and azimuthal angle θ are generated by the classical Hamiltonian

$$H_{\text{cl}} = 2Jm^2 - 2\pi h\sqrt{1-m^2}\cos\theta\ [1 + \sin(2\pi t)]. \tag{4.161}$$

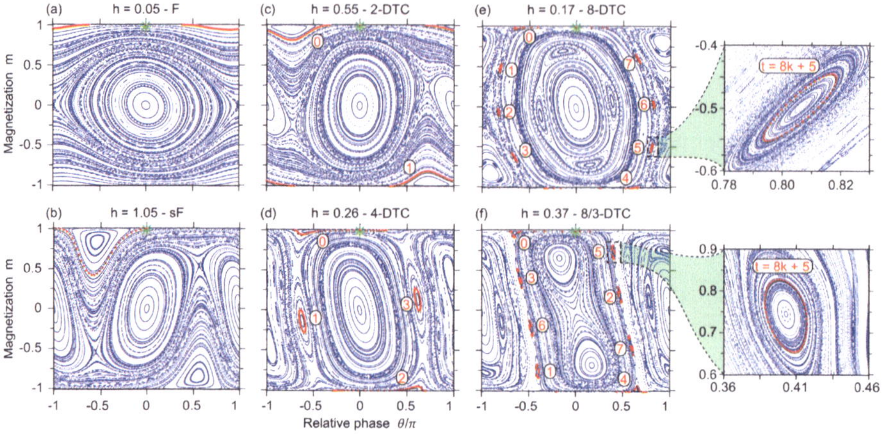

Fig. 4.18 Stroboscopic phase space portraits obtained by integration of the Hamilton equations of motion generated by the mean-field Hamiltonian (4.161) where the spin magnetization m along the z axis and the azimuthal angle θ are canonically conjugate variables. Different panels correspond to different values of the parameter h (as indicated in the plots) but to the same strength, $J = 1$, of the long-range interactions. Green asterisk represents the initial state with the magnetization $m(0) = 1$ and the azimuthal angle $\theta(0) = 0$ while red curves show stroboscopic time evolution of $m(jT)$ and $\theta(jT)$ where j is integer and $T = 1$ is the driving period. In (**a**) the magnetization always remains close to $m = 1$ because for h close to 0, the transverse magnetic field, cf. (4.158), is weak and is not able to change the direction of the spins during T—this phase is denoted as the ferromagnetic (F) phase in the panel (**a**). In (**b**) $h \approx 1$ and in each period T, the magnetization rotates nearly exactly by 2π and consequently in the stroboscopic sampling $m(jT) \approx 1$—this phase is called the stroboscopic ferromagnetic (sF) phase. In panels (**c**)–(**f**) the initial state $m(0) = 1$ is located close to the trajectories fulfilling the 2 : 1 (**c**), 4 : 1 (**d**), 8 : 1 (**e**) and 8 : 3 (**f**) resonance conditions and the system reveals n-tupling discrete time crystals (n-DTC). Reprinted figure with permission from Pizzi et al. [157]

The Hamiltonian $H_{\rm cl}$ describes a periodically driven *fictitious particle* and different classical nonlinear resonances can be expected. Such resonances are represented as elliptical islands around resonant periodic orbits in the stroboscopic pictures of the classical phase space shown in Fig. 4.18 for $J = 1$ and different values of the parameter h.

If the system is initially prepared with the maximal magnetization along the z axis, $m(0) = 1$, then for h close to 0, the driving is very weak and it is not able to change the direction of the collective spin. It is illustrated in Fig. 4.18a where the red curve shows the trajectory which remains close to $m = 1$. The Fourier transform $\tilde{m}(\nu)$ of the stroboscopic evolution of the magnetization $m(jT)$, where j is integer, results in a strong peak at zero frequency ($\nu = 0$), see Fig. 4.19.

For $h \approx 1$ and the same initial state $m(0) = 1$, the Fourier spectrum also reveals a strong peak at $\nu = 0$ but currently it is related to the fact that time-periodic transverse magnetic field rotates the collective spin nearly exactly by 2π during a single period T and consequently the system returns to its initial state, see Fig. 4.18b and Fig. 4.19.

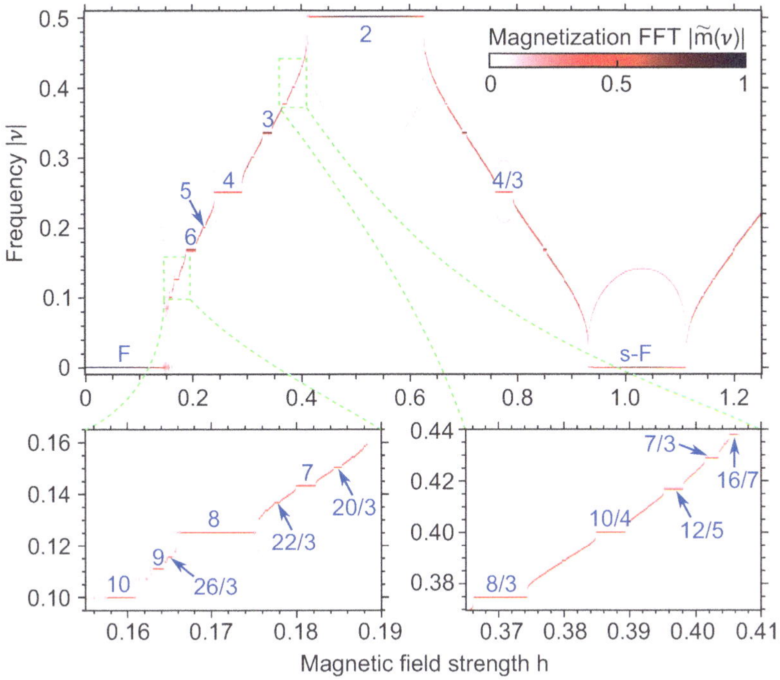

Fig. 4.19 Fourier transform $\tilde{m}(\nu)$ of stroboscopic evolution of the z component of the magnetization $m(jT)$, where j is integer, versus the strength h of the transverse magnetic field, cf. (4.158). Plateau structure indicates ranges of h where resonant elliptical islands contain the initial state with the magnetization $m(0) = 1$ and in the time evolution the system remains in the vicinity of the corresponding classical resonant trajectories. The blue numbers indicate periods of the resonant trajectories (in the units of the driving period T) which are periods of time evolution of discrete time crystals. Reprinted figure with permission from Pizzi et al. [157]

For h between 0 and 1, elliptical resonant islands corresponding to different nonlinear higher order classical resonances appear and become located around the initial point $m(0) = 1$ in the phase space as presented in Fig. 4.18. The plateau structure in Fig. 4.19 results from the fact that with increasing h, a resonant island enters the location around the initial point $m(0) = 1$ and next leaves it and then another island corresponding to a different resonance enters and subsequently leaves the location around $m(0) = 1$. Size of a plateau depends on size of the corresponding resonant island which also implies that no fine-tuning is needed because area of elliptical islands is always finite. In Fig. 4.18c–f red curves illustrate trajectories corresponding to time crystals evolving with the periods: $2T$, $4T$, $8T$ and the fractional period $8T/3$, respectively. The resonance structures in the phase space appear, thanks to the presence of the long-range interactions.

For $\alpha = 0$, $\lambda = 0$ and in the $N \to \infty$ limit, the mean-field approach predicts spontaneous breaking of the discrete time translation symmetry and time-periodic evolution of the system with periods different than the driving period and such evolution lasts forever. The quantum Lipkin–Meshkov–Glick Hamiltonian (4.159) can be also diagonalized exactly without much effort because the Hilbert space related to the total spin $S = N/2$ is N-dimensional only. Finite N results in finite lifetime of time crystals because the system can tunnel between elliptical islands belonging to the same resonance. However, the tunneling time is expected to increase exponentially quickly with the increasing N and consequently time crystals live forever in the thermodynamic limit.

Different time crystals analyzed here correspond to different size of resonant elliptical islands created in the phase space. The higher n-tupling time crystal, the size of corresponding islands is the smaller and larger values of N are required in order to trap (in the semiclassical sense) a quantum state inside an island in order to realize a time crystal. That is, $1/N$ plays a role of an effective Planck constant. For example, for the period doubling time crystal, $N \gtrsim 10$ is sufficient while for period 4-tupling we need already $N \gtrsim 100$ [157].

Due to the fact that the Lipkin–Meshkov–Glick model can be mapped to N bosons in a double well potential described within the two-mode approximation (cf. Eqs. (4.73)–(4.74) and the corresponding text), discrete time crystals proposed here can be realized in ultra-cold atom laboratories.

Finally Pizzi et al. analyze time crystal behavior for finite-range interactions ($\alpha > 0$) and in the presence of independently controlled nearest neighbor interactions ($\lambda > 0$). Then, the total spin is no longer a constant of motion and the entire Hilbert space is explored by the system. That is, the dimension of the relevant Hilbert space is not equal N but 2^N which can change lifetime of time crystals drastically. With the help of the Holstein–Primakoff transformation, Pizzi and coworkers analyze generation of spin-wave excitations [157]. They show that in order to avoid quick proliferation of spin-wave excitations and thus to avoid thermalization, the range of the interactions must be sufficiently long. For example, for the 4-tupling time crystal and for $\lambda = 0.03$ the long-range interactions must fall off slower than $\alpha < \alpha_c \approx 1.4$. If this condition is fulfilled, the time crystal is stable at least in the prethermal sense.

4.5 Discrete Time Quasi-Crystals

Quasi-crystals in condensed matter physics are spatial structures which cannot be reproduced by translation of an elementary cell [99]. Nevertheless they form diffraction patterns but with spots forbidden by crystallographic point-group symmetries [117, 182]. Discovery of quasi-crystals led to a paradigm shift within chemistry for which Dan Shechtman was awarded the Nobel Prize in 2011. Quasi-crystals are a subject of research not only in solid state physics. They are investigated in the field of optics where quasi-crystal photonic materials can be engineered [5, 107, 189, 195] and in ultra-cold atomic gases by means of a suitable arrangement of laser beams which create an effective potential for atoms of a quasi-crystal structure [21].

In the context of time crystals, quasi-crystalline structures have been also investigated. Flicker analyzed quasi-periodic behaviors in time in classical dissipative systems [63], see also [207]. Li et al. proposed spontaneous formation of time quasi-crystals by ions trapped in two rings [119] but the idea of time crystals in the ground state of systems with two-body interactions has been criticized [24, 25, 118, 199, 200]. Huang et al. investigated time quasi-crystal evolution in the time-independent Lipkin–Meshkov–Glick model and found that for a finite number N of spins, a tiny perturbation of the exact quantum ground leads to a state which can mimic spontaneous symmetry breaking and reveals quasi-periodic evolution with two incommensurate frequencies [91]. However, the frequencies scale like $1/N$ and in the thermodynamic limit the time evolution disappears. There is also research on quasi-crystal response of systems which are driven quasi-periodically in time but there the Hamiltonian does not possess the discrete time translation symmetry that could be spontaneously broken [51, 154]. Also quasi-periodic response of a periodically driven many-body system was analyzed in Ref. [127] but with no spontaneous process involved. Another idea of time quasi-crystals relies on the observation of quasi-periodic evolution of an ordinary discrete time crystal in the reference frame that rotates with a frequency which is incommensurate with respect to the driving frequency [210].

In the present section we concentrate on an isolated periodically driven quantum many-body system and show that spontaneous breaking of the discrete time translation symmetry of the Hamiltonian leads to evolution of a system that reveals a quasi-periodic pattern in time [69].

Let us start with the definition of the one-dimensional Fibonacci quasi-crystal which is a sequence of two elementary cells that we denote by R and L. The sequence is not periodic but it is also not a random pattern because it is generated by the following inflation rule $L \to LR$ and $R \to L$ [99]. Successive applications of the rule result in sequences whose lengths are equaled to the subsequent Fibonacci numbers (1, 2, 3, 5, 8, ...), i.e. $L \to LR \to LRL \to LRLLR \to LRLLRLRL \to \dots$. The long-range order of the Fibonacci quasi-crystal is reflected by its Fourier transform which reveals strong peaks [195]. The Fibonacci quasi-crystal can be also generated with the help of a line that crosses a squared lattice, see Fig. 4.20. If the gradient of the line is equaled to the golden ratio $(\sqrt{5}+1)/2$ (or inverse of the golden ratio as in Fig. 4.20), successive cuts of the line with the vertical and horizontal axes of the lattice reproduce the Fibonacci sequence. A finite fragment of the Fibonacci sequence can be obtained when the gradient of the line is given by a rational approximation of the golden ratio obtained by terminating the infinite continued fraction that represents the golden ratio. In Fig. 4.20, three lines are presented with the gradients equal the inverse of the golden ratio and its two rational approximations: 2/3 and 8/13. In the following we are going to show that the Fibonacci quasi-crystal can be revealed in dynamics of a periodically driven quantum many-body system when the discrete time translation symmetry of the Hamiltonian is spontaneously broken [69].

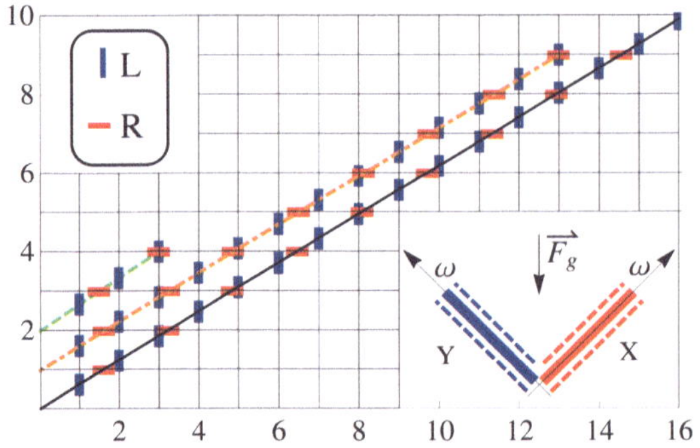

Fig. 4.20 Generation of the Fibonacci quasi-crystal. Solid black line, with the tangent equaled to the inverse of the golden ratio, crosses the squared lattice. Successive cuts of the line with the vertical and horizontal axes of the lattice, denoted by L and R, respectively, generate the Fibonacci sequence $LRLLRLRLL\ldots$. The tangents of the dashed green and dotted-dashed orange lines are rational approximation of the inverse of the golden ratio, 2/3 and 8/13, respectively. The cuts of these lines with the lattice generate finite fragments of the Fibonacci quasi-crystal, 5- and 21-element fragments, respectively. Inset presents the configuration of the two orthogonal mirrors that oscillate with the same frequency ω. The mirrors are oriented symmetrically with respect to the direction of the gravitational force $\mathbf{F}_g$. The figure reproduced from [69]

A particle bouncing resonantly on an oscillating mirror in the one-dimensional space has been described in detail in Sect. 4.2. Now let us assume that in the two-dimensional space there are two orthogonal oscillating mirrors which are oriented symmetrically side by side with respect to the direction of the gravitational force, see inset of Fig. 4.20. In the single-particle case, the system is easy to describe because the dynamics separates into two independent motions along the mirrors. The single-particle Hamiltonian (in the frame oscillating with the mirrors, cf. Eq. (4.29)) reads[17]

$$H = \frac{p_x^2 + p_y^2}{2} + x + y + \lambda_x x \cos(\omega t + \Delta\phi) + \lambda_y y \cos(\omega t), \tag{4.162}$$

where, importantly, we assume that both mirrors oscillate with the same frequency ω. The amplitudes, λ_x/ω^2 and λ_y/ω^2, of the mirrors' motions in the laboratory frame can be different and there is also a possible phase shift $\Delta\phi$ in the mirrors' oscillations. The coordinate frame is chosen so that the left mirror is located at $x = 0$, the right mirror at $y = 0$, and the gravitational force $\mathbf{F}_g$ points in the $-(\mathbf{x}+\mathbf{y})$ direction, see inset of Fig. 4.20.

In the quantum description, due to the time periodicity of the Hamiltonian $H(t + T) = H(t)$ where $T = 2\pi/\omega$, one can find Floquet states of a particle bouncing between the oscillating mirrors. Here, we focus on resonant dynamics

[17] We use the gravitational units (4.28) but assume that $g \to g/\sqrt{2}$.

where the motion of a particle along the x direction fulfills the $s_x : 1$ resonant condition with the oscillations of the left mirror (i.e., the period of particle bounces along the x directions is s_xT where s_x is integer) while the motion along the y direction satisfies the $s_y : 1$ resonant condition with the oscillations of the right mirror (the period of particle bounces along the y direction is s_yT where s_y is integer). In the quantum description of the resonant motion of a particle one can identify s_xs_y resonant Floquet states. These Floquet states are evolving with the period of the mirrors' oscillations T and they are superposition of s_xs_y localized wavepackets,

$$W_{\mathbf{i}=(i_x,i_y)}(\mathbf{r},t) = w_{i_x}(x,t)w_{i_y}(y,t). \tag{4.163}$$

The wavepackets $W_{\mathbf{i}}(\mathbf{r},t)$ are traveling in the two-dimensional space along the resonant trajectory with the period s_xs_yT. After every period T the wavepackets exchange their role so that the resonant Floquet states, which are superposition of all s_xs_y wavepackets, are periodic with the driving period T as they should—Fig. 4.21 shows an example for $s_x = 2$ and $s_y = 3$. The time periodicity of $W_{\mathbf{i}}(\mathbf{r},t+s_xs_yT) = W_{\mathbf{i}}(\mathbf{r},t)$ results from the fact that for the $s_x : 1$ resonant dynamics along the x direction, the one-dimensional localized wavepackets are periodic with the period s_xT, i.e. $w_{i_x}(x,t+s_xT) = w_{i_x}(x,t)$, while for the resonant motion along the y direction we have $w_{i_y}(y,t+s_yT) = w_{i_y}(y,t)$, see Sect. 4.2 for the detailed description of resonant dynamics in the one-dimensional case.

Now let us switch to N bosons bouncing resonantly between the two perpendicular oscillating mirrors and restrict to the resonant many-body Hilbert subspace where the many-body Floquet Hamiltonian can be approximated by the Bose–Hubbard Hamiltonian [69], cf. (4.25) and (4.95),

$$\begin{aligned} \hat{H}_F &= \frac{1}{s_xs_yT}\int_0^{s_xs_yT} dt \int d^2\mathbf{r}\, \hat{\psi}^\dagger \left(H + \frac{g_0}{2}\hat{\psi}^\dagger\hat{\psi} - i\partial_t\right)\hat{\psi} \\ &\approx -\frac{1}{2}\sum_{\langle \mathbf{i},\mathbf{j}\rangle} J_{\mathbf{ij}}\, \hat{a}_{\mathbf{i}}^\dagger \hat{a}_{\mathbf{j}} + \frac{1}{2}\sum_{\mathbf{i},\mathbf{j}} U_{\mathbf{ij}}\, \hat{a}_{\mathbf{i}}^\dagger \hat{a}_{\mathbf{j}}^\dagger \hat{a}_{\mathbf{j}}\hat{a}_{\mathbf{i}}. \end{aligned} \tag{4.164}$$

In (4.164) the field operator $\hat{\psi}$ is truncated to the sum over the bosonic annihilation operators $\hat{a}_{\mathbf{i}}$ corresponding to the localized wavepackets $W_{\mathbf{i}}$,

$$\hat{\psi}(\mathbf{r},t) \approx \sum_{\mathbf{i}} W_{\mathbf{i}}(\mathbf{r},t)\,\hat{a}_{\mathbf{i}} = \sum_{i_x=1}^{s_x}\sum_{i_y=1}^{s_y} w_{i_x}(x,t)\,w_{i_y}(y,t)\,\hat{a}_{i_x,i_y}. \tag{4.165}$$

The single-particle Hamiltonian H in (4.164) is given in Eq. (4.162) and the strength of the attractive contact interactions between bosons (we assume they are ultra-cold atoms) $g_0 < 0$. The tunneling amplitudes $J_{\mathbf{ij}}$ of atoms between the wavepackets $W_{\mathbf{i}}$ and the interaction coefficients $U_{\mathbf{ij}}$ are obtained in the analogous way like in Eqs. (4.153)–(4.154), see also Sect. 4.2. The Bose–Hubbard Hamiltonian (4.164)

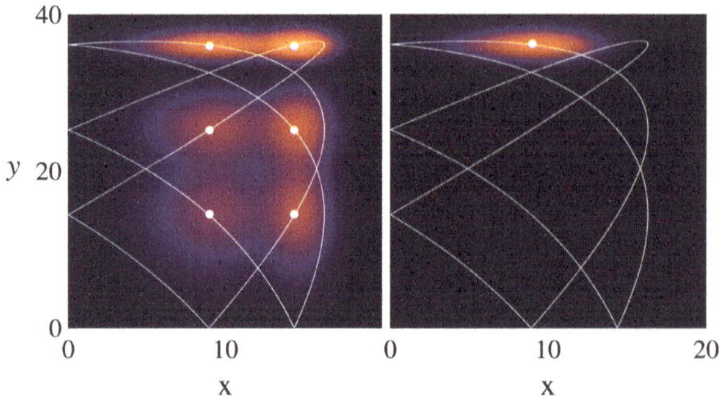

Fig. 4.21 Mean-field solutions for ultra-cold atoms bouncing resonantly between two orthogonal mirrors which oscillate with the same frequency ω. The left mirror is located at $x = 0$ while the right mirror at $y = 0$ (cf. inset of Fig. 4.20). Bouncing atoms fulfill the 3 : 1 and 2 : 1 resonance conditions with the motion of the left and right mirrors, respectively. Left panel shows density of non-interacting atoms corresponding to the lowest energy mean-field solution for the effective Bose–Hubbard model (4.164) at $t = T/3$. This solution respects the discrete time translation symmetry of the many-body Hamiltonian, i.e. it evolves with the period T and atoms are hitting the left (L) and right (R) mirrors in the alternate way forming a periodic sequence of events: $RLRLRL\ldots$. Right panel is also related to the lowest energy mean-solution for the model (4.164) but for sufficiently strong attractive interactions between atoms when the discrete time translation symmetry is spontaneously broken and the sequence of the bounces of atoms off the left and right mirrors reproduces the 5-element fragment of the Fibonacci quasi-crystal, $LRLLR$. The parameters of the system are: $\lambda_x = 0.094$, $\lambda_y = 0.030$, $\omega = 1.1$, $\Delta\phi = 2\pi/3$, $g_0 N = 0$ (left) and $g_0 N = -0.022$ (right). The latter results in $U_{\mathbf{ii}} N/J = -81$, with $J = 4.8 \times 10^{-6}$. The figure reproduced from [69]

describes actually the $s_x \times s_y$ *squared lattice* whose sites are denoted by the index $\mathbf{i} = (i_x, i_y)$. Only the dominant nearest neighbor tunneling between the *lattice sites* is taken into account and we choose the amplitudes λ_x and λ_y so that the tunneling amplitudes are the same $J \equiv J_{\mathbf{ij}}$. The on-site interaction coefficients $U_{\mathbf{ii}}$ are typically at least an order of magnitude larger than $U_{\mathbf{ij}\neq\mathbf{i}}$ which correspond to long-range interactions [69].

For negligible interactions between bosons (i.e., for $g_0 \to 0$), the lowest quasi-energy many-body Floquet state within the resonant Hilbert subspace we restrict to (or in other words the ground state of the Bose–Hubbard model (4.164)) is a Bose–Einstein condensate where all particles occupy the same single-particle Floquet state $\psi(\mathbf{r}, t) = \sum_{\mathbf{i}} W_{\mathbf{i}}(\mathbf{r}, t)/\sqrt{s_x s_y}$. However, when the attractive interactions are sufficiently strong, it becomes energetically favorable to group all bosons in one of the localized wavepacket $W_{\mathbf{i}}$.[18] Then, there is a range of the quasi-energy spectrum dominated by many-body Floquet states which have the Scrödinger cat-

[18]It should be stressed that the energy per particle corresponding to the attractive interactions we consider here is larger than the tunneling rate J but too small to induce collapse of the Bose system in the two-dimensional space. The latter would be possible if the interaction energy per particle was at least comparable to the gap between the energy of the wavepackets, $W_{\mathbf{i}}$, localized at the bottom

like structure, i.e. which are macroscopic superposition Bose–Einstein condensates. For example, for sufficiently strong attractive interactions, the lowest quasi-energy Floquet state within the resonant Hilbert subspace can be approximated by

$$\Psi_0(\mathbf{r}_1, \ldots, \mathbf{r}_N, t) \approx \frac{1}{\sqrt{s_x s_y}} \sum_{\mathbf{j}} \prod_{i=1}^{N} W_{\mathbf{j}}(\mathbf{r}_i, t). \tag{4.166}$$

The Schrödinger cat structure of Ψ_0 indicates that in the limit when $N \to \infty$,[19] spontaneous breaking of the discrete time translation symmetry of the Hamiltonian occurs and the symmetry-preserving many-body Floquet state Ψ_0 collapses to one of the Bose–Einstein condensates that form the macroscopic superposition in (4.166), i.e.

$$\Psi_0(\mathbf{r}_1, \ldots, \mathbf{r}_N, t) \to \Psi(\mathbf{r}_1, \ldots, \mathbf{r}_N, t) \approx \prod_{i=1}^{N} W_{\mathbf{j}}(\mathbf{r}_i, t), \tag{4.167}$$

and the system starts evolving with the period $s_x s_y T$. Within the Bose–Hubbard model, it can be shown that the lifetime of the symmetry broken state (4.167) goes exponentially quickly to infinity with the increasing N.

We have now all elements to analyze spontaneous emergence of quasi-crystal structures in time. Let us begin with a simple example where the $s_x : 1$ and $s_y : 1$ resonances correspond to $s_x = 2$ and $s_y = 3$. Spontaneous breaking of the time translation symmetry in the system described by the Bose–Hubbard Hamiltonian (4.164) corresponds to a quantum phase transition. While the mean-field approximation cannot describe the system at the transition point we may use the mean-field approach away from the critical point. It turns out that there are two critical values of the interaction strength where we can observe gradual breaking of the time translation symmetry. For $U_{\mathbf{ii}}N/J \to 0$, the lowest quasi-energy Floquet state within the resonant subspace is not degenerate and it describes a Bose–Einstein condensate where N atoms occupy the single-particle Floquet state $\psi(x, y, t) = \sum_{\mathbf{i}} W_{\mathbf{i}}(x, y, t)/\sqrt{s_x s_y}$ which evolves with the period T, see left panel of Fig. 4.21. If we plot the probabilities $\rho_L(t)$ and $\rho_R(t)$ for the measurement of atoms close to the left and right mirrors, respectively, where

$$\rho_L(t) = \int dy |\psi(x \approx 0, y, t)|^2, \quad \rho_R(t) = \int dx |\psi(x, y \approx 0, t)|^2, \tag{4.168}$$

of the resonance islands and the energy of the excited wavepackets in the islands, see Fig. 4.9 and the discusion in Sect. 4.2.4 and at the end of Sect. 4.2.1.

[19] Remember that in the Bose system considered here the $N \to \infty$ limit is accompanied by $g_0 \to 0$ so that $g_0 N = \text{constant}$, see Sect. 4.2.2.

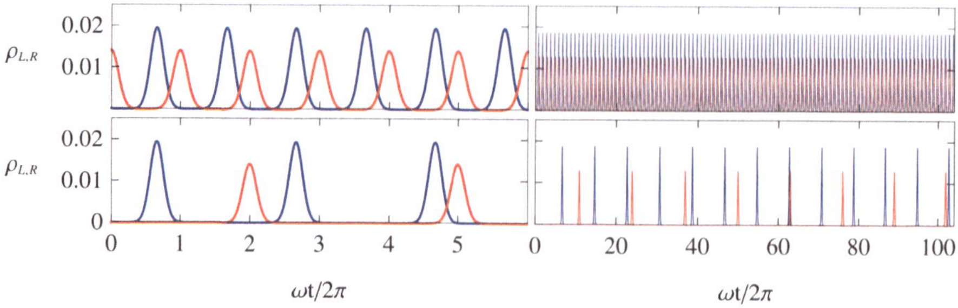

Fig. 4.22 Rescaled probabilities, $\rho_{L,R}(t)$, for hitting by atoms the left (blue) and right (red) mirrors versus time. The results correspond to the lowest energy mean-field solutions of the effective Bose–Hubbard model (4.164). Atoms are bouncing between two perpendicular mirrors that oscillate with the same frequency ω and they fulfill the $s_x : 1$ and $s_y : 1$ resonance conditions with the motions of the left and right mirrors, respectively (cf. Figs. 4.20–4.21). Left panels are related to $s_x = 2$ and $s_y = 3$ while the right panels to $s_x = 8$ and $s_y = 13$. If the interactions between atoms are negligible, then the mean-field solutions respect the discrete time translation symmetry of the many-body Hamiltonian and atoms are bouncing off the left (L) and right (R) mirrors in the alternate way forming a periodic sequence of events $RLRLRL\ldots$ (see top panels). However, if the attractive interactions between atoms are sufficiently strong, the discrete time translation symmetry is spontaneously broken and the bounces of atoms on the mirrors form the Fibonacci quasi-crystal (bottom panels). In the bottom left panel only a 5-element fragment of the Fibonacci sequence is reproduced because the chosen resonances correspond to a poor rational approximation of the golden ratio, $s_y/s_x = 3/2$. In the bottom right panel a 21-element of the Fibonacci quasi-crystal emerges spontaneously because the rational approximation is better, $s_y/s_x = 13/8$. The results shown in the left panels correspond to the same parameters as in Fig. 4.21, while in the right panels: $\lambda_x = 0.087$, $\lambda_y = 0.026$, $\omega = 1.77$, $\Delta\phi = \pi/2$, $s_y = 13$, $s_x = 8$, $g_0N = 0$ (top right panel) and $g_0N = -0.029$ (bottom right panel). The latter results in $U_{\mathbf{ii}}N/J = -80$ and $J = 2.3 \times 10^{-6}$ in (4.164). The figure reproduced from [69]

we can see that atoms bounce once off the left mirror and once off the right mirror in the alternate way (top left panel of Fig. 4.22). In other words, the probabilities $\rho_L(t)$ and $\rho_R(t)$ form a periodic sequence of events $RLRLRL\ldots$ in time.

When we increase the strength of the attractive interactions so that $-6.5 \lesssim U_{\mathbf{ii}}N/J \lesssim -4.5$, the mean-field approach shows that the lowest energy mean-field solutions for the Bose–Hubbard system (4.164) are degenerate and the corresponding wavefunctions are no longer a uniform superposition of all $s_x s_y$ localized wavepackets $W_{\mathbf{i}}(\mathbf{r}, t)$. The system prepared in one of the lowest energy mean-field states breaks the discrete time translation symmetry of the many-body Hamiltonian because it evolves with the period $s_x T$.

If the strength of the attractive interactions increases further, then we observe another critical value around $U_{\mathbf{ii}}N/J \approx -6.5$ where the system chooses spontaneously motion with the period $T = s_x s_y T$. Such a gradual breaking of the discrete time translation symmetry is the result of the presence of the long-range interactions (i.e., $U_{\mathbf{ij}\neq\mathbf{i}} \neq 0$) which in the case of $s_x = 2$ and $s_y = 3$ are sufficiently strong to influence the phase diagram of the system. For $U_{\mathbf{ii}}N/J \lesssim -25$ the symmetry broken lowest quasi-energy Floquet states within the resonant subspace reduce to single localized wavepackets $W_{\mathbf{j}}(r, t)$ with accuracy better than 99%, see right panel of Fig. 4.21. When we analyze the probabilities $\rho_L(t)$ and $\rho_R(t)$ of hitting the left and right mirrors, we find that they form in time a fragment $LRLLR$ of the Fibonacci sequence which is repeated periodically with the period $s_x s_y T = 6T$, see bottom left panel of Fig. 4.22. The fragment is short because we have chosen the resonances $s_x = 2$ and $s_y = 3$ that correspond to a very poor rational approximation $s_y/s_x = 3/2$ of the golden ratio. However, we may choose as good rational approximation as we wish and we can observe that spontaneous breaking of the discrete time translation symmetry of the Hamiltonian results in the emergence in time of the Fibonacci quasi-crystal of any length.

Let us summarize the procedure for the realization of spontaneous formation of a quasi-crystalline structure in time. Suppose we choose a rational approximation s_y/s_x of the golden ratio which allows us to generate a fragment of the Fibonacci sequence of the length $s_x + s_y$. That is, with the help of the line that crosses the squared lattice and whose gradient is given by s_x/s_y one can generate $s_x + s_y$ elements of the Fibonacci quasi-crystal, see Fig. 4.20. If we want to observe spontaneous formation of such a quasi-crystal in the time domain by ultra-cold atoms bouncing between two perpendicular oscillating mirrors, we have to load atoms to the resonant trajectory where the $s_x : 1$ and $s_y : 1$ resonant conditions with the oscillations of the left and right mirrors, respectively, are fulfilled. Then, if the interactions between atoms are very weak, the many-body Floquet states that belong to the resonant Hilbert subspace evolve with the driving period T and the probabilities of hitting the left and right mirrors form a periodic sequence of events $RLRLRLRL\ldots$. However, when the attractive interactions between atoms are sufficiently strong, there is a range of the quasi-energy spectrum in the resonant Hilbert subspace where the corresponding many-body Floquet states breaks the discrete time translation symmetry and atoms are hitting the left and right mirrors in the way that reproduces the Fibonacci quasi-crystal. Such a behavior is repeated periodically with the period $s_x s_y T$ what reflects the periodic boundary conditions of the presented quasi-crystal. The better rational approximation s_y/s_x of the golden ratio we choose, the longer Fibonacci quasi-crystal in time we can observe. Apart from the $s_y/s_x = 3/2$ case we also present in Fig. 4.22 the case corresponding to the better approximation of the golden ratio, i.e. $s_y/s_x = 13/8$.

4.6 Experimental Realizations

Research in the area of time crystals is boosted by the experiments on discrete time crystals [33, 153, 170, 186, 209] that have been performed shortly after the theoretical ideas were published [57, 102, 173]. It turned out that discrete time crystal dynamics in systems which effectively can be described by isolated periodically driven spin-1/2 models could be quite quickly realized in the laboratory. In the present section we describe the experiments starting from chronologically the first ones and try to point out differences between various demonstrations of spontaneous breaking of the discrete time translation symmetry of the Hamiltonians.

4.6.1 Periodically Driven Chain of Ions

A chain of atomic ions can be controlled and manipulated in a linear radio frequency Paul trap. In Ref. [209] Zhang et al. report the experiment where a chain of ten $^{171}Yb^+$ ions was used to realize a periodically driven spin-1/2 system with spatial disorder. Superposition of two hyperfine states of $^{171}Yb^+$ ion, i.e. the $^2S_{1/2}$ $|F = 0, m_F = 0\rangle$ and $|F = 1, m_F = 0\rangle$ states where F and m_F stand for the hyperfine and Zeeman quantum numbers, respectively, mimicked two spin-1/2 states. The periodic driving of the system that was realized in the experiment can be described by repeatable application of the Floquet unitary operator

$$U_F = e^{-iH_3t_3}e^{-iH_2t_2}e^{-iH_1t_1}, \tag{4.169}$$

which evolves the system during a single period $T = t_1 + t_2 + t_3$ where

$$H_1 = \frac{\pi}{2t_1}(1-\epsilon)\sum_i \sigma_i^x, \tag{4.170}$$

$$H_2 = J_0 \sum_{i<j} \frac{\sigma_i^z \sigma_j^z}{|i-j|^\alpha}, \tag{4.171}$$

$$H_3 = \sum_i D_i \sigma_i^z. \tag{4.172}$$

The first operator on the right-hand side of (4.169), i.e. $e^{-iH_1t_1}$, rotates spins around the x-axis by an angle $\pi(1-\epsilon)$ which in the experiment is realized by means of the optical Raman transitions. The parameter ϵ is a detuning with respect to the perfect spin flips. The middle operator, $e^{-iH_2t_2}$, evolves the system in the presence of the spin–spin interactions which are generated by spin dependent optical dipole forces that fall off algebraically with the exponent α that can be controlled experimentally—in Ref. [209] $\alpha = 1.5$ was chosen. The last part of the Floquet evolution operator, $e^{-iH_3t_3}$, introduces spatial disorder to the system. A tightly focused laser beam can be used to address individually each ion and by means of the ac Stark shift one can realize programmable spatial disorder in the ion chain. If the standard deviation W of the uniformly chosen D_i in (4.172) corresponds to $Wt_3 = \pi$, in the static (non-driven) case the many-body localization is expected even in the presence of the long-range interactions [209].

The described system closely resembles the discrete time crystal model introduced previously, see Sect. 4.3.1 for details, except the disorder is present in the field and not in the spin–spin interactions. In the experiment it is not possible to prepare a symmetry-preserving Floquet state because it is a Schrödinger cat-like state, cf. (4.111). However, it is quite easy to prepare a symmetry-broken state and investigate how it evolves in the absence and in the presence of the spin–spin interactions. The symmetry-broken state chosen in the experiment corresponded

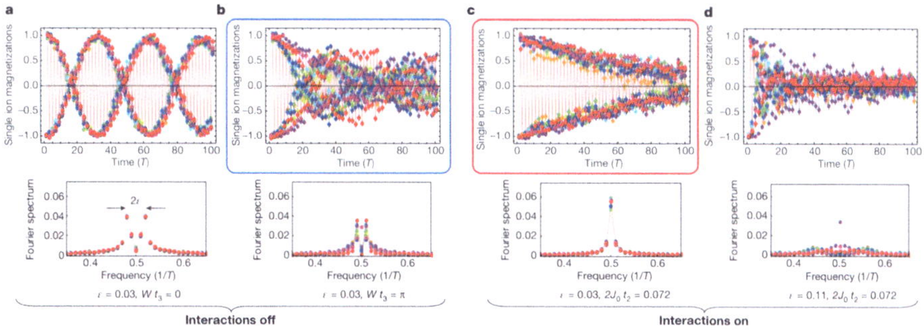

Fig. 4.23 Time evolution of single spin magnetizations along the z-axis and the corresponding Fourier spectra observed in a periodically driven chain of ten $^{171}Yb^+$ ions. Initially all spins are polarized along the z-axis. In panel (**a**) no spin–spin interactions ($J_0 = 0$) and no disorder ($W = 0$) are present and the single spin magnetizations reveal beating in time caused by a small detuning $\epsilon = 0.03$ from the perfect spin flips in (4.170). In panel (**b**) the disorder is present ($Wt_3 = \pi$) but still no spin–spin interactions and the spins precess independently with random Larmor rates. In panel (**c**) both the interactions ($2J_0t_2 = 0.072$) and disorder ($Wt_3 = \pi$) are on and despite the fact that there is the same detuning $\epsilon = 0.03$ as in the previous panels, the system self-organizes its motion and evolves with the period twice longer than the driving period demonstrating the discrete time crystal dynamics. In panel (**d**) the detuning $\epsilon = 0.11$ is too large for the system to synchronize its motion if the interaction are the same as in (**c**). Reprinted by permission from Springer Nature Customer Service Centre GmbH: Nature Springer, Zhang et al. [209], Copyright (2017)

to all spins polarized along the z-axis. Having prepared such an initial state, the periodic driving described by the Floquet unitary operator (4.169) was turned on. If the spin–spin interactions are absent ($J_0 = 0$), the spins evolve independently. Then, in the presence of a small detuning $\epsilon = 0.03$ in (4.170), the time evolution of single spin magnetizations along the z-axis reveals precession with different Larmor rates if spatial disorder is turned on, see Fig. 4.23. However, when the spin–spin interactions are present and sufficiently strong, despite the fact that there is the same detuning $\epsilon = 0.03$ as previously, the spins synchronize their precession and evolve with the period twice longer than the driving period T demonstrating spontaneous breaking of the discrete time translation symmetry of the Hamiltonian and consequently the discrete time crystal dynamics. It is illustrated by the corresponding Fourier spectra of the time evolution of the single spin magnetizations where strong peaks at half of the driving frequency are clearly visible, see Fig. 4.23.

Keeping the spin–spin interactions fixed but increasing the detuning ϵ, one observes that the system eventually enters the symmetry unbroken regime, i.e. it loses the synchronized motion. On the border between the discrete time crystal and symmetry unbroken regimes, the system, in different realizations of the spatial disorder, has difficulty to decide on which side of the phase diagram it is [204]. It corresponds to the maximal variance of the height of the peak at the half frequency of the driving in the Fourier spectra obtained in many different realizations of the experiment with randomly chosen spatial disorder. The pairs of $\{\epsilon, 2J_0t_2\}$

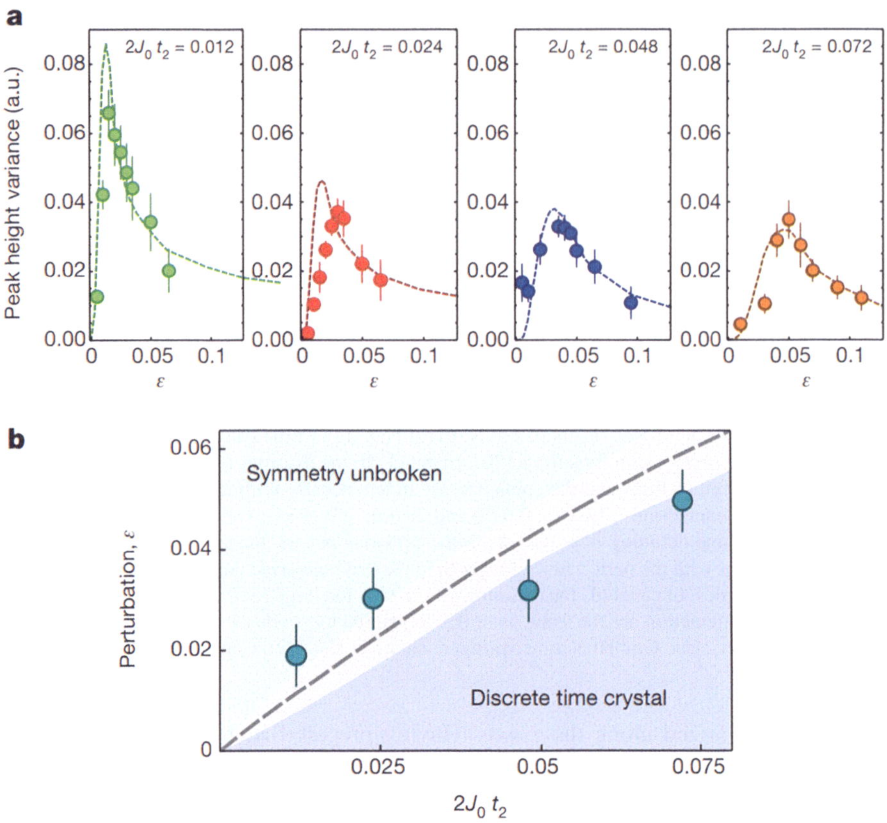

Fig. 4.24 (**a**): variance of the height of the Fourier peaks at $(2T)^{-1}$, cf. bottom row of Fig. 4.23, obtained in different experimental realizations of spatial disorder ($Wt_3 = \pi$). The variance is plotted versus the detuning ϵ. Panels from left to right correspond to the increasing strength of the spin–spin interaction. (**b**): the phase diagram—the boundary between the symmetry unbroken and discrete time crystal phases corresponds to the maximal values of the variance visible in (**a**). Reprinted by permission from Springer Nature Customer Service Centre GmbH: Nature Springer, Zhang et al. [209], Copyright (2017)

corresponding to the maximal variances allow one to determine the boundary between the time crystal and symmetry unbroken phases as depicted in Fig. 4.24—compare also Figs. 4.7 and 4.12.

The experiment we have just described demonstrates a discrete time crystal in a small system (10 ions) and in the presence of strong spatial disorder that does not come from random spin–spin interactions but from an external field along the z direction. In the periodically driven case, such a disorder does not induce many-body localization as was shown in Ref. [103]. Indeed, the leading order effective Hamiltonian that describes evolution over two driving periods becomes disorder-free and the higher order terms give too weak corrections for many-body localization to occur.

4.6.2 Periodically Driven Nitrogen-Vacancy Centers in Diamond

Choi et al. demonstrated the discrete time crystal dynamics in periodically driven nitrogen-vacancy (NV) centers in diamond at room temperature [33]. Each NV center has an electronic spin $S = 1$. Choi et al. prepared a sample with highly concentrated NV impurities (45 p.p.m.) what resulted in strong long-range dipolar interactions between the spins. The entire system consisted of 10^6 spin impurities and consequently the truly thermodynamic limit was realized in contrast to the previously described experiment with the chain of 10 ions. Both experiments were reported back to back in the same issue of Nature [33, 209]. The NV centers were a system with disorder because they were distributed randomly and also due to the presence of paramagnetic impurities in diamond. However, the system was three-dimensional and the disorder was not so strong to enforce many-body localization. Nevertheless the discrete time crystal dynamics was observed which lasted about 100 driving periods similarly as in the case of the driven chain of ions [209]. Actually in Ref. [33], two kinds of the time crystal experiments were performed. In the first case, the two electronic Zeeman states $|m_s = 0\rangle$ and $|m_s = -1\rangle$ of the spin $S = 1$ of the NV impurities were used to mimic fictitious spin-1/2 states which were periodically driven and a discrete time crystal evolving with the period twice longer than the driving period T could be realized. In the second case, all three Zeeman states of the spin $S = 1$ were employed and applying a proper driving, the experimentalists were able to observe a time crystal evolving with the period three times longer than T.

Let us begin with the period-doubling case. The periodically driven NV centers can be described by the following time-periodic Hamiltonian,

$$H(t) = \sum_i \left[\Omega_x(t)\hat{s}_i^x + D_i\hat{s}_i^y + \Omega_z(t)\hat{s}_i^z\right] + \sum_{ij} \frac{J_{ij}}{|\mathbf{r}_i - \mathbf{r}_j|^3}\left(\hat{s}_i^x\hat{s}_j^x - \hat{s}_i^y\hat{s}_j^y + \hat{s}_i^z\hat{s}_j^z\right), \tag{4.173}$$

where $\hat{s}_i^{x,y,z}$ are the operators of the fictitious spins-1/2 associated with the electronic states $|\pm 1\rangle_z \propto |m_s = 0\rangle \pm |m_s = -1\rangle$ of the NV impurities. D_i are random numbers which, in order to describe on-site random fields in the experiment, are chosen from the uniform distribution with the standard deviation $W = 2\pi \times 4\,\text{MHz}$. The average nearest neighbor separation of the spin impurities was $r_0 = 8\,\text{nm}$ and the strength of the dipolar interactions $J_{ij}/r_0^3 \approx 2\pi \times 105\,\text{kHz}$. Initially all fictitious spins-1/2 were prepared in the states polarized along the z-axis, i.e. in the $|+1\rangle_z$ states. Then, $\Omega_x(t \le \tau_1) = 0$ and the microwave field realizing $\Omega_z(t \le \tau_1) = 2\pi \times 54.6\,\text{MHz}$ was applied for duration τ_1. Next $\Omega_z(\tau_1 < t) = 0$ but a strong microwave pulse, corresponding to $\Omega_x(\tau_1 < t) = 2\pi \times 41.7\,\text{MHz}$, was turned on for duration $\tau_2 \approx 12\,\text{ns}$ which rotated the spins-1/2 around the x-axis by an angle $\theta = \Omega_x(\tau_1 < t)\tau_2 \approx \pi$ and the entire driving sequence was repeated periodically with the period $T = \tau_1 + \tau_2$.

During the time intervals τ_1 there was no rotation of the spins-1/2 but the system was subject to: the dipole interactions, spatial disorder and spin locking realized by non-zero $\Omega_z(t \leq \tau_1)$ which helped to decouple spins from the environment. The product $\Omega_z(t \leq \tau_1)\tau_1$ was always chosen to be an integer multiple of 2π in order to avoid accidental $(2T)$-periodic evolution. The latter could appear if $\Omega_z(t \leq \tau_1)\tau_1$ was an odd multiple of π because such a rotation would correct an imperfect flip of the spins and even in the absence of the spin–spin interaction, $(2T)$-periodic evolution could emerge [33]. During the periods τ_2, which were short $\tau_2 \ll \tau_1$, the system dynamics was dominated by the spin rotation realized by $\Omega_x(\tau_1 < t)$. In the experiment different interaction periods τ_1 could be chosen. If the short periods (e.g. $\tau_1 = 92$ ns) were used and there was a detuning, $\epsilon = \pi - \theta$, from the perfect rotation of the spins by π angle, e.g. $\theta = 1.034\pi$, then the spins were not able to self-organize their motion and the magnetization of the spin-1/2 along the z-axis revealed beating in time. However, when the interaction time intervals τ_1 were sufficiently long, the spins-1/2 synchronized themselves and time-periodic evolution of the magnetization with the period $2T$ was observed, i.e. similar time crystal behavior as shown in Fig. 4.23 for the driven chain of ions. For $\tau_1 < 275$ ns, the longer interaction periods τ_1, the larger deviations from the perfect π-rotation of the spins were allowed for the realization of the discrete time crystal dynamics and the phase boundary between the time crystal and symmetry unbroken phases (i.e., pairs of τ_1 and critical θ_{cr}) revealed linear increase of $|\epsilon_{cr}| = |\theta_{cr} - \pi|$ with τ_1. This behavior could be explained by means of the mean field approach [33, 34, 86]. However, for $\tau_1 > 275$ ns, the critical detuning $|\epsilon_{cr}|$ first saturated and then decreased slightly with τ_1. This regime was identified with the so-called critical time crystal behavior, see Ref. [86] for details.

In the system of periodically driven NV centers in diamond it was also possible to observe spontaneous breaking of the discrete time translation symmetry of the Hamiltonian into periodic motion with the period three times longer than the driving period T. To this end Choi et al. employed all three electronic spin states of the NV impurities and performed two short microwave nearly $\theta \approx \pi$ pulses of the duration of τ_2. The first, corresponding to the unitary transformation $U_-(\theta)$, led to the transition between the electronic states $|m_s = 0\rangle \leftrightarrow |m_s = -1\rangle$ and the other, corresponding to $U_+(\theta)$, to the transition between $|m_s = 0\rangle \leftrightarrow |m_s = +1\rangle$. If $\theta = \pi$, the microwave pulses realized the $\mathbb{Z}_3$ cyclic dynamics in the space of the three electronic spin states. That is, the application of the pulses can be described, in the basis $\{|m_s = -1\rangle, |m_s = 0\rangle, |m_s = +1\rangle\}$, by the unitary transformation

$$U_{\mathrm{rot}}(\pi) = U_+(\pi)U_-(\pi) = \begin{bmatrix} 1 & 0 & 0 \\ 0 & 0 & 1 \\ 0 & 1 & 0 \end{bmatrix} \begin{bmatrix} 0 & 1 & 0 \\ 1 & 0 & 0 \\ 0 & 0 & 1 \end{bmatrix}, \tag{4.174}$$

where, for simplicity, we have assumed that no global phase is generated during each transition. It is easy to check that $(U_{\rm rot}(\pi))^3 = 1$ and consequently any initial state returns to itself after three applications of the pair of the pulses. If $\theta \neq \pi$, the $\mathbb{Z}_3$ cyclic dynamics is broken. However, it turns out that in the presence of the sufficiently strong interactions between the spin impurities, the system can self-synchronize its motion. In the experiment, initially all NV impurities were prepared in the $|m_s = 0\rangle$ state and then the system was periodically driven with the period $T = \tau_1 + 2\tau_2$ and the time evolution of the magnetization, which was defined as the population difference between the $|m_s = 0\rangle$ and $|m_s = -1\rangle$ states, was investigated. In Fig. 4.25 the Fourier transforms of the time evolution of the magnetization are presented. If the interaction periods τ_1 were sufficiently long, a stable time evolution with the period three times longer than T was observed. Indeed, in the lower panel of Fig. 4.25c, we can see clear Fourier peaks at the frequency $1/(3T)$ and at its second harmonic frequency $2/(3T)$.

In the follow-up experiment, see Ref. [34], Choi et al. investigated in detail how discrete time crystal dynamics depended on the length of the interaction period τ_1 which was practically equaled to T because $\tau_1 \gg \tau_2$. In the short-T regime, the boundary between the discrete time crystal phase and the symmetry unbroken phase corresponded to a linear dependence of the critical detuning $|\epsilon_{\rm cr}| = |\theta_{\rm cr} - \pi|$ on T. In that regime discrete time crystal signal exhibited rapid initial decay and next slow decay with the rate Γ independent of the detuning ϵ. For longer T, discrete time crystals still existed but their stability was attributed to critically slow thermalization [34, 86]. Finally, for even longer T, the decay of the $(2T)$- or $(3T)$-periodic evolution was purely exponential with the rate which depended quadratically on the detuning, i.e. $\Gamma \approx \epsilon^2/2$. This regime was identified with dephasing of individual spins arising from coherent spin–spin interactions where the many-body system constituted its own Markovian bath. After each rotation of the spins by an angle $\theta = \pi - \epsilon$, the dephasing process results in the projection of the spins on the z axis and consequently after each driving period the magnetization along z direction becomes smaller by the factor $\cos\epsilon$. The decay rate can be obtained from a simple formula $(\cos\epsilon)^n = e^{-\Gamma n}$, where n counts how many driving periods have passed and $\Gamma = -\ln(\cos\epsilon) \approx \epsilon^2/2$. The dephasing could be also a result of coupling of the system to the external bath but the analysis showed that it would take much longer than $1/\Gamma$. In the same publication Choi et al. present results not only for the natural dipole–dipole interactions, that involve both the Ising-type of interactions and the spin-exchange transitions, but also for the purely Ising interactions between the spins [34].[20] It turns out that in the purely Ising case, the third regime dominated by the dephasing processes is reached for much longer driving period T than in the case of the dipole–dipole interactions.

[20] In the Hamiltonian (4.173) one can identify the long-range Ising interactions corresponding to the term $\sum_{ij}(-J_{ij})\hat{s}_i^y\hat{s}_j^y/|\mathbf{r}_i - \mathbf{r}_i|^3$ and the remaining spin-exchange interactions where the operators $\hat{s}_i^x\hat{s}_j^x + \hat{s}_i^z\hat{s}_j^z = (\hat{s}_i^+\hat{s}_j^- + \hat{s}_i^-\hat{s}_j^+)/2$ are responsible for flip-flop processes between the spins at i and j sites. Indeed, the operators $\hat{s}_j^\pm = \hat{s}_j^x \pm i\hat{s}_j^z$ raise and lower the spins along the y direction.

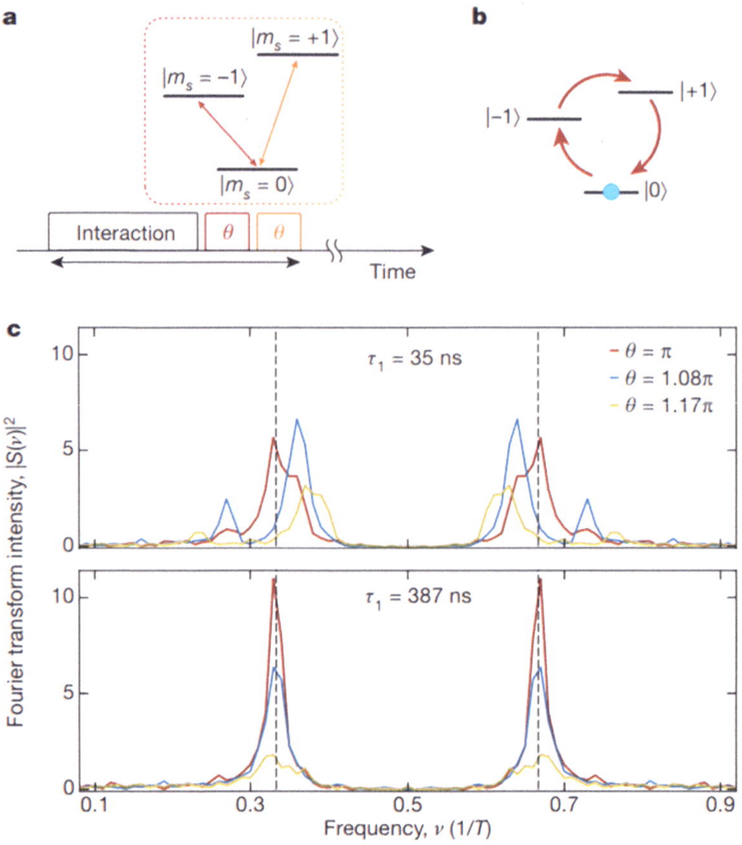

Fig. 4.25 (**a**): scheme of electronic spin $S = 1$ levels of NV impurities in diamond with the indication of the two transitions realized by means of two microwave $\theta \approx \pi$ pulses, each of the duration τ_2. In each period of the drive, $T = \tau_1 + 2\tau_2$, the pulses are preceded by the period τ_1 when the spin–spin interactions strongly influence the system dynamics. (**b**): schematic plot of the $\mathbb{Z}_3$ cyclic dynamics generated by the two microwave pulses if the angle $\theta = \pi$. (**c**): Fourier spectra of the magnetization (defined as the population imbalance of the $|m_s = 0\rangle$ and $|m_s = -1\rangle$ states) which were measured in [33] for two different interaction periods τ_1 and for different angles θ. If the interaction periods τ_1 are sufficiently long, there are strong Fourier peaks at $1/(3T)$ and $2/(3T)$ indicating the discrete time crystal dynamics with the period $3T$. Reprinted by permission from Springer Nature Customer Service Centre GmbH: Nature Springer, Choi et al. [33], Copyright (2017)

To sum up, the periodically driven 10^6 nitrogen-vacancy impurities in diamond allowed for observation of the discrete time crystal dynamics with the period twice or three times longer than the driving period. Spatial disorder was present in the system but it was not sufficiently strong to enforce many-body localization in the three-dimensional system.

4.6.3 *Periodically Driven Nuclear Magnetic Moments*

Two papers published back to back in the same issue of Physical Review Letters in 2018 reported the experiments where nuclear magnetic moments were employed to realize discrete time crystal dynamics [153, 170]. Pal et al. performed the experiments on star-shaped molecules (clusters) containing $N = 4, 10$, or 37 nuclear spins-1/2 [153] while Rovny et al. did experiments on a macroscopic number of nuclear spins-1/2 of ^{31}P in a crystal of ammonium dihydrogen phosphate. Nuclear magnetic resonance (NMR) techniques were used to realize periodic driving of the systems and to perform detection of the discrete time crystal dynamics.

In the Pal et al. experiments, periodically driven ensembles of about 10^{15} acetonitrile, trimethyl phosphite (TMP), or tetrakis(trimethylsilyl) silane (TTSS) clusters were investigated. In each case, the clusters contain the central magnetic moment associated with the spin-1/2 of the central nucleus and magnetic moments of the spins-1/2 of satellite nuclei, see Fig. 4.26. The satellite spins do not interact with each other but they do interact with the central spin. In the presence of a strong magnetic field along the z-axis and in the frame rotating at the Larmor precession frequencies of the individual nuclear spins, the resulting evolution of all spins is described by the unitary operator

$$U(t) = \exp\left(-it\frac{J}{\hbar}\hat{s}_0^z \sum_{j=1}^{N-1} \hat{s}_j^z\right), \tag{4.175}$$

where J is the strength of the Ising interactions between the central spin, described by the operator $\hat{\mathbf{s}}_0$, and the satellite spins $\hat{\mathbf{s}}_j$ where $j = 1, \ldots, N-1$. Every period T, all N spins are rotated around the x-axis by an angle θ, so that the Floquet unitary operator of the system reads

$$U_F = \exp\left(-i\theta \sum_{j=0}^{N-1} \hat{s}_j^x\right) U(T). \tag{4.176}$$

If all spins are initially polarized along the z-axis, then even if the rotation angle $\theta = \pi - \epsilon$ is slightly different from π (small ϵ), the spins synchronize their motion and the magnetization along the z-axis evolves with the period twice longer than the driving period T, see Fig. 4.26. In contrast, in the absence of the magnetic moment of the central nucleus, that is when acetonitrile contains spinless ^{12}C instead of ^{13}C, the magnetization reveals beating in time—compare panels (d) and (e) of Fig. 4.26. These two situations are intuitively illustrated with the help of the Bloch sphere representation in Fig. 4.27 which presents the results of numerical simulations [153]. In the so-called toggling frame of reference, where after each period T the basis for all spins is rotated by π around the x-axis, the periodic driving with the perfectly $\theta = \pi$ angle results in the stationary spins polarized along the z-

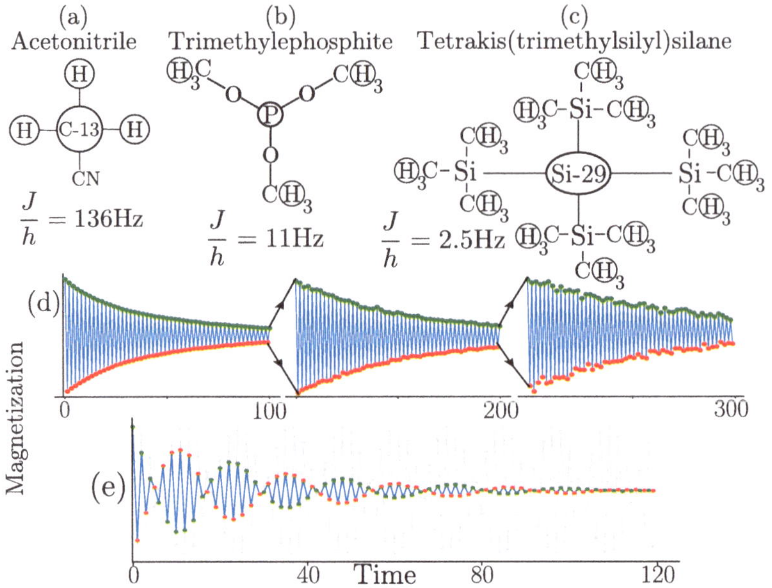

Fig. 4.26 Molecules used to demonstrate discrete time crystal dynamics in Pal et al. experiments [153]: acetonitrile (**a**), trimethyl phosphite (TMP) (**b**), and tetrakis (trimethylsilyl) silane (TTSS) (**c**). Each molecule contains the central nucleus with spin-1/2 and satellite spin-1/2 nuclei encircled in the panels which are active to nuclear magnetic resonance (NMR)—all together there are $N = 4$ (**a**), 10 (**b**), and 37 (**c**) nuclei active to NMR. The spins are initially polarized along the direction of the strong magnetic field and then subject to the periodic driving which flips the spins every period T. Despite the fact that the flips are not perfect the magnetization along the magnetic field direction evolves with the period $2T$ as illustrated in (**d**) where the experimental data corresponding to periodically driven TMP molecules are shown (for clarity the vertical scale is rescaled every 100th periods of the drive). The synchronization of the spins dynamics results from the interactions between the central and satellite spins whose strength J/h is depicted in (**a**)–(**c**). However, when the central nucleus is spinless, e.g. when acetonitrile contains not ^{13}C but ^{12}C, there are no interactions and the spins do not synchronize and the magnetization reveals beating in time—see blue curve in (**e**) that represents the experimental data, gray curve presents the theoretical prediction in the absence of the thermal bath. Reprinted figure with permission from Pal et al. [153]. Copyright (2018) by the American Physical Society

axis. However, when $\theta = \pi - \epsilon$ with $\epsilon \neq 0$, the evolution of the spins in the toggling frame reflects rotations with a small angle $-\epsilon$ about the x-axis combined with the precession resulting from the interactions between the spins, see panels (d) and (e) of Fig. 4.27. However, if the interactions are absent, there is no precession and after each period T, the spins gradually rotate about the x-axis by the full angle that in the laboratory frame corresponds to beating in time, see in Fig. 4.26e. Hence, the effect of randomization of the Larmor precession induced by the interactions is responsible for the stabilization of the $(2T)$-periodic discrete time crystal evolution. Such a semiclassical picture is valid provided the entanglement between the central and satellite spins is not maximal otherwise the magnetization of the central and

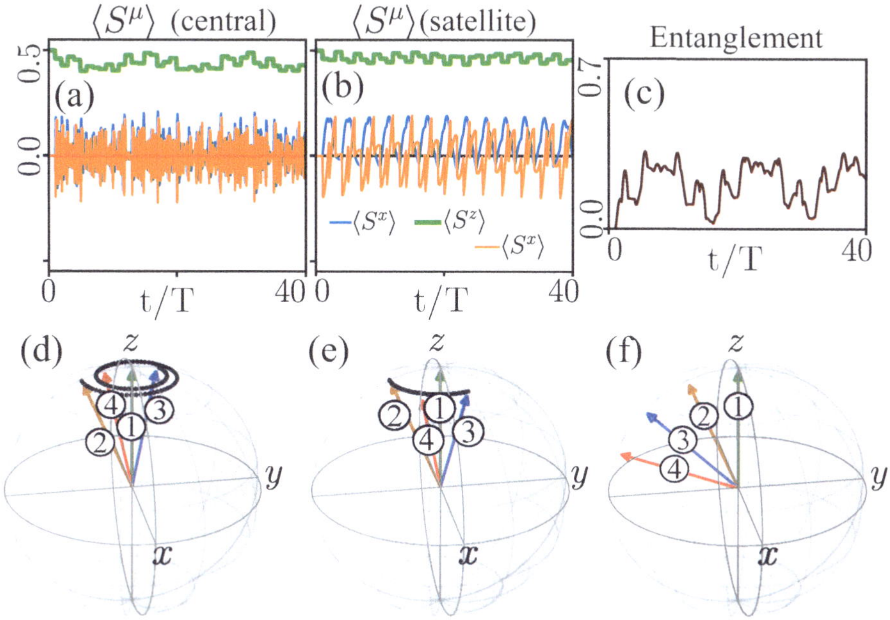

Fig. 4.27 The results of the numerical simulations of the dynamics of $N = 8$ spins described by the Floquet unitary operator (4.176) for $JT/\hbar = 4$ and $\theta = \pi - \epsilon$ where $\epsilon = 0.4$. Initially all spins are polarized along the z-axis. The toggling reference frame is used to present the results, i.e. after every period T the basis for the spins is rotated about the x-axis by π. In that frame, the perfect spin flips would correspond to the constant average values of the central and satellite spins. The imperfections of the spins flips are related to small oscillations as visible in (**a**) and (**b**). In (**d**) and (**e**) the corresponding Bloch sphere representation of the central spin and a satellite spin, respectively, are shown at $t = 0$ (number 1), just after the first flip of the spins (number 2), before the second flip (number 3) and just after the second flip (number 4). Panel (**c**) shows the entanglement entropy of the central spin which stays below the maximal value of $\ln 2$. Actually, long time numerical simulations show that the maximal entanglement between the central and satellite spins arises only for the time scale that increases exponentially with the number of spins N [153]. In (**f**) the same as in (**e**) is presented but for the non-interacting system. Reprinted figure with permission from Pal et al. [153]. Copyright (2018) by the American Physical Society

satellite spins would vanish. The numerical simulations presented in Fig. 4.27c show indeed that the von Neumann entropy stays well below its maximal value.

The theoretical description of the system based on the Floquet unitary operator (4.176) shows that the lifetime of the discrete time crystal dynamics increases exponentially with the magnetization of the satellite spins, see Ref. [153]. Indeed, the initial state, corresponding to all magnetic moments polarized along the z-axis, is a superposition of two Floquet states whose quasi-energy difference is proportional to $\epsilon^{-2|m|}$ in the toggling frame, or it is given by $\pi/T + O(\epsilon^{-2|m|})$ in the laboratory frame, where m is an eigenvalue of the total spin operator $\sum_{j=1}^{N-1} \hat{s}_j^z$ of all satellite nucleus. In other words, for large $|m|$, these Floquet states become Schrödinger cat like states, $|m_0, m\rangle \pm |-m_0, -m\rangle$ where m_0 is an eigenvalue of the central spin

$\hat{s}_0^z$. However, in the experiments, the discrete time crystal dynamics decays faster than expected from the theoretical prediction (cf. Fig. 4.26) which is attributed to the contact of the system with thermal bath.

While in the Pal et al. experiment the discrete time crystals were demonstrated in rather small systems of nuclear magnetic moments, in the Rovny et al. experiments [169, 170] a macroscopic number of nuclear spins were involved. A crystal of ammonium dihydrogen phosphate (the chemical formula $NH_4H_2PO_4$) was subject to a strong magnetic field and with the help of radio-frequency pulses the magnetic moments of the spins-1/2 of the ^{31}P nuclei were driven periodically. The spin magnetization of the ^{31}P nuclei achieved in the strong magnetic field at room temperature was periodically rotated about an axis perpendicular to the magnetic field direction by an angle $\theta = \pi - \epsilon$ by means of the short radio-frequency pulses of the duration τ_p. Between the successive radio-frequency pulses, the spins of the ^{31}P nuclei evolved for the time periods $\tau = T - \tau_p$ in the presence of the magnetic field and experienced dipole interactions between themselves, dipole interactions with the spins-1/2 of the ^{1}H nuclei and with the spins-1 of the ^{14}N nuclei described by the parts H_{int}^{PP}, H_{int}^{PH} and H_{int}^{PN} of the system Hamiltonian, respectively. The experiment could be performed in the presence of the coupling H_{int}^{PH} or when this coupling was turned off with the help of the high-power continuous-wave electromagnetic radiation. In the latter case substantial heating of the spins system was observed if the experiments lasted sufficiently long.

Qualitatively the $(2T)$-periodic dynamics observed by Rovny et al. was similar like in the Pal et al. case and like in the previously described experiments, see Fig. 4.26 and Sects. 4.6.1 and 4.6.2. Rovny et al. performed analysis which allowed them to shed light on causes of finite lifetime of the observed dynamics. They noticed that the envelopes of the magnetization signals closely followed a simple formula $(\cos\epsilon)^n$ where $\epsilon = \pi - \theta$ is deviation with respect to the perfect rotation of the spins by π and n is a number of periods after which the spin magnetization was detected. It suggested that after each rotation of the spins by the angle θ, the interactions described by the Hamiltonian $H_{\text{int}} = H_{\text{int}}^{PP} + H_{\text{int}}^{PH} + H_{\text{int}}^{PN}$ resulted in dephasing of the spins precession around the magnetic field vector and only the spins components along the magnetic field direction survived. In other words, after each period T of the driving, the magnetization got smaller by the factor $\cos\epsilon$. Such a dephasing resembles the similar dephasing observed by Choi et al. [33, 34] for nitrogen-vacancy centers in diamond in the limit of the long driving period. The contribution to the dephasing process coming from the part H_{int}^{PP} of the Hamiltonian H_{int} could be verified experimentally. Having turned off the part H_{int}^{PH} by means of the high-power continuous-wave, Rovny et al. were able to reverse the forward evolution of the driven system. That is, at the end of $n \leq 10$ periods of the driving, they applied a sequence of the rotations by $-\theta$ each followed by the evolution according to $-H_{\text{int}}^{PP}$. The resulting echo was maximal if a number n' of the backwards steps was close to the number n of the driving periods of the forward evolution. The echo signals exceeded the envelop imposed by the

$(\cos\epsilon)^{n+n'}$ formula indicating that the dephasing induced by $H_{\rm int}^{PP}$ was coherent to large extent and could be reversed.

If the rotation angle $\theta = \pi$, one may expect no decay of the $(2T)$-periodic dynamics provided each such a flip of the magnetic moments in the course of the periodic driving of the system is instantaneous [170]. However, the experimental data show that even if $\epsilon = 0$, the signals diminish. Rovny et al. investigated the experimental conditions and ruled out the imperfections of the radio-frequency pulses as a reason for the observed decay. Then, they turned to an analysis of effects of the finite time periods needed to rotate the spins by an angle $\theta = \pi$ which were $\tau_p \approx 7.5$ μs.[21] It turned out that the dipolar interactions present during the rotation periods τ_p played an important role in the observed decay of the signals. If the rotation angle $\theta \neq \pi$, the influence of the interactions during the rotation periods τ_p can be expected to be even greater [170].

The results of the numerical simulations performed by Pal et al. (see Fig. 4.27) show that the presence of the Ising spin–spin interactions does not cause the decay of the time crystal signals even if the spins rotation angle $\theta \neq \pi$ provided the magnetization of the satellite spins is large. In the Rovny et al. case, the dephasing related to the interactions $H_{\rm int}^{PP}$ contributes to the decay of the time crystal dynamics [170]. There are differences between these two experiments. In the Pal et al. numerical simulations, the rotations of the spins are instantaneous while in the Rovny et al. experiment the rotations last finite time periods and the interactions during these periods definitely influence the lifetime of the $(2T)$-periodic dynamics.

In the experiment on discrete time crystals in a chain of ions [209], there was strong disorder realized by an external field but there was no disorder in spin–spin interactions. Consequently, in the periodically driven case, many-body localization was not realized [103]. In the case of the discrete time crystal in periodically driven nitrogen-vacancy centers in diamond, disorder was present but it was too weak to ensure many-body localization phenomena. The systems of driven nuclear magnetic moments, which are described in the current section, contain little or even no spatial disorder.

4.6.4 Space-Time Crystals in an Atomic Bose–Einstein Condensate

In the Smits et al. experiment [186], a very different system, as compared to the experiments with periodically driven spin systems described in the previous sections, was employed to demonstrate discrete time crystal dynamics. Bose–

[21]The total driving period T ranged from 20 μs to even 1 s while the strengths of the dipolar interactions of the nuclear magnetic moment of ^{31}P with other ^{31}P and with ^{14}N, defined as the root mean square angular frequencies, were estimated to be 508 Hz and 97 Hz, respectively [170]. The interactions with ^{1}H were experimentally turned off in this analysis.

Einstein condensate of $N \approx 5 \times 10^7$ ultra-cold ^{23}Na atoms was prepared in the elongated cylindrical harmonic potential with the frequencies $\omega_\perp = 2\pi \times 52.7$ Hz and $\omega_z = 2\pi \times 1.43$ Hz along the radial (transverse) and axial (longitudinal) directions, respectively. The condensate fraction was greater than 0.9. The realization of the periodic driving and the method for visualization of the discrete time crystal dynamics were quite unique in this experiment.

The Floquet systems are systems which are periodically driven by external force and dynamics of equipment that provides force is not included in the Floquet Hamiltonian of the driven systems under consideration. In the Smits et al. experiment, the situation was different. The longitudinal degrees of freedom of the Bose–Einstein condensate corresponded to the system which was periodically driven while a periodically evolving transverse degree of freedom provided periodic driving force for the longitudinal modes. Due to the fact that the back-action from the longitudinal modes to the transverse mode was negligible, the dumping of the transverse oscillations was weak and the driving force for the longitudinal degrees of freedom was approximately periodic.

Usually in order to visualize discrete time crystal dynamics, many realizations of the same experiment are prepared and in each realization a certain quantity is measured. By changing time moment when this quantity is detected one obtains information about the time evolution of a system. However, Smits et al. could detect the discrete time crystal evolution in a single realization of the experiment. By means of the phase contrast imaging technique, they were able to get information about density profile of the Bose–Einstein condensate [186]. The information was so weak that the quantum state of the Bose system was not disturbed too much and it could evolve further and the same phase imaging technique could be applied many times in the course of the time evolution of the system. It was like in a standard situation for ordinary space crystals where we can look at a crystal many times and always see the same crystal without necessity of preparing a crystal again after each detection.

The experiment started with the preparation of the Bose–Einstein condensate in the elongated harmonic trap. Then, the frequency $\omega_\perp$ of the trap along the transverse directions was periodically modulated in time by means of three consecutive pulses with the modulation depth 0.125 that lasted 50 ms. This perturbation excited the radial breathing mode of the Bose–Einstein condensate with the frequency $\omega \approx 2\omega_\perp$, i.e. the atomic cloud started breathing with the frequency ω along the transverse directions. Due to the strong trap asymmetry, the energy of the radial mode $\hbar\omega$ was much greater than the energy $\hbar\omega_z$ corresponding to the axial direction and coupling of the radial mode to many longitudinal modes could be expected. After 500 ms starting from the onset of the experiment, Smits et al. observed a periodic structure in space in the longitudinal direction which oscillated in time with the period $2T$ where $T = 2\pi/\omega$ is the driving period related to the radial breathing mode oscillations, see Fig. 4.28. The periodic oscillations of the longitudinal pattern in time formed a two-dimensional periodic space-time structure which, when Fourier transformed, resulted in the peaks in the momentum-frequency

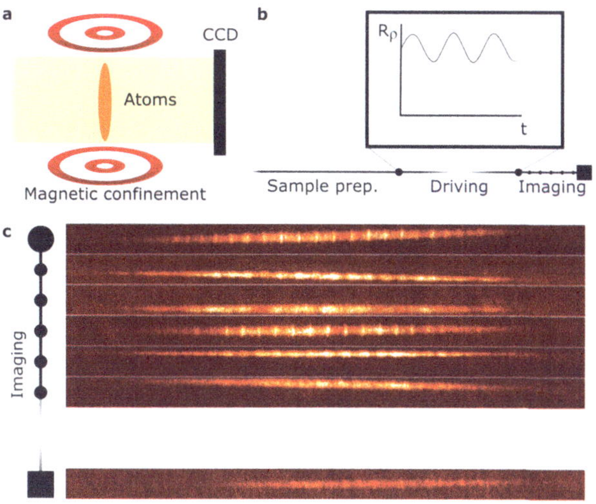

Fig. 4.28 Geometry of the Smits et al. experiment (**a**) and an illustration of the experimental sequence (**b**). Bose–Einstein condensate of ultra-cold ^{23}Na atoms was prepared in an elongated cylindrical harmonic trap with the frequencies along the radial and longitudinal directions fulfilling $\omega_\perp \gg \omega_z$. Then, the radial breathing mode of the atomic cloud was excited by modulating the trap frequency $\omega_\perp$ with the help of three pulses of 50 ms duration. The oscillating breathing mode provided periodic driving, with the frequency $\omega \approx 2\omega_\perp$, for longitudinal modes of the condensate. The excited longitudinal modes revealed a periodic structure in space which oscillated with the frequency $\omega/2$. (**c**) shows the first six images and the last image obtained in a single realization of the experiment where 50 images were taken every 3.28 ms. The images were obtained by means of the phase contrast method where atoms were illuminated with proper laser beams and signal was detected with a CCD camera, see (**a**). Reprinted figure with permission from Smits et al. [186]. Copyright (2018) by the American Physical Society

domain that confirmed actually the existence of a space-time crystal, see Fig. 4.29. As we will see in the following there are two degenerate $(2T)$-periodic mean field solutions for the longitudinal motion which are shifted in the time evolution with respect to each other by the time period T. Which of them is observed in a given realization of the experiment is determined spontaneously in the process of the experimental preparation.

In order to describe the system one can apply the mean-field approach [121, 186]. The initial Bose–Einstein condensate and the dynamics of the system when the space-time crystal is formed can be well described by the Gross–Pitaevskii equation where all atoms occupy the same single particle wavefunction $\psi(\mathbf{r}, t)$ (see background information in Sect. 3.1.1). If we parametrize the wavefunction by the atomic density $n(\mathbf{r}, t)$ and the phase $\phi(\mathbf{r}, t)$, i.e. $\psi = \sqrt{n}e^{i\phi}$, we obtain the

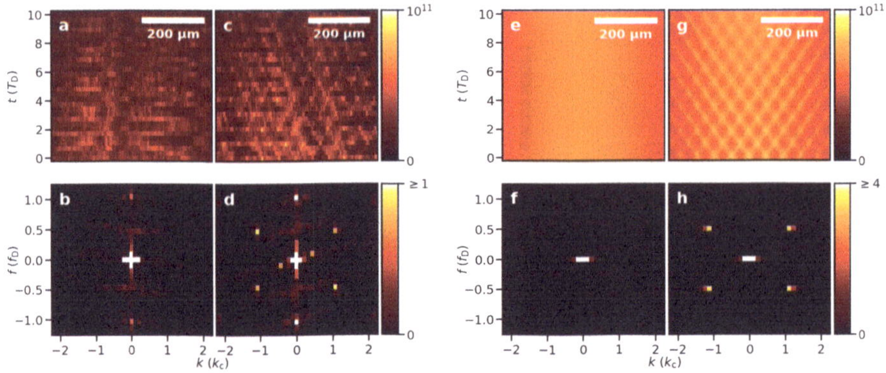

Fig. 4.29 (**a**) and (**c**) present the experimental data that illustrate how the density of atoms along the axial direction of the harmonic trapping potential, $n(0, 0, z, t)$, evolves in time before the onset of the space-time crystal formation (**a**) and after the crystal is formed (**c**). The time t in the vertical axes is given in the driving period units $T_D = T = 2\pi/\omega$ while the length scale in the horizontal directions, corresponding to the position z, is depicted by the 200 μm bars. Panels (**b**) and (**d**) show the Fourier transforms of the data presented in (**a**) and (**c**), respectively. The Fourier transforms switch from the space-time space to the momentum-frequency space. The frequency is given in the units of $f_D = \omega/(2\pi)$ while the momentum is presented in the units of $k_c = k_{19}$ where k_{19} is the momentum corresponding to the dominant longitudinal mode of the Bose–Einstein condensate. Note that the Fourier peaks in (**d**) confirm the formation of the space-time crystalline structure. Especially the peaks located at $f = f_D/2$ indicate that the system evolves with the period twice longer than the driving $T_D = T$. Panels (**e**)–(**h**) show the same as (**a**)–(**d**) but the results are obtained in the numerical integration of the Gross–Pitaevskii equation. Reprinted figure with permission from Smits et al. [186]. Copyright (2018) by the American Physical Society

hydrodynamic formulation of the Gross–Pitaevskii equation where the gradient of the phase $\nabla\phi$ plays the role of the velocity flow of atoms[22] [155].

The initial Bose–Einstein condensate is well described by the Thomas–Fermi profile, i.e. for a large atomic cloud with repulsive interactions (positive s-wave scattering length a_s of atoms) the density of atoms in the harmonic trap is given by a parabolic shape [155], cf. (4.178). When the radial breathing mode is excited, the cloud starts breathing what in turns excites axial modes of the condensate. Then, the density and phase of the condensate wavefunction can be described (provided the axial perturbation is small) by

$$n(\mathbf{r}, t) = n_0(\mathbf{r}, t) + \delta n(\mathbf{r}, t), \qquad \phi(\mathbf{r}, t) = \phi_0(\mathbf{r}, t) + \delta\phi(\mathbf{r}, t), \quad (4.177)$$

where δn and $\delta\phi$ describe the axial modes, which will be determined in the following, and

[22] Note that we assume here that the wavefunction is normalized not to the unity but to the total number of atoms in the system, i.e. $\langle\psi|\psi\rangle = N$.

$$n_0(\mathbf{r}, t) = \frac{m\omega_\perp^2 R_\perp^2}{2g_0 b_\perp^2(t) b_z(t)} \left[1 - \left(\frac{x}{b_\perp(t)R_\perp}\right)^2 - \left(\frac{y}{b_\perp(t)R_\perp}\right)^2 - \left(\frac{z}{b_z(t)R_z}\right)^2\right], \tag{4.178}$$

$$\nabla\phi_0(\mathbf{r}, t) = \frac{mR_\perp}{\hbar}\left(\dot{b}_\perp(t)\frac{x\,\mathbf{e}_x + y\,\mathbf{e}_y}{b_\perp(t)R_\perp} + \dot{b}_z(t)\frac{\omega_\perp z\,\mathbf{e}_z}{\omega_z b_z(t)R_z}\right), \tag{4.179}$$

where $R_\perp$ and R_z are the Thomas–Fermi radii of the initial unperturbed condensate in the harmonic trap, m is the mass of an atom and $g_0 = 4\pi\hbar^2 a_s/m$ with the s-wave scattering length $a_s > 0$. The parameters $b_z(t)$ and $b_\perp(t)$ describe the time evolution of the radial breathing mode and in the absence of the back-action from the axial modes, what we assume here, their time dependence is known [121]. We also assume that the breathing mode is weakly excited, i.e. $b_z(t) = 1$ and $b_\perp(t) = 1 + A\cos\omega t$ with $A \ll 1$ and $\omega \approx 2\omega_\perp$.

In the one-dimensional approximation, the axial perturbation can be expanded in the basis of $L_j(z/R_z) = P_{4j+2}(z/R_z) - P_{4j}(z/R_z)$ where $P_j(z)$ are Legendre polynomials.[23] The following ansatz is assumed to describe the time evolution of the axial perturbation [121]

$$\delta n(z, t) = -\sum_j \dot{\kappa}_j(t)\, L_j\left(\frac{z}{R_z}\right), \tag{4.180}$$

$$\delta\phi(z, t) = \frac{g_0}{\hbar}\sum_j \kappa_j(t)\, L_j\left(\frac{z}{R_z}\right), \tag{4.181}$$

where $\kappa_j(t)$ contains the entire information about the time evolution of a given j-mode. In the frame rotating with the frequency $\omega/2$, i.e.

$$\kappa_j(t) = \tilde{\kappa}_j(t)e^{-i\omega t/2} + \tilde{\kappa}_j^*(t)e^{i\omega t/2}, \tag{4.182}$$

when one neglects coupling between different axial modes and retains the terms up to the first order in A and employs the rotating wave approximation, the slowly varying amplitudes $\tilde{\kappa}_j(t)$ can be described by the linear equations [121]

$$i\frac{d}{dt}\begin{bmatrix}\tilde{\kappa}_j(t)\\ \tilde{\kappa}_j^*(t)\end{bmatrix} = \begin{bmatrix}-\delta_j & A\omega/4\\ -A\omega/4 & \delta_j\end{bmatrix}\begin{bmatrix}\tilde{\kappa}_j(t)\\ \tilde{\kappa}_j^*(t)\end{bmatrix}, \tag{4.183}$$

where for large j

[23] The Legendre polynomials form the basis suitable for description of excitations of a one-dimensional Bose–Einstein condensate in a harmonic trap in the Thomas-Fermi regime [121]. Note that only even Legendre polynomials are involved because due to the parity conservation, the radial breathing mode is coupled to axial modes of even parity only [121].

$$\delta_j \approx \frac{\omega}{2} - \left(2j + \frac{3}{4}\right)\omega_z. \tag{4.184}$$

Terms proportional to $\ddot{\kappa}_j(t)$ are neglected in the derivation of (4.183). Solutions of (4.183) evolve harmonically in time with the frequencies $\pm\Delta\Omega_j$, where

$$\Delta\Omega_j = \sqrt{\delta_j^2 - \left(\frac{\omega A}{4}\right)^2}, \tag{4.185}$$

provided the amplitude of the radial breathing mode $A \to 0$. Thus, if the driving force is weak and a given j-mode in the expansions (4.180)–(4.181) is initially populated, it evolves with the frequency $\omega/2 \pm \Delta\Omega_j$ and its amplitude does not change in time. However, when the amplitude A of the driving force increases, at some point, the frequency $\Delta\Omega_j$ related to the minimal value of $|\delta_j|$ becomes a purely imaginary complex number. Then, the contribution of the corresponding longitudinal mode to the axial perturbation δn and $\delta\phi$ increases in time and actually it diverges exponentially quickly. This is the onset of the formation of the space-time crystal. If a few frequencies $\Delta\Omega_j$ become imaginary complex numbers, then the mode with the greatest imaginary value will eventually dominate the perturbation of the system along the axial direction.

In the Smits et al. experiment $\omega \approx 2\omega_\perp$ and increasing A, the first longitudinal mode which starts growing exponentially in time corresponds to $j \approx 18$ according to (4.184) (a more accurate estimation results in $j = 19$ [121, 186]). This mode dominates the evolution of the experimentally measured atomic density. That is, for sufficiently large A, changes of the density along the longitudinal direction follow a periodic pattern in space with the spatial period $2\pi/k_{19} \approx 54$ μm and this pattern oscillates in time with the frequency $\omega/2$ and with the amplitude that initially increases exponentially with time with the rate $\mathrm{Im}(\Delta\Omega_{19})$. The exponential growth of $\tilde{\kappa}_{19}(t)$ is valid for a short time period when $\tilde{\kappa}_{19}(t)$ is small because it is obtained with the help of Eqs. (4.183) which are the result of a low-order expansion of the hydrodynamic equations in δn and $\delta\phi$ [121]. For the description of long time evolution of the system, one has to refer to the full numerical integration of the Gross–Pitaevskii equation which was performed in Refs. [121, 186] and the results are presented in Fig. 4.29.

In order to analyze spontaneous breaking of the discrete time translation symmetry we have to consider the quantum many-body Hamiltonian for the longitudinal degrees of freedom. The easiest way to obtain the desired Hamiltonian is to quantize the energy corresponding to the dynamical systems defined by Eqs. (4.183). Starting with the Lagrangian that generates the equations of motion (4.183), we can obtain the energy and then substitute $\tilde{\kappa}_j \to q_j\hat{a}_j$ and $\tilde{\kappa}_j^* \to q_j\hat{a}_j^\dagger$ where q_j's are normalization factors and the bosonic operators $\hat{a}_j$ and $\hat{a}_j^\dagger$ fulfill $[\hat{a}_j, \hat{a}_j^\dagger] = 1$. It results in the following quantum Hamiltonian

$$\hat{H} = \sum_j \left[-\hbar\delta_j \hat{a}_j^\dagger \hat{a}_j + \frac{\hbar\omega A}{8} \left(\hat{a}_j \hat{a}_j + \hat{a}_j^\dagger \hat{a}_j^\dagger \right) \right], \tag{4.186}$$

which describes quantum-mechanically the longitudinal degrees of freedom of the Bose–Einstein condensate [121, 186]. The U(1) symmetry, related to the unitary transformations $e^{iq\hat{N}_a}$ where q is a real parameter, is broken because the Hamiltonian $\hat{H}$ does not commute with the operator of the number of excitations of the longitudinal modes $\hat{N}_a = \sum_j \hat{a}_j^\dagger \hat{a}_j$. The Hamiltonian $\hat{H}$ possesses the $\mathbb{Z}_2$ symmetry, i.e. the transformation $\hat{a}_j \to -\hat{a}_j$ does not change the Hamiltonian. This symmetry is spontaneously broken when a given mode has a large occupation. Then, $\tilde{\kappa}_j \approx \pm q_j \langle \hat{a}_j \rangle \neq 0$ where the sign is chosen spontaneously in each realization of the experiment.

On the mean field level, and for the parameters of the Smits et al. experiment, there are two degenerate solutions $\pm\tilde{\kappa}_{19}(t)$ of the Eqs. (4.183) corresponding to the longitudinal mode $L_{19}(z/R_z)$, cf. (4.180) and (4.181), which dominates the axial perturbation in the experiment. Each of these two solutions leads to different evolution of the density pattern of the Bose–Einstein condensate along the axial direction, i.e.

$$n_\pm(0,0,z,t) \approx n_0(0,0,z,t) \pm \frac{i\omega}{2} \left[\tilde{\kappa}_{19}(t) e^{-i\omega t/2} - \tilde{\kappa}_{19}^*(t) e^{i\omega t/2} \right] L_{19}\left(\frac{z}{R_z} \right), \tag{4.187}$$

where the terms $\dot{\tilde{\kappa}}_{19}(t)$ and $\dot{\tilde{\kappa}}_{19}^*(t)$ are omitted because $\tilde{\kappa}_{19}(t)$ changes slowly as compared to the driving period $T = 2\pi/\omega$. The two densities $n_\pm$ evolve with the period $2T$ (assuming $\tilde{\kappa}_{19}(t) \approx$ constant) and one goes into the other after every period T, i.e. $n_+(0,0,z,t) = n_-(0,0,z,t+T)$. Which of these two densities is realized in the experiment is determined randomly in the process of the experimental preparation.

The lifetime of the space-time crystal observed in the experiment is finite due to the presence of a thermal could of atoms [121, 186]. The resulting dumping of the space-time crystal has not been included in the analysis presented here.

4.7 Dissipative Systems

The previous sections of the present chapter were devoted to the concept of discrete time crystals in isolated periodically driven quantum many-body systems. Spontaneous breaking of the discrete time translation symmetry of time-periodic Hamiltonians and the resulting stable and long-lived subharmonic evolution of such systems is non-trivial because a generic isolated quantum many-body system is expected to be heated to the infinite temperature state under a time-periodic drive. The lack of the heating is basically attributed to integrability of the systems. For example, due to the presence of local integrals of motion in systems which reveal many-body localization or due to underlying integrable classical dynamics, the discrete time crystals are resistant to the heating.

For a given discrete time crystal in an isolated system, the question arises if such a crystalline behavior survives when a system is coupled to an environment? Lazarides and Moessner analyzed this problem in Ref. [114]. They focused on periodically driven spin systems with strong disorder where in the many-body localization regime any short-range correlated state evolves with the period twice longer than the driving period and such evolution is expected to last forever, see Sect. 4.3.1. It turns out that if there is coupling between such a system and an environment, discrete time crystal dynamics can only survive if an environment is very fine-tuned. For example, it is the case for an environment which performs *measurement* on a system by projecting a state of a system on the time-dependent basis that evolves according to the Floquet operator of a system itself. However, for a typical, non-fine-tuned, environment, there is no chance for discrete time crystals that have been proposed in many-body localized systems and they have to die [114].

Can an environment be our ally and an essential contributor to the formation of a steady state of a system that evolves periodically in time? It is known from the systematic studies of Prigogine and others that open systems that are out of thermodynamic equilibrium can possess steady states that break spontaneously time translation symmetry and reveal periodic evolution in time [144]. Examples of such dissipative structures range from hydrodynamics, chemical reactions, and biological organisms to quantum systems. In his Nobel lecture Ilya Prigogine writes [161]:

> There is a striking similarity between the ferromagnetic system and the case of oscillating chemical reactions. When we increase the distance from equilibrium, the system begins to oscillate. It will move along the limit cycle. The phase on the limit cycle is determined by the initial fluctuation and plays the same role as the direction of magnetization. If the system is finite, fluctuations will progressively take over and perturb the rotation. However, if the system is infinite, then we may obtain a long-range temporal order very similar to the long-range space order in the ferromagnetic system. We see, therefore, that the appearance of a periodic reaction is a process that breaks time symmetry exactly as ferromagnetism is a process that breaks space symmetry.

In the present monograph it is impossible to cover even in part the field of dissipative systems. We restrict to the description of only a few examples investigated recently in the context of time crystals. The considered examples concern periodically driven systems but also time-independent systems that can develop steady-state periodic motion.

4.7.1 Dicke Time Crystals in Driven Dissipative Systems

There are already articles that investigate subharmonic response of a dissipative system to a periodic drive in the context of the discrete time crystal dynamics [65, 71, 116, 151, 211]. Among them the first paper was published by Gong et al. where the authors considered a periodically modulated open Dicke model and presented a variety of different discrete time crystal orders [71]. Before we switch to the description of the driven case we first remind a reader basic information concerning the original Dicke model which describes superradiance of atoms in an optical cavity.

It should be mentioned that quantum optics is a perfect playground for investigation of non-equilibrium phenomena. Exact quantum solution for subharmonic generation in nonlinear optics, that is related to the discrete time crystal dynamics in dissipative systems, was provided long time ago [49, 50] (see also [190, 191] and the references therein).

Dicke Model

In 1954 Dicke considered spontaneous emission of light by N atoms which are prepared in their excited state and which are located close to each other on a distance of a fraction of the wavelength of the spontaneously emitted light [46, 105]. It turns out that emission of a single photon triggers a chain reaction leading to the decay of all atoms and emission of N photons in free space. This phenomenon was dubbed superradiance [46]. Later it was shown that steady state superradiance can be observed in an ensemble of atoms collectively coupled to a quantized mode of the electromagnetic cavity [84, 197]. When the coupling between atoms and the electromagnetic field reaches a critical value, a continuous quantum phase transition occurs and the number of photons in the cavity becomes proportional to N.

Let us consider N two-level atoms (each described by the Hamiltonian $\Omega_0 \hat{s}_i^z$ where $\hat{s}_i^z$ are z-component of the spin-1/2 operator and Ω_0 is the energy splitting of the two atomic levels) which are collectively coupled to a single electromagnetic mode of the cavity. The energy of the electromagnetic field is described by $\Omega \hat{a}^\dagger \hat{a}$ where $\hat{a}$ and $\hat{a}^\dagger$ are standard bosonic operators that annihilate and create a photon of the cavity mode of the frequency Ω. The entire Hamiltonian of the system reads

$$H = \Omega \hat{a}^\dagger \hat{a} + \Omega_0 \sum_{i=1}^{N} \hat{s}_i^z + \frac{2\lambda}{\sqrt{N}} (\hat{a} + \hat{a}^\dagger) \sum_{i=1}^{N} \hat{s}_i^x . \tag{4.188}$$

The last term in (4.188) describes the collective coupling between atoms and the cavity field with the strength proportional to $1/\sqrt{N}$ which guarantees a well-defined thermodynamic limit. Taking the advantage of the collective coupling we can define the total spin operators $\hat{S}^{x,y,z} = \sum_{i=1}^{N} \hat{s}_i^{x,y,z}$ and write the Hamiltonian in the form

$$H = \Omega \hat{a}^\dagger \hat{a} + \Omega_0 \hat{S}^z + \frac{2\lambda}{\sqrt{N}} (\hat{a} + \hat{a}^\dagger) \hat{S}^x . \tag{4.189}$$

(continued)

The system possesses the $\mathbb{Z}_2$ symmetry, i.e. the Hamiltonian commutes with the parity operator $P = e^{i\pi(\hat{a}^\dagger\hat{a}+\hat{S}_z+N/2)}$ which is related to the simultaneous change of $\hat{a} \to -\hat{a}$, $\hat{S}_x \to -\hat{S}_x$ and $\hat{S}_y \to -\hat{S}_y$. When $\lambda \to 0$, the ground state of the system corresponds to all atoms in the lowest energy level and no photons in the cavity. However, when λ reaches a critical value, the $\mathbb{Z}_2$ symmetry becomes spontaneously broken in the thermodynamic limit and there are two degenerate symmetry broken ground states with the average number of photons in the cavity proportional to N.

The analysis of the quantum phase transition can be easily performed within the mean field approach but before we apply it let us make the problem more general and assume that photons can leak out of the cavity due to coupling to a bath. Assuming that the bath is Markovian and there are no photons in the bath, the density matrix of the atom and cavity mode system can be described by the quantum master equation

$$\dot{\rho} = \mathcal{L}(\rho) = -i[H, \rho] + \frac{\kappa}{2}\left(2\hat{a}\rho\hat{a}^\dagger - \hat{a}^\dagger\hat{a}\rho - \rho\hat{a}^\dagger\hat{a}\right), \tag{4.190}$$

where $\mathcal{L}$ is the Lindblad generator [71]. In the Heisenberg picture, the corresponding time evolution of an operator $\hat{O}$ is described by the adjoint Lindblad generator $\mathcal{L}^\dagger$ [23],

$$\frac{d\langle\hat{O}\rangle}{dt} = \left\langle\mathcal{L}^\dagger(\hat{O})\right\rangle = \left\langle i[H, \hat{O}] + \frac{\kappa}{2}\left(2\hat{a}^\dagger\hat{O}\hat{a} - \hat{a}^\dagger\hat{a}\hat{O} - \hat{O}\hat{a}^\dagger\hat{a}\right)\right\rangle, \tag{4.191}$$

where it is assumed that $\hat{O}$ is explicitly time-independent. For $\hat{O} = \hat{a}$ and $\hat{O} = \hat{S}^{x,y,z}$ we get

$$\begin{aligned}
\frac{d\langle\hat{a}\rangle}{dt} &= -\left(i\Omega + \frac{\kappa}{2}\right)\langle\hat{a}\rangle - \frac{i2\lambda}{\sqrt{N}}\langle\hat{S}^x\rangle, \\
\frac{d\langle\hat{S}^x\rangle}{dt} &= -\Omega_0\langle\hat{S}^y\rangle, \\
\frac{d\langle\hat{S}^y\rangle}{dt} &= \Omega_0\langle\hat{S}^x\rangle - \frac{2\lambda}{\sqrt{N}}\left\langle(\hat{a}+\hat{a}^\dagger)\hat{S}^z\right\rangle, \\
\frac{d\langle\hat{S}^z\rangle}{dt} &= \frac{2\lambda}{\sqrt{N}}\left\langle(\hat{a}+\hat{a}^\dagger)\hat{S}^y\right\rangle.
\end{aligned} \tag{4.192}$$

In the large N limit, in order to find steady states of the system we may employ the mean field approximation for the electromagnetic mode, i.e. $\hat{a} \approx \langle\hat{a}\rangle \equiv a$, and assume that this mode is coupled to the classical spin degrees of freedom

(continued)

$S^{x,y,z} = \langle \hat{S}^{x,y,z} \rangle$ where the length of the spin vector $S = N/2$ tells us how many atoms are in the cavity. In the mean field approximation we may $\left\langle \hat{a}\hat{S}^{y,z} \right\rangle \approx \langle \hat{a} \rangle \langle \hat{S}^{y,z} \rangle$ because in the limit of $N \to \infty$, quantum fluctuations around the average values are negligible. Then, for $\lambda > \lambda_c$, where

$$\lambda_c = \frac{1}{2}\sqrt{\frac{\Omega_0}{\Omega}\left(\Omega^2 + \frac{\kappa^2}{4}\right)}, \tag{4.193}$$

there are two steady state solutions of (4.192) which indicate a quantum phase transition to the superradiant phase where the $\mathbb{Z}_2$ symmetry is broken,

$$\begin{aligned} S^x &= \pm \frac{N}{2}\sqrt{1 - \frac{\lambda_c^4}{\lambda^4}}, \\ S^y &= 0, \\ S^z &= -\frac{N}{2}\frac{\lambda_c^2}{\lambda^2}, \\ a &= \mp\sqrt{N}\frac{\lambda}{\Omega - i\frac{\kappa}{2}}\sqrt{1 - \frac{\lambda_c^4}{\lambda^4}}. \end{aligned} \tag{4.194}$$

Indeed, the average number of photons in the cavity $\langle \hat{a}^\dagger \hat{a} \rangle = |a|^2 \propto N$ and the steady states are not invariant under the exchange of $a \to -a$ and $S_x \to -S_x$.

In the case of the closed system (i.e., for $\kappa = 0$), mean field ground states can be obtained by the substitution $\hat{a} \to a$ and $\hat{S}^{x,y,z} \to S^{x,y,z}$ in (4.189) and minimization of the resulting energy that leads to the two degenerate solutions (4.194) with $\kappa = 0$.

Let us consider a driven Dicke model where we assume that the coupling constant λ in (4.188) is modulated periodically in time with the frequency ω in the following way [71]

$$\lambda(t) = \lambda(t+T) = \begin{cases} \lambda_0 & \text{for} \quad 0 \le t < \frac{T}{2}, \\ 0 & \text{for} \quad \frac{T}{2} \le t < T, \end{cases} \tag{4.195}$$

where $T = 2\pi/\omega$ and $\lambda_0 > \lambda_c$.

For $\Omega = \Omega_0 = \omega$ and $\kappa = 0$, the free evolution during the second part of the period T corresponds to the parity operator (up to an unimportant global phase). Indeed, the time evolution during $T/2 \le t < T$, generated by the

Hamiltonian (4.188), leads to the unitary transformation

$$e^{-iTH/2} = e^{iT(\Omega\hat{a}^\dagger\hat{a}+\Omega_0\hat{S}^z)/2} = e^{-i\pi(\hat{a}^\dagger\hat{a}+\hat{S}^z)}, \tag{4.196}$$

that changes $\hat{a} \to -\hat{a}$ and $\hat{S}^x \to -\hat{S}^x$. Hence, if at $t = T/2$ the system is in one of the steady states (4.194), at $t = T$ it ends up in the other steady state. Thus, if we start with one of the steady states, the modulated Dicke model reveals $(2T)$-periodic evolution where spontaneous breaking of the discrete time translation symmetry of the time-periodic Hamiltonian corresponds to the spontaneous breaking of the $\mathbb{Z}_2$ symmetry in the static Dicke model.

In the presence of the dissipation ($\kappa \neq 0$), the period-doubling evolution is robust even if there is a small detuning between the cavity frequency Ω and the resonant frequency Ω_0 of the two-level atoms, i.e. the detuning that results in imperfect switching between the two steady states during the evolution in the second part of the driving periods. The robustness is due to the collective behavior built during the first part of the driving periods, cf. (4.195), and the contractive nature of the dissipation which drives the system to one of the steady states.

In Ref. [71] the detuning ϵ is defined via the relations $\Omega = (1-\epsilon)\omega$ and $\Omega_0 = (1+\epsilon)\omega$ and for $\kappa = 0.05$, $\lambda_0 = 1$ and $\omega = 1$, the phase diagram is obtained, see Fig. 4.30. Because of the nonlinearity of the mean field equations that describe

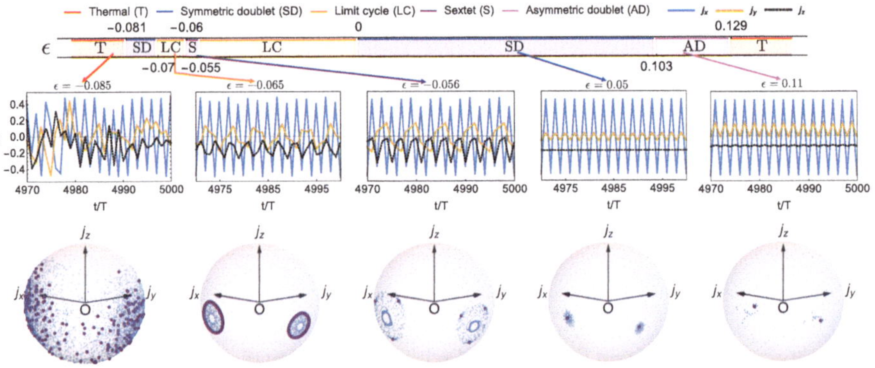

Fig. 4.30 Periodically driven Dicke model. Top row shows the phase diagram vs. the detuning ϵ for $\lambda_0 = 1$ and $\kappa = 0.05$. For increasing positive values of ϵ, one observes first the symmetric period-doubling phase (SD), then the asymmetric period-doubling order (AD) and finally irregular thermal trajectories (T). For $\epsilon < 0$ and increasing negative values of ϵ, first a pair of two limit cycles (LC) emerges, then the period-sixtupling order (S), the limit cycles again, the symmetric period-doubling order and finally the thermal phase. Middle row presents examples of evolution of the spin components for different phases—the plotted spin components are renormalized, i.e. $j^{x,y,z} = S^{x,y,z}/\sqrt{N}$. The bottom row shows the same trajectories as in the middle row but in the Bloch sphere representation. In the middle row the last 30 periods of the time evolution obtained numerically is presented only. In the bottom row full trajectories (sampled every period T) and the last 200 points are presented with blue and purple dots, respectively. Reprinted figure with permission from Gong et al. [71]. Copyright (2018) by the American Physical Society

evolution of a and $S^{x,y,z}$, there is variety of different steady state behavior of the periodically driven system for different values of the system parameters. For small and positive ϵ, the symmetric period-doubling is observed where after each period $S^{x,y} \to -S^{x,y}$. When ϵ becomes greater than about 0.103, the period-doubling becomes asymmetric, i.e. $S^y \to -S^y$ is no longer true, while for $\epsilon \gtrsim 0.129$ the system enters a thermal phase with irregular trajectories. For small negative ϵ one observes that the system tends to the steady state where the spin evolution forms a pair of symmetric loops on the Bloch sphere (i.e., a pair of limit cycles). If ϵ becomes more and more negative, a period-sixtupling, again limit cycle and symmetric period-doubling and finally thermal steady states are subsequently observed, see Fig. 4.30.

In the $N \to \infty$ limit, the mean field approach describes accurately the Dicke model. Such a mean-field solvability is due to the lack of interactions between atoms and the resulting collective character of the model. In Ref. [211] Zhu and coworkers analyzed if the Dicke time crystal is stable in the presence of short-range atom–atom interactions. They considered the Hamiltonian (4.188) supplemented with an interaction term, i.e.

$$H = \Omega \hat{a}^\dagger \hat{a} + \Omega_0 \sum_{i=1}^{N} \hat{s}_i^z + \frac{2\lambda}{\sqrt{N}}(\hat{a} + \hat{a}^\dagger) \sum_{i=1}^{N} \hat{s}_i^x + J \sum_{i=1}^{N} \hat{s}_i^x \hat{s}_{i+1}^x, \tag{4.197}$$

where J determines the strength of the interactions. Taking into account the dissipation and employing the time-dependent spin-wave approach, Zhu et al. showed that for weak interactions, i.e. for $J/\omega \ll 1$ where $\omega = (\Omega + \Omega_0)/2$, the Dicke time crystal survives and is stable despite a small detuning $(\Omega_0 - \Omega)/\omega = -0.12$, see Fig. 4.31a. Strong interactions create substantial quantum fluctuations and in the presence of the time-periodic driving the system suffers from heating. For moderate interactions ($J/\omega \lesssim 1$), the heating can be suppressed by sufficiently strong dissipation ($\kappa/\omega \gtrsim 1$) which works as a contractor guiding the system to the desired steady states (Fig. 4.31b). However, if dissipation is very strong the dynamics becomes irregular or atoms are cooled to their ground state and the Dicke time crystal is destroyed, see Fig. 4.31. The results presented in this figure correspond to the antiferromagnetic atom–atom interactions ($J > 0$). It turns out that for ferromagnetic interactions ($J < 0$), time crystalline behavior can be also observed and even when the interaction strength is beyond the perturbative regime [211].

Finally we would like to mention that driven dissipative atom-cavity systems allow for realization of various steady states that break the time translation symmetry and can evolve even with a period that is incommensurate multiple of the driving period [35].

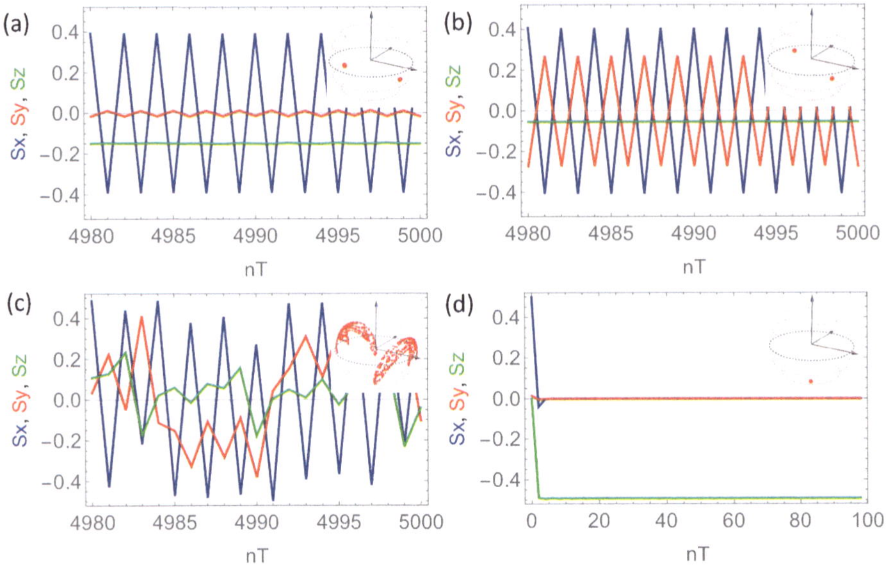

Fig. 4.31 Periodically driven Dicke model with the antiferromagnetic atom–atom interactions. Long time dynamics is presented for the detuning $(\Omega_0 - \Omega)/\omega = -0.12$. Despite the presence of the interactions, stable discrete time crystal dynamics is observed provided the dissipation is properly chosen: $J/\omega = 0.056$ and $\kappa/\omega = 0.375$ in (**a**) and $J/\omega = 0.2$ and $\kappa/\omega = 1.45$ in (**b**). In (**c**) there is no interaction ($J = 0$) and too strong dissipation ($\kappa/\omega = 0.9$) makes the dynamics irregular. In (**d**), which corresponds to $J/\omega = 0.04$ and $\kappa/\omega = 3.2$, the dissipation is so strong that it cools atoms to their ground states and no dynamics is observed. The presented spin components are normalized so that the length of the spin vector $S = 1/2$. Note that in (**a**)–(**c**), the last 20 periods of the time evolution obtained numerically is presented only. Reprinted figure from Zhu et al. [211]

4.7.2 *Discrete Time Crystals in Dissipative Systems in the Absence of Manifest Symmetries*

Let us assume for a moment that we have an open time-independent quantum many-body system which possesses the $\mathbb{Z}_2$ symmetry. If, in the thermodynamic limit, this symmetry is spontaneously broken, two steady state solutions emerge which do not respect the $\mathbb{Z}_2$ symmetry. Then applying a time-periodic driving that in each period transforms a system between the two steady states, we can observe a discrete time crystal dynamics where a system evolves with the period twice longer than the driving period. We may expect that dissipation can be our ally because an imperfect transfer between the steady states should be corrected by the contractive dissipative dynamics which pushes a system towards a desire steady state. This scenario is illustrated in Fig. 4.32 and its signatures can be identified in the Dicke time crystal—see the previous section.

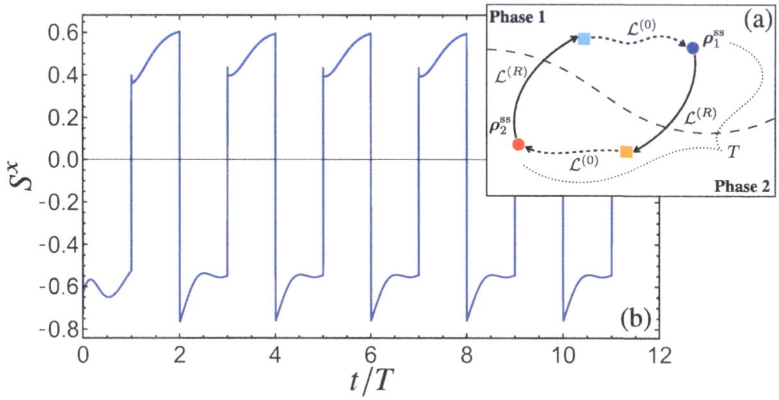

Fig. 4.32 Panel (**a**): suppose there are two steady states, ρ_1^{ss} (blue dot) and ρ_2^{ss} (red dot), surrounded by separate basins of attraction in the phase space. If there is an appropriate combination of transformation $\mathcal{L}^{(R)}$ and dissipative evolution $\mathcal{L}^{(0)}$, then one steady state is transformed into the basin of attraction of the other one and the subsequent contractive dynamics $\mathcal{L}^{(0)}$ pushes it towards the desired steady state. If the transformation $\mathcal{L}^{(R)}$ is applied periodically, we can observe time evolution of a system with the period twice longer than the driving period. Panel (**b**): mean field time evolution of the spin magnetization along the x axis demonstrating discrete time crystal dynamics in a driven dissipative gas of Rydberg atoms considered in Ref. [65]. Dark and bright areas in both panels indicate the basins of attraction of the steady states ρ_1^{ss} and ρ_2^{ss}, respectively. Reprinted figure with permission from Gambetta et al. [65]. Copyright (2019) by the American Physical Society

It turns out that similar mechanism can be also realized even if there is no manifest symmetry that can be spontaneously broken. Gambetta and coworkers showed that metastable states of open time-independent systems are also suitable for discrete time crystal dynamics if a system is appropriately driven [65], see also [198]. Before we sketch their idea we need some basic information about spectral properties of Lindblad generators [23].

Spectral Properties of Lindblad Generators

Let us consider a quantum system coupled to a Markovian bath [23, 113]. Time evolution of a density matrix ρ of a system can be approximated by the quantum master equation which can be written the following general form

$$\frac{d\rho}{dt} = \mathcal{L}(\rho) = -i[H, \rho] + \sum_j \left(\hat{\ell}_j \rho \hat{\ell}_j^\dagger - \frac{1}{2}\left\{ \hat{\ell}_j^\dagger \hat{\ell}_j, \rho \right\} \right) \tag{4.198}$$

(continued)

where $\mathcal{L}$ is the Lindblad generator (often called a superoperator because it acts in the space of density matrices, which are operators themselves, acting in the Hilbert space), H is the Hamiltonian of a system, and $\{\cdot, \cdot\}$ stands for the anticommutator.[24] The so-called jump operators $\hat{\ell}_j$ are responsible for the coupling between a system and a bath via $\sum_j \hat{\ell}_j \hat{\Gamma}_j$ where $\hat{\Gamma}_j$ are operators defined in the Hilbert space of a bath. The Lindblad form of the evolution equation (4.198) guarantees that $\rho(t)$ is positively defined (i.e., eigenvalues of $\rho(t)$ are non-negative) and $\mathrm{Tr}[\rho(t)] = 1$. It means that $\rho(t)$ properly describes a quantum state of a system.

Density matrices form a vector space and it is convenient to consider a superoperator $\mathcal{L}$ as a matrix which multiplies a vector made of elements of a density matrix [113]. Let us write a density matrix in the vector form by means of the Choi–Jamiolkowski isomorphism, $|i\rangle\langle j| \to |i\rangle|j\rangle$,

$$\rho = \sum_{i,j} \rho_{ij} |i\rangle\langle j| \to \mathrm{vec}(\rho) = \sum_{i,j} \rho_{ij} |i\rangle|j\rangle. \tag{4.199}$$

Then, the action of the Lindblad generator, Eq. (4.198), can be written in the form,

$$\mathcal{L}(\rho) \to \hat{\mathcal{L}}\mathrm{vec}(\rho), \tag{4.200}$$

where $\hat{\mathcal{L}}$ is a matrix by which the vector $\mathrm{vec}(\rho)$ is multiplied. Solutions of the master equation can be thus written as

$$\mathrm{vec}(\rho(t)) = e^{\hat{\mathcal{L}}t}\mathrm{vec}(\rho(0)). \tag{4.201}$$

If $\hat{\mathcal{L}}$ is diagonalizable (which in general is not guaranteed), then solutions (4.201) can be expanded in the eigenvector basis. Suppose $\hat{\mathcal{L}}$ is diagonalizable. Because it is not a Hermitian matrix, its eigenvalues λ_n need not be real and right $\mathrm{vec}(R_n)$ and left $\mathrm{vec}(L_n)$ eigenvectors corresponding to the same eigenvalue need not be the same,

$$\hat{\mathcal{L}}\mathrm{vec}(R_n) = \lambda_n \mathrm{vec}(R_n), \qquad \mathrm{vec}(L_n)^\dagger \mathcal{L} = \lambda_n \mathrm{vec}(L_n)^\dagger, \tag{4.202}$$

where, in general, the matrices $L_n \neq R_n$. The diagonal decomposition of $\hat{\mathcal{L}}$ can be written in the form

$$\hat{\mathcal{L}} = \sum_n \lambda_n \mathrm{vec}(R_n)\mathrm{vec}(L_n)^\dagger, \tag{4.203}$$

(continued)

[24] Explicitly, for two operators $\hat{a}$ and $\hat{b}$, the anticommutator is $\{\hat{a}, \hat{b}\} = \hat{a}\hat{b} + \hat{b}\hat{a}$.

with

$$\text{vec}(L_n)^\dagger \text{vec}(R_{n'}) = \delta_{nn'}. \tag{4.204}$$

The scalar product of $\text{vec}(L_n)$ and $\text{vec}(R_{n'})$ is actually the Schmidt–Hilbert inner product of the corresponding matrices, i.e. $\text{Tr}[L_n^\dagger R_{n'}] = \text{vec}(L_n)^\dagger \text{vec}(R_{n'})$.

The trace of a density matrix does not change in the time evolution

$$\begin{aligned}\frac{d}{dt}\text{Tr}[\rho(t)] = \frac{d}{dt}\text{Tr}[I^\dagger \rho(t)] = \frac{d}{dt}\text{vec}(I)^\dagger \text{vec}(\rho(t)) \\ = \text{vec}(I)^\dagger \hat{\mathcal{L}}\text{vec}(\rho(t)) = 0,\end{aligned} \tag{4.205}$$

which implies that $\text{vec}(I)$, where I is the identity matrix, is the left eigenvector of $\hat{\mathcal{L}}$ corresponding to the zero eigenvalue $\lambda_1 = 0$. If ρ_{ss} is the steady state solution of the master equation (which, let us assume, is unique), then

$$\frac{d}{dt}\text{vec}(\rho_{ss}) = \hat{\mathcal{L}}\text{vec}(\rho_{ss}) = 0, \tag{4.206}$$

and it means that $\text{vec}(\rho_{ss})$ is the right eigenvector of $\hat{\mathcal{L}}$ corresponding to the eigenvalue $\lambda_1 = 0$. We can now decompose the time evolution of the density matrix in the eigenvector basis

$$\text{vec}(\rho(t)) = \text{vec}(\rho_{ss}) + \sum_{n=2} c_n e^{\lambda_n t}\text{vec}(R_n), \tag{4.207}$$

or in the matrix form

$$\rho(t) = \rho_{ss} + \sum_{n=2} c_n e^{\lambda_n t} R_n, \tag{4.208}$$

where $c_n = \text{Tr}[L_n^\dagger \rho(0)]$ (note that $c_1 = \text{Tr}[I^\dagger \rho(0)] = 1$). Because $\text{Tr}[\rho_{ss}] = 1$, we have necessarily $\text{Tr}[R_n] = 0$ for $n \geq 2$ and such R_n matrices alone cannot be a density matrix. Moreover, real parts of all non-zero eigenvalues must be negative, i.e. $\text{Re}(\lambda_n) < 0$ for $n \geq 2$, otherwise we would have unphysical exponential divergence of time evolving components of solutions of the master equation.

Finally let us show that complex eigenvalues always come in conjugate pairs λ_n and λ_n^*. Indeed, on the one hand we have

(continued)

$$(\mathcal{L}(R_n))^\dagger = (\lambda_n R_n)^\dagger = \lambda_n^* R_n^\dagger, \tag{4.209}$$

and on the other hand it can be checked by means of (4.198) that the Hermitian conjugate of the result of $\mathcal{L}(R_n)$ reduces to

$$(\mathcal{L}(R_n))^\dagger = \mathcal{L}(R_n^\dagger). \tag{4.210}$$

Thus, $\mathcal{L}(R_n^\dagger) = \lambda_n^* R_n^\dagger$ and consequently $R_n^\dagger$ is the right eigensolution corresponding to the eigenvalue λ_n^*.

Now we can come back to the discrete time crystal proposed by Gambetta et al. [65]. Let us assume that an open time-independent quantum system consisting of, e.g., N spins is described by the master equation with the Lindblad generator $\mathcal{L}^{(0)}$. We also assume that for $N < \infty$ there is a unique stationary state ρ^{ss} which is the eigenstate (or rather eigenmatrix) of $\mathcal{L}^{(0)}$ corresponding to the non-degenerate eigenvalue $\lambda_1 = 0$, i.e. $\mathcal{L}^{(0)}(\rho^{ss}) = 0$. All other eigenvalues λ_m of $\mathcal{L}^{(0)}$ have non-vanishing negative real parts which we order as follows: $\mathrm{Re}(\lambda_{m+1}) \leq \mathrm{Re}(\lambda_m) < 0$. Finally we assume that when $N \to \infty$, the second eigenvalue λ_2 is real-valued and $\lambda_2 \to 0$ while $\mathrm{Re}(\lambda_3) < 0$. The vanishing gap between λ_2 and $\lambda_1 = 0$ implies the existence of a metastable regime [131, 168]. That is, for $t \gg \tau_2$ (where $\tau_2 = 1/\lambda_2$) the system reaches the stationary state ρ^{ss}, however, when $\tau_3 \ll t \ll \tau_2$ (where $\tau_3 = 1/\mathrm{Re}(\lambda_3)$) the dynamics is effectively reduced to the subspace spanned by the right eigenmatrices $R_1 = \rho^{ss}$ and R_2 of $\mathcal{L}^{(0)}$ corresponding to the eigenvalues $\lambda_1 = 0$ and λ_2, respectively. The eigenmatrix R_2 cannot be considered as a density matrix of a system because $\mathrm{Tr}[R_2] = 0$. Therefore, it is more convenient to define the following extreme density matrices as a basis in the metastable subspace [131, 168]

$$\tilde{\rho}_1 = \rho^{ss} + c_2^{\max} R_2, \qquad \tilde{\rho}_2 = \rho^{ss} + c_2^{\min} R_2, \tag{4.211}$$

where $c_2^{\min}$ and $c_2^{\max}$ are extreme values of the coefficient $c_2 = \mathrm{Tr}[L_2 \rho(0)]$ which are equal to the minimal and maximal eigenvalues of the left eigenmatrix L_2 of $\mathcal{L}^{(0)}$, respectively.[25] In the metastable subspace any density matrix of a system can be written as $p_1\tilde{\rho}_1 + p_2\tilde{\rho}_2$, where $p_1 + p_2 = 1$ and $p_{1,2} \geq 0$, due to the convexity of the metastable subspace inherited from the convexity of the full space of density matrices. In the limit $N \to \infty$ the two extreme density matrices $\tilde{\rho}_{1,2}$ become two effective stationary states with their own basins of attraction. Note that in general there is no $\mathbb{Z}_2$ symmetry of a system which could transform one of these two effective stationary states into the other.

[25] For a real eigenvalue λ_m of a Lindblad generator, the corresponding left and right eigenmatrices L_m and R_m can be chosen as Hermitian matrices.

In order to realize a discrete time crystal with the help of the effective stationary states $\tilde{\rho}_{1,2}$ it is necessary to invent another Lindblad operator $\mathcal{L}^{(R)}$ which generates the dynamics $e^{\hat{\mathcal{L}}^{(R)}t}$ that transforms one of the stationary states (4.211) into the basin of attraction of the other one. All together we need a time-periodic Lindblad generator

$$\mathcal{L}(t) = \mathcal{L}(t+T) = \begin{cases} \mathcal{L}^{(R)} & \text{for} \quad 0 \le t < t_R, \\ \mathcal{L}^{(0)} & \text{for} \quad t_R \le t < T, \end{cases} \tag{4.212}$$

where $T - t_R \gg \tau_3$. During the time intervals t_R, the evolution operator $e^{\hat{\mathcal{L}}^{(R)}t}$ results in the transfer of a system from one of the effective steady states (4.211) to the basin of attraction of the other one. Next, during the periods $T - t_R$, the contractive dynamics corresponding to the Lindblad generator $\mathcal{L}^{(0)}$ pushes a system to a desired steady state and the whole procedure is repeated periodically, see Fig. 4.32. The resulting periodic evolution of a system takes place with a period twice longer than the driving period demonstrating discrete time crystal dynamics.

Let us describe a concrete example analyzed by Gambetta and coworkers [65]. They consider N Rydberg atoms on a lattice whose ground and Rydberg energy levels correspond to the eigenvalues $\pm 1/2$ of the spin operators $\hat{s}_i^z = \sigma_i^z/2$, where $\sigma_i^{x,y,z}$ are the Pauli matrices, which describe Rydberg atoms located in different lattice sites (i is the lattice site index). There is long-range interaction between atoms but only if they are in the Rydberg state which can be described by $\sum_{i \neq j} V_{ij}\hat{n}_i\hat{n}_j$ where $\hat{n}_i = I/2 + \hat{s}_i^z$, with the identity matrix I, and V_{ij} determines the strength of the interaction between i and j atoms. The two atomic energy states are coupled by laser pulses with the Rabi frequencies $\Omega_{x,y}(t+T) = \Omega_{x,y}(t)$ and with the detuning $\Delta(t+T) = \Delta(t)$. The laser pulses are applied periodically with the period T. The entire Hamiltonian reads

$$H = \sum_{i=1}^{N} \left[2\Omega_x(t)\hat{s}_i^x + 2\Omega_y(t)\hat{s}_i^y + \Delta(t)\hat{n}_i\right] + \sum_{i \neq j}^{N} V_{ij}\hat{n}_i\hat{n}_j. \tag{4.213}$$

The finite lifetime of the Rydberg states due to spontaneous emission, with the rate Γ, is described by the master equation with the time-periodic Lindblad generator $\mathcal{L}(t+T) = \mathcal{L}(t)$,

$$\frac{d\rho}{dt} = \mathcal{L}(t)(\rho) = -i[H(t), \rho] + \Gamma \sum_{j=1}^{N} \left(\sigma_j^- \rho \sigma_j^+ - \frac{1}{2}\left\{\sigma_j^+\sigma_j^-, \rho\right\} \right), \tag{4.214}$$

where the jump operators $\sigma_j^{\pm} = (\sigma_j^x \pm i\sigma_j^y)/2$. During the first part of the driving period, i.e. for $0 \le t < t_R$, the Lindblad generator $\mathcal{L}(t) = \mathcal{L}^{(R)}$, with the parameters $(\Omega_x(t), \Omega_y(t), \Delta(t)) = (\Omega_x^{(R)}, \Omega_y^{(R)}, \Delta^{(R)})$, is used while for $t_R \le t < T$ the generator $\mathcal{L}(t) = \mathcal{L}^{(0)}$ is applied with the parameters $(\Omega_x(t), \Omega_y(t), \Delta(t)) = (\Omega_x^{(0)}, 0, \Delta^{(0)})$.

Let us first analyze stationary properties of the Lindblad generator $\mathcal{L}^{(0)}$. In the limit when $N \to \infty$, which is the limit where we expect the existence of the metastable regime, we may employ the mean-field approximation by assuming that all spin operators can be substituted by the same expectation value $\hat{s}_i^{x,y,z} \to \langle \hat{s}_i^{x,y,z} \rangle \equiv S^{x,y,z}$ and similarly $\hat{n}_i \to \langle \hat{n}_i \rangle \equiv n$. Depending on values of the parameters $\Delta^{(0)}/\Gamma$, $\Omega_x^{(0)}/\Gamma$ and V/Γ (where $V = 2N^{-1}\sum_{i \neq j} V_{ij}$) there is one or two stable stationary mean field solutions, $S^{x,y,z}$ and n, of the master equation (compare Eq. (4.192) and the related text in Sect. 4.7.1)—the entire phase diagram is analyzed in detail in Refs. [65, 132, 201]. We focus on the regime where there are two stable mean-field stationary solutions that we denote by the vectors $\mathbf{M}_{1,2}^{ss} = (S_{1,2}^x, S_{1,2}^y, S_{1,2}^z)$ and which correspond to the two metastable states (4.211) described in the quantum many-body description when $N < \infty$. Note that there is no $\mathbb{Z}_2$ symmetry of the system which can transform one of the mean-field stationary solutions into the other one.

Finally we have to find the Lindblad generator $\mathcal{L}^{(R)}$ which drives the system between two basins of attraction of the two stationary states. The idea is to perform, in the spin space, the global rotation by π around the bisecant of the two unit vectors $\mathbf{m}_{1,2}^{ss} = \mathbf{M}_{1,2}^{ss}/|\mathbf{M}_{1,2}^{ss}|$, i.e. around the direction of

$$\mathbf{d} = (d_x, d_y, d_z) = \frac{\mathbf{m}_1^{ss} + \mathbf{m}_2^{ss}}{|\mathbf{m}_1^{ss} + \mathbf{m}_2^{ss}|}. \tag{4.215}$$

That is, we need the dynamics which during time intervals $0 \leq t < t_R$ realizes the rotation $e^{-i\pi \sum_j \hat{\mathbf{s}}_j \cdot \mathbf{d}}$. If we forget about the interactions and the spontaneous emission of atoms, such a dynamics is realized by the Hamiltonian (4.213) with the parameters

$$\Omega_x^{(R)} = \frac{\pi d_x}{2t_R}, \qquad \Omega_y^{(R)} = \frac{\pi d_y}{2t_R}, \qquad \Delta^{(R)} = \frac{\pi d_z}{t_R}. \tag{4.216}$$

Influence of the interactions and dissipation, which are present during the rotation periods τ_R, can be reduced if t_R is chosen sufficiently short.

In Fig. 4.32b discrete time crystal dynamics is illustrated where the mean field time evolution of S^x (under the periodic application of the evolution operators with the Lindblad generators $\mathcal{L}^{(R)}$ and $\mathcal{L}^{(0)}$) is presented. The transfer of the system between the two basins of attraction of the stationary states and subsequent dissipative dynamics lead to time-periodic evolution with the period $2T$ and it is robust even if one modifies slightly the global rotation (4.216). Gambetta et al. analyzed also quantum dynamics beyond the mean field approximation for a finite system ($N = 12, 20, 28$) and in the case of the uniform all to all spin–spin interactions, i.e. when $V_{ij} = V_0/N$ for any i and j. They found that the lifetime of the discrete time crystal dynamics increased algebraically with the number of Rydberg atoms like N^α where $\alpha \approx 0.5$.

4.7.3 Analysis of Discrete Time Crystals in Open Systems

In the previous section we have described an idea for realization of discrete time crystals in open systems which is summarized in Fig. 4.32a. This proposal can be applied to a system either with a $\mathbb{Z}_2$ symmetry that is spontaneously broken or without such a symmetry but when effective stationary states emerge in the limit of large number of degrees of freedom of a system [65]. In the proposal an environment coupled to a system drains energy from a system which otherwise could be heated to infinite temperature by a periodic drive. However, as pointed out by Lazarides and coworkers [116], coupling to an environment not only introduces the desired cooling but can also generate noise which can eventually lead to loss of coherence and decay of time crystal dynamics. In the present section we describe the results of the Lazarides et al. work [116].

Let us assume that the Hilbert space of an open time-independent system can be divided into two sectors and each sector possesses a steady state, e.g., resulting from spontaneous breaking of a $\mathbb{Z}_2$ symmetry of a system. Each steady state is characterized by a different average value of an observable $\hat{M}$ (which can be the magnetization of a quantum spin system). Having a system with such properties, let us turn on periodic driving (with the period $T = t_R + t_C$) which is similar to the driving considered by Gambetta et al. [65], see Fig. 4.32a. That is, during the first part of the driving period (i.e., for $0 \leq t < t_R$), there is a global *rotation* which in the absence of the dissipation would map a neighborhood of one of the symmetry broken ground states to states belonging to the other sector. In the second part of the driving period ($t_R \leq t < T = t_R + t_C$), which is a cooling period, a system evolves in the presence of the dissipation and presumably is driven towards a desired steady state and the whole procedure is repeated periodically.

Let us consider realization of such a protocol in a quantum spin system. Assume that we start with the state of the system in the Hilbert space sector characterized by the positive magnetization $m(0) = \mathrm{Tr}[\rho(0)\hat{M}] > 0$ where ρ is the density matrix of the system. We also require that the initial state corresponds to the probability distribution of the observable $\hat{M}$ which is negligibly small in the other sector where the magnetization is opposite. After the application of the global rotation, the magnetization becomes reversed, $m(t_R) < 0$, and we may expect that after the subsequent cooling period the system will end up in the steady state of the sector characterized by the negative magnetization. If after each driving period the same scenario is observed, we will deal with $(2T)$-periodic evolution of the magnetization and consequently with the discrete time crystal dynamics, see Fig. 4.33a. However, if the global rotation is not instantaneous, the system is traveling through excited states during t_R and it is at the same time being cooled by the dissipation. As a result, before the global rotation is completed, some part of the population can be cooled down and trapped in the wrong sector. It increases the width of the probability distribution of the observable $\hat{M}$ and such a generation of noise can eventually kill $(2T)$-periodic evolution of $m(t)$, see Fig. 4.33b.

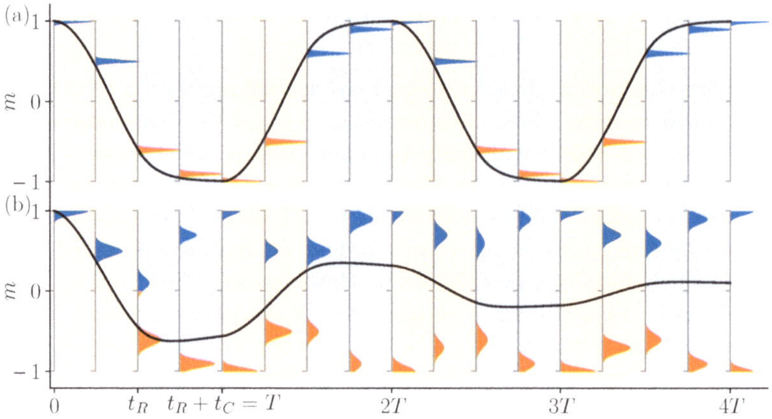

Fig. 4.33 Time evolution of the magnetization $m = \langle \hat{M} \rangle$ of a dissipative quantum spin system. It is assumed that due to spontaneous breaking of a $\mathbb{Z}_2$ symmetry of a system, there are two steady states corresponding to the opposite directions of the magnetization. Each steady state belongs to one of the two different sectors of the Hilbert space which are characterized by positive and negative values of m. Moreover, it is assumed that the global rotation of all spins is performed periodically, with the period $T = t_R + t_C$, which lasts t_R and which reverses the direction of the magnetization. During the entire evolution the dissipation is present. Panel (**a**) illustrates a situation when in the course of the global rotations, when a state of a system is not transferred yet to the desired Hilbert space sector, the cooling process does not significantly populate a wrong sector and the $(2T)$-periodic evolution of the magnetization survives. In panel (**b**) the situation is different. That is, before a system lands in a desired sector, the cooling is able to populate a wrong sector sufficiently strongly so that in the course of time evolution a mixture of two steady states is formed and the time crystal dynamics of the magnetization dies. Reprinted figure with permission from Lazarides et al. [116]

Lazarides et al. analyze concrete examples of discrete time crystal dynamics in open systems [116]. They begin with the ferromagnetic Ising model in one-dimensional space described by the Hamiltonian,

$$H_0 = -\sum_i \sigma_i^z \sigma_{i+1}^z + g \sum_i \sigma_i^x, \tag{4.217}$$

where $\sigma_i^{x,z}$ are the Pauli matrices. In the ferromagnetic phase ($|g| < 1$) and in the thermodynamic limit, there are two $\mathbb{Z}_2$ symmetry broken ground states which for $g \to 0$ correspond to all spins up $|\Uparrow\rangle = |\uparrow, \uparrow, \ldots\rangle$ or all spins down $|\Downarrow\rangle = |\downarrow, \downarrow, \ldots\rangle$. The time-periodic driving of the system is introduced by assuming that the Hamiltonian is time-periodic $H(t + T) = H(t)$ and during the first part of the period ($0 \leq t < t_R$) a perfect rotation of all spins around the x direction by an angle θ is realized, i.e.

$$H(t) = H(t+T) = \begin{cases} \frac{\theta}{t_R}\sum_i \sigma_i^x & \text{for} \quad 0 \le t < t_R, \\ H_0 & \text{for} \quad t_R \le t < T = t_R + t_C. \end{cases} \tag{4.218}$$

Moreover, it is assumed that the system is coupled to a Markovian bath and its time evolution is described by the master equation

$$\frac{d\rho}{dt} = -i[H(t), \rho] + \sum_\alpha \left(\hat{\ell}_\alpha \rho \hat{\ell}_\alpha^\dagger - \frac{1}{2}\{\hat{\ell}_\alpha^\dagger \hat{\ell}_\alpha, \rho\} \right), \tag{4.219}$$

with the time-independent quantum jump operators,

$$\hat{\ell}_\alpha = \sqrt{\gamma}\,[\Theta(m_\alpha)| \Uparrow \rangle\langle\alpha| + \Theta(-m_\alpha)| \Downarrow \rangle\langle\alpha|], \tag{4.220}$$

where γ determines the effective decay rate, $\Theta(m_\alpha)$ is the Heaviside step function, $m_\alpha = \langle\alpha|\hat{M}|\alpha\rangle$ is the magnetization in a product spin state $|\alpha\rangle$ (i.e., a state of the form $|\uparrow, \downarrow, \downarrow, \ldots\rangle$) and $\hat{M} = \sum_i \sigma_i^z$. A jump operator $\hat{\ell}_\alpha$ projects a state of the system on a product state $|\alpha\rangle$ and, depending on if $|\alpha\rangle$ belongs to the positive ($m_\alpha > 0$) or negative ($m_\alpha < 0$) magnetization sectors, a jump operator $\hat{\ell}_\alpha$ transfers a system state directly to the corresponding ground state $| \Uparrow \rangle$ or $| \Downarrow \rangle$. Such a coupling to the environment ensures a very efficient cooling but on the other hand it may also leave some population in a ground state of a wrong sector. Numerical simulations show that the latter effect is fatal to the discrete time crystal. That is, the system tends to a mixture of the two ground states related to the opposite magnetization and the $(2T)$-periodic evolution of the magnetization decays exponentially with time, see Fig. 4.34. If we consider an observable whose average values are the same for both ground states, then it should be finite in a statistical mixture of them. In Fig. 4.34, the $\mathbb{Z}_2$ invariant correlator,

$$C(t) = \frac{1}{N^2}\sum_{i \neq j} \mathrm{Tr}[\rho(t)\sigma_i^z \sigma_j^z], \tag{4.221}$$

is shown which still reveals T-periodic evolution even if the $(2T)$-periodic oscillations of the magnetization decay in time like $e^{-t/\tau}$.[26] The lifetime τ increases exponentially with γt_C (for fixed γt_R) but it decreases polynomially with γt_R (for fixed γt_C). To conclude the $(2T)$-periodic evolution of the magnetization always decays exponentially with time and consequently the considered time crystal is doomed to the destruction.

In Ref. [116] also another kind of a coupling to an environment is analyzed where the jump operators in the master equation can move domain walls in the

[26]It is an indication that the system evolves towards a state that is fundamentally different from the discrete time crystal considered in an isolated periodically driven spin system with strong spatial disorder where both $C(t)$ and $m(t)$ show up persistent oscillations, see Ref. [102].

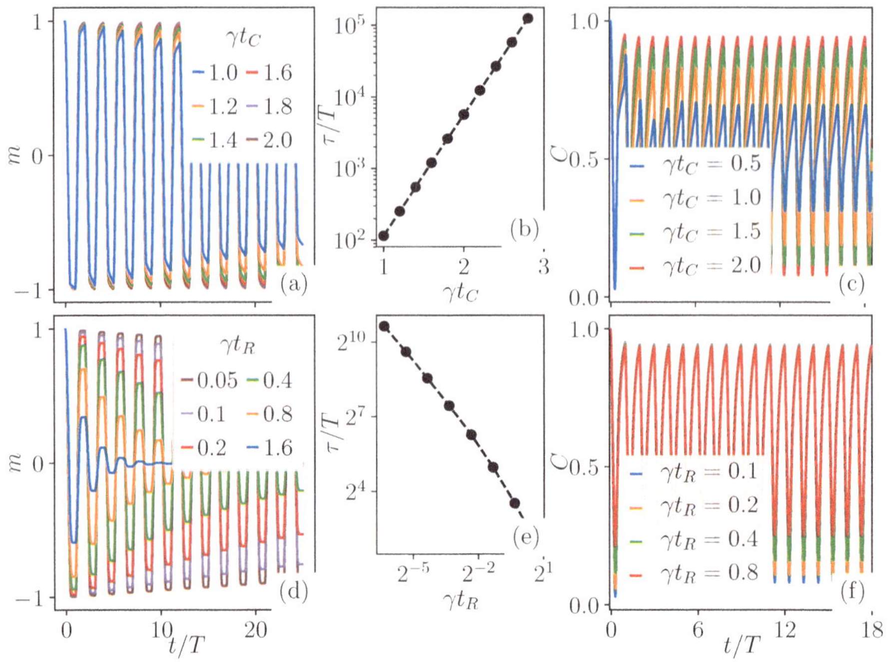

Fig. 4.34 Periodically driven ferromagnetic Ising model in one-dimensional space in the presence of the dissipation. The system is described by the master equation (4.219) with the quantum jump operators (4.220). In panel (**a**) time evolution of the magnetization $m = \mathrm{Tr}[\rho(t)\hat{M}]$ is presented for fixed $\gamma t_R = 0.1$ and different values of γt_C while in panel (**d**) $\gamma t_C = 10$ is fixed and different γt_R are chosen. In all cases the $(2T)$-periodic oscillations of the magnetization decay in time like $e^{-t/\tau}$. For fixed γt_R the lifetime τ increases exponentially with γt_C, see panel (**b**), while for fixed γt_C, τ decreases polynomially with γt_R, see panel (**e**). The $\mathbb{Z}_2$ symmetric correlator $C(t)$, Eq. (4.221), for fixed values of $\gamma t_R = 0.1$ or $\gamma t_C = 2$ is presented in panels (**c**) and (**f**), respectively. Periodic oscillations of $C(t)$ do not decay but its amplitude depends on γt_R and γt_C. The number of spins $N = 51$ and the angle of the global rotation: $\theta = 0.45\pi$ (**a**)–(**b**), $\theta = \pi/2$ (**d**)–(**e**) and $\theta = 0.4\pi$ (**c**)–(**f**), see (4.218). Reprinted figure with permission from Lazarides et al. [116]

ferromagnetic Ising model. The jump operators are not able to annihilate a single domain wall but two domain walls can annihilate if they move towards each other. Numerical simulations show that in the one-dimensional case the discrete time crystal dynamics decays exponentially with time similarly as in the case of the jump operators (4.220). However, it turns out that in the two-dimensional case, the $(2T)$-periodic oscillations of the magnetization are stable and discrete time crystal dynamics is not killed.

In Sects. 4.7.1 and 4.7.2 discrete time crystals in dissipative systems are considered where the protocol of the time-periodic driving of the systems also relies on switching between two distinct sectors of the Hilbert space. In these cases if fully connected models (i.e., Hamiltonians with all to all interactions) are used, then in the limit of a large number of degrees of freedom, $N \to \infty$, the noise that we analyze in the present section is suppressed and the master equation solutions for the density matrix result in no broadening of the probability distributions [17, 66, 116].

4.7.4 Breaking of Continuous Time Translation Symmetry in Dissipative Systems

In the previous sections we have provided examples of periodically driven dissipative systems that reveal time-periodic dynamics in a steady state which breaks the discrete time translation symmetry dictated by an external periodic drive. It is also known that open systems are able to perform time-periodic dynamics even if they are not periodically driven [144, 161]. Indeed, in the presence of non-conservative forces, a time-independent dynamical system can be attracted not only to a stable fixed point, which is a time-independent configuration, but also to its one-dimensional version which is a limit cycle—a non-trivial periodic trajectory. Such systems are time independent, therefore, the time-periodic evolution in a steady state breaks actually continuous time translation symmetry of a system. In the present section we describe an example of periodic evolution which arises in a time-independent open system and which has been recently analyzed in the context of time crystals.

Iemini et al. considered a time-independent system that can be divided into two parts: a boundary consisting of N_b degrees of freedom and a bulk with N_B degrees of freedom [97], see Fig. 4.35. If the boundary is a macroscopic subsystem, i.e. $N_b \to \infty$, but still much smaller than the bulk, i.e. $N_B \to \infty$ and $N_b/N_B \to 0$, the boundary can be described by the density matrix ρ_b reduced to the N_b degrees of freedom which fulfills the master equation provided the bulk can be modeled by a Markovian bath. For example, let us focus on a collection of N_b spins-1/2 which play a role of the boundary degrees of freedom and which can be described by the Hamiltonian $H_b = \omega_0 \hat{S}^x$ with the x component of the collective spin operator $\hat{S}^x = \sum_{j=1}^{N_b} \hat{s}_j^x$. We assume that the reduced density matrix ρ_b fulfills the master equation

$$\frac{d\rho_b}{dt} = \mathcal{L}(\rho_b) = -i[\omega_0 \hat{S}^x, \rho_b] + \frac{\kappa}{S}\left(\hat{S}_- \rho_b \hat{S}_+ - \frac{1}{2}\left\{\hat{S}_+ \hat{S}_-, \rho_b\right\}\right), \tag{4.222}$$

where κ is the effective decay rate, $S = N_b/2$ is the total spin and the quantum jump operator $\hat{S}_- = \hat{S}_+^\dagger = \hat{S}^x - i\hat{S}^y$ [97]. This model was used to describe cooperative emission of atoms in electromagnetic cavities [47, 48, 77, 196]. The model fulfills the requirements needed to observe time-periodic evolution in the thermodynamic limit. That is, for $N_b \to \infty$ the spectrum of the Lindblad generator $\mathcal{L}$ has a vanishing gap between the zero eigenvalue and real parts of complex eigenvalues with non-vanishing imaginary parts (see background information in Sect. 4.7.2 for basic properties of $\mathcal{L}$). The vanishing real parts of the complex

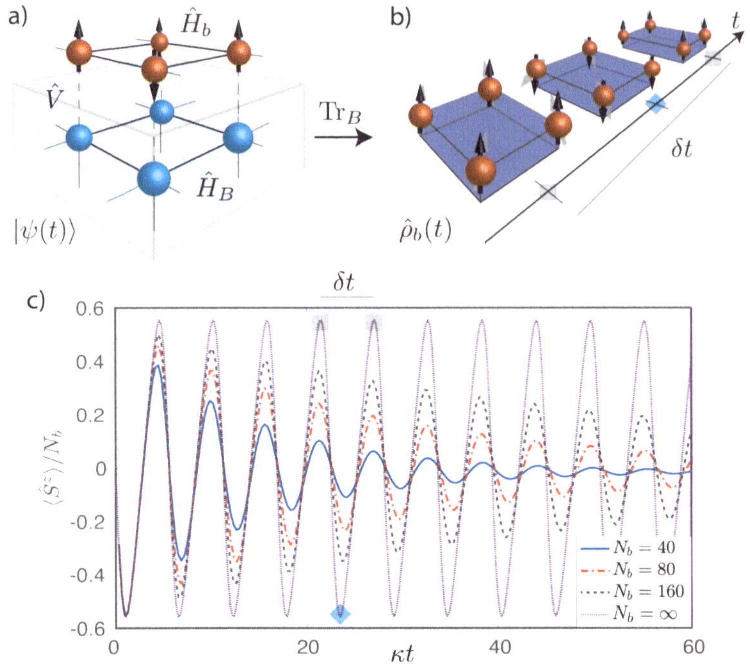

Fig. 4.35 Panels (**a**) and (**b**) show schematic plots of a system where a boundary and a bulk are identified. The boundary and the bulk are described by the Hamiltonians H_b and H_B, respectively, and they are coupled via an interaction term V. In the thermodynamic limit the numbers of degrees of freedom of the boundary, N_b, and the bulk, N_B, go to infinity but $N_b/N_B \to 0$. In that limit the boundary can be described by the reduced density matrix, $\rho_b(t) = \mathrm{Tr}_B[|\psi(t)\rangle\langle\psi(t)|]$, which fulfills the quantum master equation provided the bulk can be modeled by a Markovian bath. The system is time-independent but in the time crystal phase, the boundary evolves periodically in time as depicted in (**c**) in the case when the boundary is represented by a collection of spins. The lifetime of the periodic oscillations of the z component of the collective spin increases with the number of degrees of freedom N_b. The results presented in (**c**) correspond to the system initialized in the pure state with all spins pointed along the x direction. Reprinted figure with permission from Iemini et al. [97]. Copyright (2018) by the American Physical Society

eigenvalues ensure that there are metastable solutions of the master equation while the non-zero imaginary parts imply that these solutions should reveal non-trivial oscillations.[27] A fragment of the spectrum of $\mathcal{L}$ that is close to the zero eigenvalue $\lambda_0 = 0$ is shown in Fig. 4.36. The results correspond to the weak dissipation regime ($\omega_0/\kappa > 1$) which we focus on here. The real parts of the eigenvalues approach algebraically zero with $N_b \to \infty$ and the corresponding imaginary parts form bands separated by a fundamental frequency $\Gamma_{\omega_0/\kappa}$ which depends on ω_0/κ. These properties of the spectrum are responsible for the appearance of the time-

[27]Compare also Sect. 4.7.2 where the metastable subspace corresponding to purely real eigenvalues is explored to realize a discrete time crystal in a driven dissipative system [65].

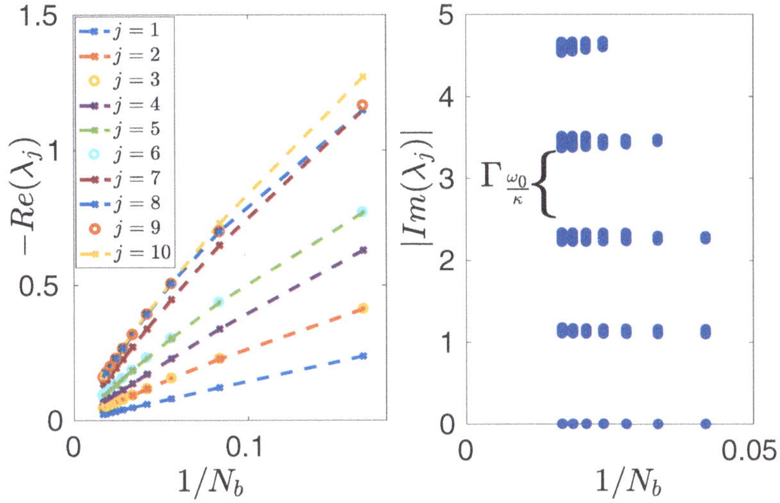

Fig. 4.36 Left panel: the smallest absolute values of 10 real parts of the eigenvalues of the Lindblad generator in (4.222) as a function of $1/N_b$, where N_b is the number of spins in the boundary, in the case of the weak dissipation, $\omega_0/\kappa > 1$, (the zero eigenvalue $\lambda_0 = 0$ is not shown in the plot). The eigenvalues are ordered according to $|\mathrm{Re}(\lambda_{j+1})| \geq |\mathrm{Re}(\lambda_j)|$. The presented real parts tend to zero algebraically with N_b. Right panel: imaginary parts of the eigenvalues λ_j corresponding to $j^2 \leq 0.025 N_b$. The imaginary parts form bands whose widths remain finite in the thermodynamic limit and which are separated by $\Gamma_{\omega_0/\kappa}$. Reprinted figure with permission from Iemini et al. [97]. Copyright (2018) by the American Physical Society

periodic evolution that lasts infinitely long in the thermodynamic limit and does not require specific choice of the initial conditions. In Fig. 4.35 such evolution is demonstrated by periodic oscillations of the average value of the z component of the collective spin, $\langle \hat{S}^z(t) \rangle = \mathrm{Tr}[\rho_b(t)\hat{S}^z]$, for the system initialized with all spins aligned along the x direction. The lifetime of the oscillations increases algebraically with N_b and for $N_b \to \infty$ the system can be described by means of the mean field approximation. That is, following the mean field approach presented in the background information in Sect. 4.7.1, we obtain the classical equations of motion for the average values of the spin component operators,

$$
\begin{aligned}
\frac{dS^x}{dt} &= \frac{\kappa}{S} S^x S^z, \\
\frac{dS^y}{dt} &= -\omega_0 S^z + \frac{\kappa}{S} S^y S^z, \\
\frac{dS^z}{dt} &= \omega_0 S^y - \frac{\kappa}{S}\left[\left(S^x\right)^2 + \left(S^y\right)^2 \right],
\end{aligned}
\tag{4.223}
$$

where $S^{x,y,z} = \langle \hat{S}^{x,y,z,} \rangle$ and $S = N_b/2$.

In the strong dissipation regime ($\omega_0/\kappa < 1$), there are two steady state solutions of (4.223) corresponding to the collective spin pointed along the z direction, i.e.

$$(S^x, S^y, S^z) = S\left(0, \frac{\omega_0}{\kappa}, \pm\sqrt{1-\left(\frac{\omega_0}{\kappa}\right)^2}\right). \tag{4.224}$$

Despite the fact that these two solutions can be transformed one to the other by changing $S^z \to -S^z$, they are not equivalent because the first solution is unstable while the other one stable [77].

In the weak dissipation regime, which we are interested in here, there are also two steady state solutions with the spin nearly along the x axis,

$$(S^x, S^y, S^z) = S\left(\pm\sqrt{1-\left(\frac{\kappa}{\omega_0}\right)^2}, \frac{\kappa}{\omega_0}, 0\right), \tag{4.225}$$

but they represent neither stable nor unstable fixed point. The stability analysis shows that the eigenvalues of the Jacobian corresponding to these solutions are purely imaginary [77]. Thus, passing the critical point at $\omega_0/\kappa = 1$ we observe neither the standard pitchfork bifurcation (where a stable fixed point becomes unstable and two stable fixed points are born) nor the Hopf bifurcation (where a fixed point becomes a limit cycle attracting a system to a periodic orbit). The phase space consists of two families of periodic trajectories circulating around each of the steady state solutions (4.225) [97]. In the thermodynamic limit, time-periodic evolution of the system we analyze in the present section is related to the motion along these trajectories, see Fig. 4.35.

Time-independent dissipative models can also arise in periodically driven open systems when a specific frame of reference is chosen—for example, in atomic systems coupled to electromagnetic modes of a cavity. In the context of time crystals these models are considered in Refs. [100, 194] where appropriate choice of parameters leads to a steady state that is a limit cycle, see also [156].

Finally we would like to mention that Buča et al. identified a class of dissipative time-independent systems which reveal time crystal dynamics in the long time limit [26]. They showed that if there exists a non-trivial operator A that commutes with jump operators of a master equation and fulfills $[H, A] = -\lambda A$, where λ is real, then $\rho_{nm} = A^n\rho_{ss}(A^\dagger)^m$, with integer $n, m > 0$ and $\mathcal{L}\rho_{ss} = 0$, fulfills $\mathcal{L}\rho_{nm} = i(m-n)\lambda\rho_{nm}$. Thus, all initial states that contain ρ_{nm}, with $n \neq m$, will continuously oscillate in the long time limit.

4.7.5 *Imaginary Time Crystals*

We have seen quite a few examples illustrating that crystalline structures can emerge not only in space but also in the time domain. Time is a real parameter in quantum mechanics, however, Wilczek in his first paper on quantum time crystals discussed also a possibility for realization of crystalline structure not in real time but in imaginary time [203]. This idea was followed by Cai et al. in Ref. [28] and we are going to describe their work in the present section. The existence of the so-called imaginary time crystals is related to the fact that the inverse temperature in the path integral formulation of statistical mechanics plays a role of the imaginary time. Before we describe the finding of Cai et al. we have to provide basic information about the path integral approach to statistical mechanics.

Path Integral Formulation of Statistical Mechanics

The central quantity of statistical mechanics is the partition function which for a weak coupling of a system with a thermal bath reduces to

$$Z = \mathrm{Tr}\left[e^{-\beta\hat{H}}\right], \tag{4.226}$$

where $\hat{H}$ is the quantum Hamiltonian of a system and $\beta = 1/T$ is the inverse temperature where we have assumed the Boltzmann constant $k_B = 1$.

On the other hand, for an isolated system, time evolution of a quantum state of a system can be written in the form

$$\psi(z,t) = \int_0^t dy\; G(z,t;y)\,\psi(y,0), \tag{4.227}$$

with the kernel of the evolution operator,

$$G(z,t;y) = \left\langle z \left| e^{-it\hat{H}} \right| y \right\rangle, \tag{4.228}$$

where we have assumed a single-particle system in one-dimensional space with the quantum Hamiltonian $\hat{H} = \hat{p}^2/2m + V(\hat{x})$ and $\hbar = 1$. Comparing (4.226) and (4.228) one may conclude that the partition function reduces to the trace of the kernel if we consider the inverse temperature as the imaginary time parameter. Indeed, substituting $t = -i\beta$ in the kernel (more precisely the substitution can be considered as an analytical continuation—the Wick rotation) we obtain

$$Z = \int dz\; G(z,-i\beta;z). \tag{4.229}$$

(continued)

In order to calculate the kernel (4.228) of the evolution operator let us divide the time period t into small N pieces, $\varepsilon = t/N$,

$$G(z,t;y) = \left\langle z \left| \left(e^{-i\varepsilon\hat{H}}\right)^N \right| y \right\rangle \approx \left\langle z \left| \left(e^{-i\varepsilon\hat{p}^2/2m} e^{-i\varepsilon V(\hat{x})}\right)^N \right| y \right\rangle, \tag{4.230}$$

where the right-hand side expression becomes exact when $N \to \infty$. Inserting the identity operators $1 = \int dx_j |x_j\rangle\langle x_j|$ and $1 = \int dp_k |p_k\rangle\langle p_k|$ between every product of the exponential functions of the operators $e^{-i\varepsilon\hat{p}^2/2m}e^{-i\varepsilon V(\hat{x})}$, one can calculate the action of the operators because the first is diagonal in the momentum representation, $e^{-i\varepsilon\hat{p}^2/2m}|p_k\rangle = e^{-i\varepsilon p_k^2/2m}|p_k\rangle$, while the other in the position representation, $e^{-i\varepsilon V(\hat{x})}|x_j\rangle = e^{-i\varepsilon V(x_j)}|x_j\rangle$. Moreover, due to the fact that $\langle p_k|x_j\rangle = e^{-ip_kx_j}/\sqrt{2\pi}$ and the kinetic energy of a particle depends quadratically on the momentum, all integration over the momenta can be performed exactly leading to the final formula

$$G(z,t;y) = \lim_{N\to\infty} \int dx_1 \ldots dx_{N-1} \left(\frac{m}{i2\pi\varepsilon}\right)^{N/2} e^{i\sum_{j=0}^{N-1}\left[\frac{m}{2}\left(\frac{x_{j+1}-x_j}{\varepsilon}\right)^2 - V(x_j)\right]\varepsilon} \tag{4.231}$$

$$= \int_{x(0)=y}^{x(t)=z} e^{iS[\mathbf{x},\dot{\mathbf{x}}]}\, \mathcal{D}\mathbf{x}, \tag{4.232}$$

where the sum in the exponent in (4.231) is actually a Riemann sum over a trajectory defined by a set of points $\mathbf{x} = \{y, x_1, x_2, \ldots, x_{N-1}, z\}$ and in the limit when $N \to \infty$ it approaches the classical action,

$$S[\mathbf{x},\dot{\mathbf{x}}] = \int_0^t dt' \left[\frac{m\dot{x}^2(t')}{2} - V(x(t'))\right], \tag{4.233}$$

calculated for a trajectory which fulfills $x(0) = y$ and $x(t) = z$. The integration $\int dx_1 \ldots dx_{N-1}$ in (4.231) means that the kernel is a sum of all possible sets $\mathbf{x} = \{y, x_1, x_2, \ldots, x_{N-1}, z\}$, or in other words it is the integration over all possible (not necessarily classical) trajectories, that connect the initial point $x(0) = y$ and the final point $x(t) = z$, with the weight given by e^{iS}. The integration over all trajectories is denoted by $\int_{x(0)=y}^{x(t)=z} \mathcal{D}\mathbf{x}$ in (4.232) and its definition is given in (4.231).

Now let us substitute $t = -i\beta$ and $t' = -i\tau$ in (4.233), or equivalently $\varepsilon = -i\epsilon$ in (4.231), and choose the final point to be the same as the initial point $z = y$. Then, the trace of $G(z, -i\beta, z)$ reduces to the partition function where the final imaginary time determines the inverse temperature $t = -i\beta$,

(continued)

i.e.

$$Z = \int dz\, G(z, -i\beta, z) = \int dz \int_{x(0)=z}^{x(t)=z} e^{-S_E[\mathbf{x},\dot{\mathbf{x}}]}\, \mathcal{D}\mathbf{x}, \tag{4.234}$$

where

$$S_E[\mathbf{x}, \dot{\mathbf{x}}] = \int_0^\beta d\tau \left[\frac{m\dot{x}^2(\tau)}{2} + V(x(\tau))\right], \tag{4.235}$$

is the so-called Euclidean action, i.e. an action when in the context of quantum field theory we change the space geometry from Lorentzian to Euclidean because the kinetic energy part has the same sign as the potential energy, compare (4.235) and (4.233).

To summarize, we have shown that thermodynamic description of a system coupled to a thermal bath can be obtained by integration over all possible trajectories of a system that evolve not in real but in the imaginary time $t' = -i\tau$.

Now we are able to return to the work by Cai et al. on the imaginary time crystals [28]. We have seen that the thermodynamic description of a time-independent system can be obtained if we collect all possible trajectories of a system and for each trajectory perform integration of the classical action (4.233) but along the imaginary time between 0 and $-i\beta$ where $\beta = 1/T$ is the inverse temperature. With the help of the substitution $t' = -i\tau$ in the classical action we arrive at the action (4.235) defined in the Euclidean space and finally at the partition function (4.234).

The idea of the imaginary time crystals (iTime cystals) is related to the question if dominant contributions to the partition function come from trajectories evolving periodically in the imaginary time with the same period? In other words if, for example, density-density correlation functions of a system can reveal periodic behavior versus the imaginary time. It should be stressed that such periodic paths in the imaginary time cannot be directly measured. However, signatures of their existence should be visible in oscillatory behavior of observables of a system versus the inverse temperature $\beta = 1/T$ [28, 203].

For weak interaction between a system and a thermal bath, the density matrix of a system at the thermal equilibrium reads

$$\rho = \frac{1}{Z} e^{-\beta H}, \tag{4.236}$$

where H is the system Hamiltonian and $Z = \mathrm{Tr}[e^{-\beta H}]$ is the partition function. However, such a Boltzmann–Gibbs ensemble requires corrections when the energy of the interaction between a system and a bath is comparable with typical energy scales of a system.

Let us consider bosons in one-dimensional lattice of L sites and in the half-filling case (i.e., the total number of particles N fulfills $N/L = 1/2$). We assume hard-core bosons what means that no more than one boson can occupy a single lattice site. The Hamiltonian of the system reads

$$H = -J \sum_{i=1}^{L} \left(\hat{a}_{i+1}^{\dagger} \hat{a}_i + \hat{a}_i^{\dagger} \hat{a}_{i+1} \right), \tag{4.237}$$

with the hard-core bosons constraint $n_i = \langle \hat{a}_i^{\dagger} \hat{a}_i \rangle \leq 1$. If each site of the lattice is coupled to one of L independent thermal baths formed by fermionic quasi-particles, then starting with full description of the hard-core bosons system and the baths one can eliminate the baths by calculating the trace over the baths' degrees of freedom. It leads to the density matrix of the hard-core bosons

$$\rho = \frac{1}{Z} e^{-\beta H - S_R}, \tag{4.238}$$

where

$$S_R = -\alpha \int_0^{\beta} d\tau \int_0^{\beta} d\tau' \sum_{i=1}^{L} \left(n_i(\tau) - \frac{1}{2} \right) D(\tau - \tau') \left(n_i(\tau') - \frac{1}{2} \right), \tag{4.239}$$

describes effective action of bath-induced onsite interactions between bosons with the coupling constant $\alpha \geq 0$ [28]. The kernel function in (4.239) is

$$D(\tau - \tau') = F(\tau - \tau') + F(\beta - |\tau - \tau'|), \tag{4.240}$$

where

$$F(\tau - \tau') = e^{-\omega_d |\tau - \tau'|} \cos[2\pi \omega_c (\tau - \tau')]. \tag{4.241}$$

In terms of the imaginary time, the interactions are retarded, i.e. they depend on different moments of iTime, cf. (4.239). If the coupling constant between the system and the baths $\alpha \to 0$, we recover the standard Boltzmann–Gibbs ensemble. However, when the coupling is significant, Eq. (4.239) shows that the baths introduce the retarded interactions between bosons in each lattice site. The kernel function (4.241) chosen by Cai et al. indicates that the retarded interactions reveal oscillations versus $\tau - \tau'$ which decay exponentially with $|\tau - \tau'|$. The oscillations in the kernel are not a necessary condition for the observation of the imaginary time crystals. It is sufficient if the kernel possesses at least one minimum at some value of $|\tau - \tau'|$ and such a minimum determines the time lattice constant in iTime crystals [28].

Cai et al. applied path integral quantum Monte Carlo simulations to analyze properties of the ground state of the system[28] [28]. For $\alpha = 0$, the ground state corresponds to a quasi-superfluid state of hard-core bosons with algebraic decay of the density–density correlation functions both in space and in the imaginary time (remember we deal with a one-dimensional Bose system in the half-filling case). However, when $\alpha \neq 0$ and it is sufficiently large, trajectories evolving periodically in the imaginary time and with periodic behavior in space have the largest weight in the description of the system within the path integral approach because their energies are the lowest. The competition between hopping of bosons on lattice sites and the retarded interactions leads to novel quantum phases which can be characterized by the density–density correlation functions in iTime and in space,[29]

$$S(r,\tau) = \frac{1}{L}\sum_i \left\langle \left(n_{i+r}(\tau) - \frac{1}{2}\right)\left(n_i(0) - \frac{1}{2}\right)\right\rangle, \tag{4.242}$$

and by the order parameter

$$m_\tau^2 = \frac{1}{\beta}\int_0^\beta d\tau\, S(0,\tau)\cos(2\pi\omega_c\tau), \tag{4.243}$$

which is able to indicate spontaneous breaking of the translation symmetry in the imaginary time. The presence of crystalline behavior in space can be also identified by the structure factor $\tilde{S}(q)$ which is the Fourier transform of the equal-iTime correlation function $S(r,0)$, i.e.

$$\tilde{S}(q) = \frac{1}{L}\sum_r e^{iqr} S(r,0). \tag{4.244}$$

Figure 4.37 illustrates crucial results obtained by Cai et al. The onsite density–density correlation function $S(0,\tau)$ decays algebraically unless the coupling constant α is greater than a critical value. In the latter case periodic oscillations of $S(0,\tau)$ in the imaginary time τ with the period dictated by the period of the oscillatory part of the kernel function (4.241) are observed and consequently an iTime crystal emerges. The structure factor $m_s^2 = \tilde{S}(q = \pi)$ and the equal-iTime correlation function $S(r,0)$, which are shown in Fig. 4.37, indicate that a crystalline structure is not only present in iTime but also appears in space and it has the periodicity twice longer than the lattice constant and corresponds to a charge density wave with an alternate occupations of lattice sites (i.e., $n_i \approx 1, n_{i+1} \approx 0, n_{i+2} \approx 1$

---

[28]A finite size scaling was used in Ref. [28] where the inverse temperature was determined by the size of the system $\beta = L$. It means that the zero temperature limit is achieved for $L \to \infty$.

[29]Note the same symbol used for the density-density correlation function in (4.242) and for the action (4.233) in the background information - the symbol is the same but the quantities are totally different.

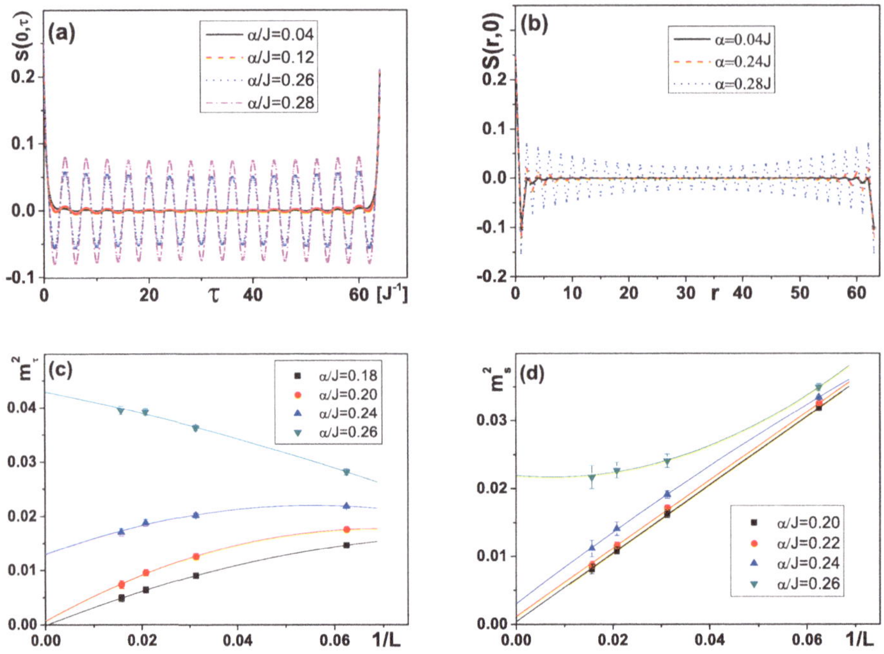

Fig. 4.37 Panel (**a**): the onsite density–density correlation function $S(0, \tau)$ versus the imaginary time τ, see Eq. (4.242). For small values of the coupling strength α, the correlation function $S(0, \tau)$ decays algebraically with τ. However, for a sufficiently large α the long-range order emerges indicated by oscillations of $S(0, \tau)$ with the characteristic period $1/\omega_c = 4/J$. Such spontaneous breaking of the imaginary time translation symmetry is reflected by the behavior of the order parameter m_τ, Eq. (4.243), which is depicted in panel (**c**). That is, for α smaller than a critical value, the order parameter decays with the increasing size L of the system otherwise it tends to a finite value. Panels (**b**) and (**d**) present the equal-iTime density–density correlation function $S(r, 0)$ and its Fourier transform, i.e. the structure factor $m_s^2 = \tilde{S}(q = \pi)$, see (4.244). For a sufficiently large α the hard-core bosons system reveals a charge density wave with the spatial period twice longer than the lattice constant. The temperature of the system corresponds to $\beta = L$, and the parameters of the kernel function (4.241) are the following $\omega_c = J/4$ and $\omega_d = 0.1J$. Reprinted figure with permission from Cai et al. [28]

and so on). The retarded onsite interactions are able to develop periodic hopping of hard-core bosons in the imaginary time (i.e., periodic modulation of the densities $n_i(\tau)$ in iTime) if bosons have a room to do it. In the half-filling case we consider here, the optimal spatial configuration to realize hopping in the imaginary time and thus the iTime crystal corresponds to a charge density wave with the spatial period twice longer than the lattice constant. Charge density waves can appear in a time-independent lattice system when there are nearest neighbor repulsive interactions. The present results indicate that the retarded onsite interactions induced by the baths play a similar role as nearest neighbor repulsive interactions between bosons in space—the latter are of course absent in our case. The analysis of the correlation length in iTime and the correlation length in space performed in Ref. [28] shows

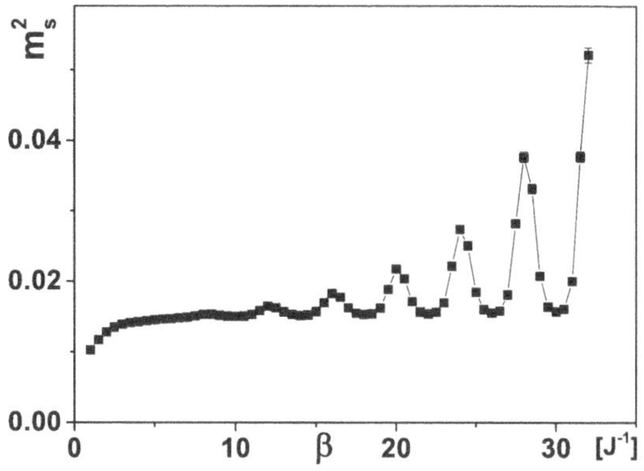

Fig. 4.38 Structure factor $m_s^2 = \tilde{S}(q = \pi)$ as a function of the inverse temperature $\beta = 1/T$ for the hard-core bosons system in the iTime crystal phase. The periodic appearance of the peaks is a signature of the periodic structure in the imaginary time. That is, when the inverse temperature β is an integer multiple of the iTime crystal period, dictated by the kernel function (4.241) of the bath-induced retarded interactions, charge density waves in space appear together with the crystalline structure in iTime. Reprinted figure with permission from Cai et al. [28]

that the critical value for the emergence of the iTime crystal is $\alpha/J = 0.23(1)$ and it is slightly smaller than the critical value for the emergence of the crystalline structure in space which happens for $\alpha/J = 0.245(5)$.

The imaginary time is not accessible experimentally. However, experimental signatures of the existence of iTime crystals can be observed if the charge density waves, which are observable in space, appear periodically when we are decreasing the temperature of the system. In Fig. 4.38 the structure factor $m_s^2 = \tilde{S}(q = \pi)$ is presented which reveals peaks that appear periodically as a function of the inverse temperature β with the period very similar to the iTime crystal period. Thus, if the temperature is sufficiently low, further decrease of the temperature should show up signatures of the iTime crystals.

4.8 Further Reading

It is not possible to describe the whole research performed in the field of discrete time crystals and related phenomena. A reader is encouraged to refer to the scientific literature. Some of the articles, certainly an incomplete collection, one can find in the following list: [7, 8, 12, 16, 20, 32, 36, 39, 44, 52, 54, 55, 64, 73, 76, 78–81, 83, 92, 93, 96, 98, 104, 120, 122, 125, 128, 129, 133, 135, 137, 142, 145–147, 150, 158, 172, 176, 178, 184, 205].

References

1. Abanin, D.A., De Roeck, W., Huveneers, F.: Exponentially slow heating in periodically driven many-body systems. Phys. Rev. Lett. **115**, 256803 (2015). https://doi.org/10.1103/PhysRevLett.115.256803
2. Abanin, D., De Roeck, W., Ho, W.W., Huveneers, F.: A rigorous theory of many-body prethermalization for periodically driven and closed quantum systems. Commun. Math. Phys. **354**(3), 809–827 (2017). https://doi.org/10.1007/s00220-017-2930-x
3. Abanin, D.A., De Roeck, W., Ho, W.W., Huveneers, F.M.C.: Effective Hamiltonians, prethermalization, and slow energy absorption in periodically driven many-body systems. Phys. Rev. B **95**, 014112 (2017). https://doi.org/10.1103/PhysRevB.95.014112
4. Agarwal, K., Gopalakrishnan, S., Knap, M., Müller, M., Demler, E.: Anomalous diffusion and Griffiths effects near the many-body localization transition. Phys. Rev. Lett. **114**, 160401 (2015). https://doi.org/10.1103/PhysRevLett.114.160401
5. Albuquerque, E., Cottam, M.: Theory of elementary excitations in quasiperiodic structures. Phys. Rep. **376**(4), 225–337 (2003). https://doi.org/10.1016/S0370-1573(02)00559-8. http://www.sciencedirect.com/science/article/pii/S0370157302005598
6. Altman, E., Vosk, R.: Universal dynamics and renormalization in many-body-localized systems. Annu. Rev. Condens. Matter Phys. **6**, 383 (2015). https://doi.org/10.1146/annurev-conmatphys-031214-014701
7. Autti, S., Heikkinen, P.J., Mäkinen, J.T., Volovik, G.E., Zavjalov, V.V., Eltsov, V.B.: AC Josephson effect between two superfluid time crystals (2020). e-prints arXiv:2003.06313
8. Balatsky, A.V., Sukhachov, P.O., Band yopadhyay, S.: Quantum pairing time orders. Annalen der Physik **532**(2), 1900529 (2020). https://doi.org/10.1002/andp.201900529
9. Bar Lev, Y., Reichman, D.R.: Dynamics of many-body localization. Phys. Rev. B **89**, 220201 (2014). https://doi.org/10.1103/PhysRevB.89.220201
10. Bar Lev, Y., Cohen, G., Reichman, D.R.: Absence of diffusion in an interacting system of spinless fermions on a one-dimensional disordered lattice. Phys. Rev. Lett. **114**, 100601 (2015). https://doi.org/10.1103/PhysRevLett.114.100601
11. Bardarson, J.H., Pollmann, F., Moore, J.E.: Unbounded growth of entanglement in models of many-body localization. Phys. Rev. Lett. **109**, 017202 (2012). https://doi.org/10.1103/PhysRevLett.109.017202
12. Barfknecht, R.E., Rasmussen, S.E., Foerster, A., Zinner, N.T.: Realizing time crystals in discrete quantum few-body systems. Phys. Rev. B **99**, 144304 (2019). https://doi.org/10.1103/PhysRevB.99.144304
13. Barišić, O.S., Prelovšek, P.: Conductivity in a disordered one-dimensional system of interacting fermions. Phys. Rev. B **82**, 161106 (2010). https://doi.org/10.1103/PhysRevB.82.161106
14. Barišić, O.S., Kokalj, J., Balog, I., Prelovšek, P.: Dynamical conductivity and its fluctuations along the crossover to many-body localization. Phys. Rev. B **94**, 045126 (2016). https://doi.org/10.1103/PhysRevB.94.045126
15. Basko, D., Aleiner, I., Altschuler, B.: Metal-insulator transition in a weakly interacting many-electron system with localized single-particle states. Ann. Phys. **321**, 1126 (2006)
16. Bello, L., Calvanese Strinati, M., Dalla Torre, E.G., Pe'er, A.: Persistent coherent beating in coupled parametric oscillators. Phys. Rev. Lett. **123**, 083901 (2019). https://doi.org/10.1103/PhysRevLett.123.083901
17. Benatti, F., Carollo, F., Floreanini, R., Narnhofer, H.: Quantum spin chain dissipative mean-field dynamics. J. Phys. A Math. Theoret. **51**(32), 325001 (2018). https://doi.org/10.1088/1751-8121/aacbdb
18. Berkelbach, T.C., Reichman, D.R.: Conductivity of disordered quantum lattice models at infinite temperature: many-body localization. Phys. Rev. B **81**, 224429 (2010). https://doi.org/10.1103/PhysRevB.81.224429

19. Berman, G., Zaslavsky, G.: Theory of quantum nonlinear resonance. Phys. Lett. A **61**(5), 295–296 (1977). https://doi.org/10.1016/0375-9601(77)90618-1. http://www.sciencedirect.com/science/article/pii/0375960177906181
20. Bomantara, R.W., Gong, J.: Simulation of non-Abelian braiding in Majorana time crystals. Phys. Rev. Lett. **120**, 230405 (2018). https://doi.org/10.1103/PhysRevLett.120.230405
21. Bordia, P., Luschen, H., Schneider, U., Knap, M., Bloch, I.: Periodically driving a many-body localized quantum system. Nat. Phys. **13**, 460–464 (2017). Article. **Advance online publication**. http://dx.doi.org/10.1038/nphys4020
22. Botzung, T., Hagenmüller, D., Masella, G., Dubail, J., Defenu, N., Trombettoni, A., Pupillo, G.: Effects of energy extensivity on the quantum phases of long-range interacting systems (2019). e-prints arXiv:1909.12105
23. Breuer, H.P., Petruccione, F.: The Theory of Open Quantum Systems. Oxford University Press, Oxford (2002)
24. Bruno, P.: Comment on "Space-Time Crystals of Trapped Ions". Phys. Rev. Lett. **111**, 029301 (2013). https://doi.org/10.1103/PhysRevLett.111.029301
25. Bruno, P.: Impossibility of spontaneously rotating time crystals: a no-go theorem. Phys. Rev. Lett. **111**, 070402 (2013). https://doi.org/10.1103/PhysRevLett.111.070402
26. Buča, B., Tindall, J., Jaksch, D.: Complex coherent quantum many-body dynamics through dissipation. Nat. Commun. **10**, 1730 (2019). https://doi.org/10.1038/s41467-019-09757-y
27. Buchleitner, A., Delande, D., Zakrzewski, J.: Non-dispersive wave packets in periodically driven quantum systems. Phys. Rep. **368**(5), 409–547 (2002). http://www.sciencedirect.com/science/article/pii/S0370157302002703
28. Cai, Z., huang, Y., Liu, W.V.: Imaginary time crystal of thermal quantum matter (2019). e-prints arXiv:1902.09747
29. Castin, Y.: Bose-Einstein condensates in atomic gases: Simple theoretical results. In: Kaiser, R., Westbrook, C., David, F. (eds.) Coherent Atomic Matter Waves, pp. 1–136. Springer, Berlin (2001)
30. Castin, Y., Dum, R.: Low-temperature Bose-Einstein condensates in time-dependent traps: Beyond the $u(1)$ symmetry-breaking approach. Phys. Rev. A **57**, 3008–3021 (1998). https://doi.org/10.1103/PhysRevA.57.3008
31. Chandran, A., Kim, I.H., Vidal, G., Abanin, D.A.: Constructing local integrals of motion in the many-body localized phase. Phys. Rev. B **91**, 085425 (2015). https://doi.org/10.1103/PhysRevB.91.085425
32. Chinzei, K., Ikeda, T.N.: Time crystals protected by floquet dynamical symmetry in Hubbard models (2020). e-prints arXiv:2003.13315
33. Choi, S., Choi, J., Landig, R., Kucsko, G., Zhou, H., Isoya, J., Jelezko, F., Onoda, S., Sumiya, H., Khemani, V., von Keyserlingk, C., Yao, N.Y., Demler, E., Lukin, M.D.: Observation of discrete time-crystalline order in a disordered dipolar many-body system. Nature **543**(7644), 221–225 (2017). Letter. https://doi.org/10.1038/nature21426
34. Choi, J., Zhou, H., Choi, S., Landig, R., Ho, W.W., Isoya, J., Jelezko, F., Onoda, S., Sumiya, H., Abanin, D.A., Lukin, M.D.: Probing quantum thermalization of a disordered dipolar spin ensemble with discrete time-crystalline order. Phys. Rev. Lett. **122**, 043603 (2019). https://doi.org/10.1103/PhysRevLett.122.043603
35. Cosme, J.G., Skulte, J., Mathey, L.: Time crystals in a shaken atom-cavity system. Phys. Rev. A **100**(5), 053615 (2019). https://doi.org/10.1103/PhysRevA.100.053615
36. Dai, C.M., Gu, Z.C., Yi, X.X.: Discrete time-crystalline order in Bose–Hubbard model with dissipation. New J. Phys. **22**(2), 023026 (2020). https://doi.org/10.1088/1367-2630/ab6ead
37. D'Alessio, L., Rigol, M.: Long-time behavior of isolated periodically driven interacting lattice systems. Phys. Rev. X **4**, 041048 (2014). https://doi.org/10.1103/PhysRevX.4.041048
38. D'Alessio, L., Kafri, Y., Polkovnikov, A., Rigol, M.: From quantum chaos and eigenstate thermalization to statistical mechanics and thermodynamics. Adv. Phys. **65**(3), 239–362 (2016). https://doi.org/10.1080/00018732.2016.1198134
39. Dai, J., Peng, X., Niemi, A.J.: Topologically enhanced time crystals and molecular knots (2019). e-prints arXiv:1910.13787

40. Dalton, B.: Phase space theory of Bose-Einstein condensates and time-dependent modes. Ann. Phys. **327**(10), 2432–2490 (2012). https://doi.org/10.1016/j.aop.2012.06.005. http://www.sciencedirect.com/science/article/pii/S0003491612001017
41. Dalton, B., Ghanbari, S.: Two mode theory of Bose-Einstein condensates: interferometry and the Josephson model. J. Modern Opt. **59**(4), 287–353 (2012). https://doi.org/10.1080/09500340.2011.632100
42. De Luca, A., Scardicchio, A.: Ergodicity breaking in a model showing many-body localization. Europhys. Lett. **101**, 37003 (2013). http://stacks.iop.org/0295-5075/101/i=3/a=37003
43. De Luca, A., Altshuler, B.L., Kravtsov, V.E., Scardicchio, A.: Anderson localization on the Bethe lattice: nonergodicity of extended states. Phys. Rev. Lett. **113**, 046806 (2014). https://doi.org/10.1103/PhysRevLett.113.046806
44. De Martino, D., Barato, A.C.: Oscillations in feedback-driven systems: Thermodynamics and noise. Phys. Rev. E **100**(6), 062123 (2019). https://doi.org/10.1103/PhysRevE.100.062123
45. Deutsch, J.M.: Quantum statistical mechanics in a closed system. Phys. Rev. A **43**, 2046–2049 (1991). https://doi.org/10.1103/PhysRevA.43.2046
46. Dicke, R.H.: Coherence in spontaneous radiation processes. Phys. Rev. **93**, 99–110 (1954). https://doi.org/10.1103/PhysRev.93.99
47. Drummond, P.D.: Observables and moments of cooperative resonance fluorescence. Phys. Rev. A **22**, 1179–1184 (1980). https://doi.org/10.1103/PhysRevA.22.1179
48. Drummond, P., Carmichael, H.: Volterra cycles and the cooperative fluorescence critical point. Opt. Commun. **27**(1), 160–164 (1978). https://doi.org/10.1016/0030-4018(78)90198-0. http://www.sciencedirect.com/science/article/pii/0030401878901980
49. Drummond, P., McNeil, K., Walls, D.: Non-equilibrium transitions in sub/second harmonic generation. Opt. Acta Int. J. Opt. **27**(3), 321–335 (1980). https://doi.org/10.1080/713820226
50. Drummond, P., McNeil, K., Walls, D.: Non-equilibrium transitions in sub/second harmonic generation. Opt. Acta **28**, 211 (1981). https://doi.org/10.1080/713820531
51. Dumitrescu, P.T., Vasseur, R., Potter, A.C.: Logarithmically slow relaxation in quasiperiodically driven random spin chains. Phys. Rev. Lett. **120**, 070602 (2018). https://doi.org/10.1103/PhysRevLett.120.070602
52. Dykman, M.I.: Coherent multiple-period states of periodically modulated qubits. Phys. Rev. A **100**(4), 042101 (2019). https://doi.org/10.1103/PhysRevA.100.042101
53. Dziarmaga, J., Sacha, K.: Images of a Bose–Einstein condensate: diagonal dynamical Bogoliubov vacuum. J. Phys. B Atom. Mol. Opt. Phys. **39**(1), 57–68 (2005). https://doi.org/10.1088/0953-4075/39/1/007
54. Efetov, K.B.: Mean-field thermodynamic quantum time-space crystal: spontaneous breaking of time-translation symmetry in a macroscopic fermion system. Phys. Rev. B **100**, 245128 (2019). https://doi.org/10.1103/PhysRevB.100.245128
55. Efetov, K.B.: Thermodynamic quantum time crystals (2019). e-prints arXiv:1902.07520
56. Eisert, J., Friesdorf, M., Gogolin, C.: Quantum many-body systems out of equilibrium. Nat. Phys. **11**(2), 124–130 (2015). http://dx.doi.org/10.1038/nphys3215
57. Else, D.V., Bauer, B., Nayak, C.: Floquet time crystals. Phys. Rev. Lett. **117**, 090402 (2016). https://doi.org/10.1103/PhysRevLett.117.090402
58. Else, D.V., Bauer, B., Nayak, C.: Prethermal phases of matter protected by time-translation symmetry. Phys. Rev. X **7**, 011026 (2017). https://doi.org/10.1103/PhysRevX.7.011026
59. Else, D.V., Monroe, C., Nayak, C., Yao, N.Y.: Discrete time crystals. Ann. Review of Condens. Matter Phys. **11**(1), 467–499 (2020). https://doi.org/10.1146/annurev-conmatphys-031119-050658
60. Estarellas, M.P., Osada, T., Bastidas, V.M., Renoust, B., Sanaka, K., Munro, W.J., Nemoto, K.: Simulating complex quantum networks with time crystals (2019). e-prints arXiv:1907.13146
61. Fan, C.h., Rossini, D., Zhang, H.X., Wu, J.H., Artoni, M., La Rocca, G.C.: Discrete time crystal in a finite chain of Rydberg atoms without disorder. Phys. Rev. A **101**(1), 013417 (2020). https://doi.org/10.1103/PhysRevA.101.013417
62. Flatté, M.E., Holthaus, M.: Classical and quantum dynamics of a periodically driven particle in a triangular well. Ann. Phys. **245**(1), 113–146 (1996). https://doi.org/10.1006/aphy.1996.0005. http://www.sciencedirect.com/science/article/pii/S0003491696900056

63. Flicker, F.: Time quasilattices in dissipative dynamical systems. SciPost Phys. **5**, 1 (2018). https://doi.org/10.21468/SciPostPhys.5.1.001. https://scipost.org/10.21468/SciPostPhys.5.1.001
64. Gambetta, F.M., Carollo, F., Lazarides, A., Lesanovsky, I., Garrahan, J.P.: Classical stochastic discrete time crystals. Phys. Rev. E **100**, 060105 (2019). https://doi.org/10.1103/PhysRevE.100.060105
65. Gambetta, F.M., Carollo, F., Marcuzzi, M., Garrahan, J.P., Lesanovsky, I.: Discrete time crystals in the absence of manifest symmetries or disorder in open quantum systems. Phys. Rev. Lett. **122**, 015701 (2019). https://doi.org/10.1103/PhysRevLett.122.015701
66. Gelhausen, J., Buchhold, M.: Dissipative Dicke model with collective atomic decay: bistability, noise-driven activation, and the nonthermal first-order superradiance transition. Phys. Rev. A **97**, 023807 (2018). https://doi.org/10.1103/PhysRevA.97.023807
67. Gerry, C., Knight, P.: Introductory Quantum Optics. Cambridge University Press, New York (2005)
68. Giergiel, K., Kosior, A., Hannaford, P., Sacha, K.: Time crystals: analysis of experimental conditions. Phys. Rev. A **98**, 013613 (2018). https://doi.org/10.1103/PhysRevA.98.013613
69. Giergiel, K., Kuroś, A., Sacha, K.: Discrete time quasicrystals. Phys. Rev. B **99**, 220303 (2019). https://doi.org/10.1103/PhysRevB.99.220303
70. Giergiel, K., Tran, T., Zaheer, A., Singh, A., Sidorov, A., Sacha, K., Hannaford, P.: Creating big time crystals with ultracold atoms. New Journal of Physics. IOP Publishing. **22**(8), 085004 (2020). https://doi.org/10.1088/1367-2630/aba3e6
71. Gong, Z., Hamazaki, R., Ueda, M.: Discrete time-crystalline order in cavity and circuit QED systems. Phys. Rev. Lett. **120**, 040404 (2018). https://doi.org/10.1103/PhysRevLett.120.040404
72. Gornyi, I.V., Mirlin, A.D., Polyakov, D.G.: Interacting electrons in disordered wires: Anderson localization and low-t transport. Phys. Rev. Lett. **95**, 206603 (2005). https://doi.org/10.1103/PhysRevLett.95.206603
73. Gosner, J., Kubala, B., Ankerhold, J.: Relaxation dynamics and dissipative phase transition in quantum oscillators with period tripling. Phys. Rev. B **101**, 054501 (2020). https://doi.org/10.1103/PhysRevB.101.054501
74. Gramsch, C., Rigol, M.: Quenches in a quasidisordered integrable lattice system: dynamics and statistical description of observables after relaxation. Phys. Rev. A **86**, 053615 (2012). https://doi.org/10.1103/PhysRevA.86.053615
75. Haake, F.: Quantum Signatures of Chaos. Physics and Astronomy Online Library. Springer, Berlin (2001). https://books.google.pl/books?id=Orv0BXoorFEC
76. Hahn, W., Dobrovitski, V.V.: Long-lived coherence in driven spin systems: From two- to infinite spatial dimensions (2019). e-prints arXiv:1911.06272
77. Hannukainen, J., Larson, J.: Dissipation-driven quantum phase transitions and symmetry breaking. Phys. Rev. A **98**, 042113 (2018). https://doi.org/10.1103/PhysRevA.98.042113
78. Harper, F., Roy, R., Rudner, M.S., Sondhi, S.L.: Topology and broken symmetry in floquet systems (2019). e-prints arXiv:1905.01317
79. Hayata, T., Hidaka, Y.: Diffusive Nambu-Goldstone modes in quantum time-crystals (2018). e-prints arXiv:1808.07636
80. Heissenberg, C., Strocchi, F.: Generalized criteria of symmetry breaking. a strategy for quantum time crystals. Ann. Phys. **415**, 168097 (2020). https://doi.org/10.1016/j.aop.2020.168097
81. Helmrich, S., Arias, A., Lochead, G., Wintermantel, T.M., Buchhold, M., Diehl, S., Whitlock, S.: Signatures of self-organized criticality in an ultracold atomic gas. Nature **577**(7791), 481–486 (2020). https://doi.org/10.1038/s41586-019-1908-6
82. Henkel, J., Holthaus, M.: Classical resonances in quantum mechanics. Phys. Rev. A **45**, 1978–1986 (1992). https://doi.org/10.1103/PhysRevA.45.1978
83. Heo, M.S., Kim, Y., Kim, K., Moon, G., Lee, J., Noh, H.R., Dykman, M.I., Jhe, W.: Ideal mean-field transition in a modulated cold atom system. Phys. Rev. E **82**, 031134 (2010). https://doi.org/10.1103/PhysRevE.82.031134

84. Hepp, K., Lieb, E.H.: On the superradiant phase transition for molecules in a quantized radiation field: the Dicke maser model. Ann. Phys. **76**(2), 360–404 (1973). https://doi.org/10.1016/0003-4916(73)90039-0
85. Heyl, M.: Dynamical quantum phase transitions: a review. Rep. Progr. Phys. **81**(5), 054001 (2018). https://doi.org/10.1088/1361-6633/aaaf9a
86. Ho, W.W., Choi, S., Lukin, M.D., Abanin, D.A.: Critical time crystals in dipolar systems. Phys. Rev. Lett. **119**, 010602 (2017). https://doi.org/10.1103/PhysRevLett.119.010602
87. Holthaus, M.: On the classical-quantum correspondence for periodically time dependent systems. Chaos Soliton. Fract. **5**(7), 1143–1167 (1995). Quantum Chaos: Present and Future. https://doi.org/10.1016/0960-0779(94)E0059-X. http://www.sciencedirect.com/science/article/pii/096007799 4E0059X
88. Holthaus, M.: Floquet engineering with quasienergy bands of periodically driven optical lattices. J. Phys. B Atom. Mol. Opt. Phys. **49**(1), 013001 (2015). https://doi.org/10.1088/0953-4075/49/1/013001
89. Holthaus, M., Flatté, M.E.: Subharmonic generation in quantum systems. Phys. Lett. A **187**(2), 151–156 (1994). https://doi.org/10.1016/0375-9601(94)90053-1. http://www.sciencedirect.com/science/article/pii/0375960194900531
90. Huang, B., Wu, Y.H., Liu, W.V.: Clean floquet time crystals: models and realizations in cold atoms. Phys. Rev. Lett. **120**, 110603 (2018). https://doi.org/10.1103/PhysRevLett.120.110603
91. Huang, Y., Li, T., Yin, Z.q.: Symmetry-breaking dynamics of the finite-size Lipkin-Meshkov-Glick model near ground state. Phys. Rev. A **97**, 012115 (2018). https://doi.org/10.1103/PhysRevA.97.012115
92. Huang, Y., Guo, Q., Xiong, A., Li, T., Yin, Z.q.: Time crystal based on a levitated charged nanoparticle (2020). e-prints arXiv:2001.10187
93. Hurtado-Gutiérrez, R., Carollo, F., Pérez-Espigares, C., Hurtado, P.I.: Building continuous time crystals from rare events (2019). e-prints arXiv:1912.02733
94. Huse, D.A., Nandkishore, R., Oganesyan, V., Pal, A., Sondhi, S.L.: Localization-protected quantum order. Phys. Rev. B **88**, 014206 (2013). https://doi.org/10.1103/PhysRevB.88.014206
95. Huse, D.A., Nandkishore, R., Oganesyan, V.: Phenomenology of fully many-body-localized systems. Phys. Rev. B **90**, 174202 (2014). https://doi.org/10.1103/PhysRevB.90.174202
96. Iadecola, T., Schecter, M., Xu, S.: Quantum many-body scars from Magnon condensation. Phys. Rev. B **100**, 184312 (2019). https://doi.org/10.1103/PhysRevB.100.184312
97. Iemini, F., Russomanno, A., Keeling, J., Schirò, M., Dalmonte, M., Fazio, R.: Boundary time crystals. Phys. Rev. Lett. **121**, 035301 (2018). https://doi.org/10.1103/PhysRevLett.121.035301
98. Ikeda, T.N., Sato, M.: General description for nonequilibrium steady states in periodically driven dissipative quantum systems (2020). e-prints arXiv:2003.02876
99. Janot, C.: Quasicrystals: A Primer. Oxford University Press, Oxford (1994)
100. Keßler, H., Cosme, J.G., Hemmerling, M., Mathey, L., Hemmerich, A.: Emergent limit cycles and time crystal dynamics in an atom-cavity system. Phys. Rev. A **99**, 053605 (2019). https://doi.org/10.1103/PhysRevA.99.053605
101. Khemani, V., Nandkishore, R., Sondhi, S.L.: Nonlocal adiabatic response of a localized system to local manipulations. Nat. Phys. **11**, 560 (2015). https://doi.org/10.1038/nphys3344
102. Khemani, V., Lazarides, A., Moessner, R., Sondhi, S.L.: Phase structure of driven quantum systems. Phys. Rev. Lett. **116**, 250401 (2016). https://doi.org/10.1103/PhysRevLett.116.250401
103. Khemani, V., Moessner, R., Sondhi, S.L.: A brief history of time crystals (2019). e-prints arXiv:1910.10745
104. Kim, K., Heo, M.S., Lee, K.H., Jang, K., Noh, H.R., Kim, D., Jhe, W.: Spontaneous symmetry breaking of population in a nonadiabatically driven atomic trap: an Ising-class phase transition. Phys. Rev. Lett. **96**, 150601 (2006). https://doi.org/10.1103/PhysRevLett.96.150601

105. Kirton, P., Roses, M.M., Keeling, J., Dalla Torre, E.G.: Introduction to the Dicke model: from equilibrium to nonequilibrium, and vice versa. Adv. Quantum Technol. **2**(1–2), 1800043 (2019). https://doi.org/10.1002/qute.201800043. https://onlinelibrary.wiley.com/doi/abs/10.1002/qute.201800043
106. Kjäll, J.A., Bardarson, J.H., Pollmann, F.: Many-body localization in a disordered quantum ising chain. Phys. Rev. Lett. **113**, 107204 (2014). https://doi.org/10.1103/PhysRevLett.113.107204
107. Kohmoto, M., Sutherland, B., Iguchi, K.: Localization of optics: quasiperiodic media. Phys. Rev. Lett. **58**, 2436–2438 (1987). https://doi.org/10.1103/PhysRevLett.58.2436
108. Kosior, A., Sacha, K.: Dynamical quantum phase transitions in discrete time crystals. Phys. Rev. A **97**, 053621 (2018). https://doi.org/10.1103/PhysRevA.97.053621
109. Kosior, A., Syrwid, A., Sacha, K.: Dynamical quantum phase transitions in systems with broken continuous time and space translation symmetries. Phys. Rev. A **98**, 023612 (2018). https://doi.org/10.1103/PhysRevA.98.023612
110. Kozarzewski, M., Prelovšek, P., Mierzejewski, M.: Distinctive response of many-body localized systems to a strong electric field. Phys. Rev. B **93**, 235151 (2016). https://doi.org/10.1103/PhysRevB.93.235151
111. Kuroś, A., Mukherjee, R., Golletz, W., Sauvage, F., Giergiel, K., Mintert, F., Sacha, K.: Phase diagram and optimal control for n-tupling discrete time crystal. Submitted to New J. Phys. (2020). https://ui.adsabs.harvard.edu/abs/2020arXiv200414982K
112. Kuwahara, T., Mori, T., Saito, K.: Floquet-Magnus theory and generic transient dynamics in periodically driven many-body quantum systems. Ann. Phys. **367**, 96–124 (2016). http://doi.org/10.1016/j.aop.2016.01.012. http://www.sciencedirect.com/science/article/pii/S0003491616000142
113. Landi, G.T.: Lecture notes on Quantum Information and Quantum Noise. Unpublished http://www.fmt.if.usp.br/~gtlandi/quantum-information-and.html
114. Lazarides, A., Moessner, R.: Fate of a discrete time crystal in an open system. Phys. Rev. B **95**, 195135 (2017). https://doi.org/10.1103/PhysRevB.95.195135
115. Lazarides, A., Das, A., Moessner, R.: Equilibrium states of generic quantum systems subject to periodic driving. Phys. Rev. E **90**, 012110 (2014). https://doi.org/10.1103/PhysRevE.90.012110
116. Lazarides, A., Roy, S., Piazza, F., Moessner, R.: Time crystallinity in dissipative floquet systems. Phys. Rev. Res. **2**, 022002 (2020). https://doi.org/10.1103/PhysRevResearch.2.022002
117. Levine, D., Steinhardt, P.J.: Quasicrystals: A new class of ordered structures. Phys. Rev. Lett. **53**, 2477–2480 (1984). https://doi.org/10.1103/PhysRevLett.53.2477
118. Li, T., Gong, Z.X., Yin, Z.Q., Quan, H.T., Yin, X., Zhang, P., Duan, L.M., Zhang, X.: Reply to Comment on "Space-Time Crystals of Trapped Ions" (2012). e-prints arXiv:1212.6959
119. Li, T., Gong, Z.X., Yin, Z.Q., Quan, H.T., Yin, X., Zhang, P., Duan, L.M., Zhang, X.: Space-time crystals of trapped ions. Phys. Rev. Lett. **109**, 163001 (2012). https://doi.org/10.1103/PhysRevLett.109.163001
120. Li, X., Chai, J., Zhu, H., Wang, P.: Discrete Lorentz symmetry and discrete spacetime translational symmetry in two- and three-dimensional crystals. J. Phys. Conden. Matt. **32**(14), 145402 (2020). https://doi.org/10.1088/1361-648X/ab6464
121. Liao, L., Smits, J., van der Straten, P., Stoof, H.T.C.: Dynamics of a space-time crystal in an atomic Bose-Einstein condensate. Phys. Rev. A **99**, 013625 (2019). https://doi.org/10.1103/PhysRevA.99.013625
122. Libal, A., Balazs, T., Reichhardt, C., Reichhardt, C.J.O.: Colloidal dynamics on a choreographic time crystal (2019). e-prints arXiv:1912.04824
123. Lichtenberg, A., Lieberman, M.: Regular and Chaotic Dynamics. Applied Mathematical Sciences. Springer, Berlin (1992). https://books.google.pl/books?id=2ssPAQAAMAAJ
124. Lieb, E.H., Seiringer, R., Yngvason, J.: Bosons in a trap: a rigorous derivation of the Gross-Pitaevskii energy functional. Phys. Rev. A **61**, 043602 (2000). https://doi.org/10.1103/PhysRevA.61.043602

125. Lledó, C., Mavrogordatos, T.K., Szymańska, M.H.: driven Bose-Hubbard dimer under nonlocal dissipation: a bistable time crystal. Phys. Rev. B **100**, 054303 (2019). https://doi.org/10.1103/PhysRevB.100.054303
126. Luitz, D.J., Laflorencie, N., Alet, F.: Extended slow dynamical regime prefiguring the many-body localization transition. Phys. Rev. B **93**, 060201 (2015). https://doi.org/10.1103/PhysRevB.93.060201
127. Luitz, D.J., Lazarides, A., Bar Lev, Y.: Periodic and quasiperiodic revivals in periodically driven interacting quantum systems. Phys. Rev. B **97**, 020303 (2018). https://doi.org/10.1103/PhysRevB.97.020303
128. Lupo, C., Weber, C.: Nanoscopic time crystal obtained by nonergodic spin dynamics. Phys. Rev. B **100**(19), 195431 (2019). https://doi.org/10.1103/PhysRevB.100.195431
129. Lyu, C., Choudhury, S., Lv, C., Yan, Y., Zhou, Q.: An eternal discrete time crystal beating the Heisenberg limit (2019). e-prints arXiv:1907.00474
130. Machado, F., Else, D.V., Kahanamoku-Meyer, G.D., Nayak, C., Yao, N.Y.: Long-range prethermal phases of nonequilibrium matter. Phys. Rev. X **10**, 011043 (2020). https://doi.org/10.1103/PhysRevX.10.011043
131. Macieszczak, K., Guţă, M.u.u.u.u., Lesanovsky, I., Garrahan, J.P.: Towards a theory of metastability in open quantum dynamics. Phys. Rev. Lett. **116**, 240404 (2016). https://doi.org/10.1103/PhysRevLett.116.240404
132. Marcuzzi, M., Levi, E., Diehl, S., Garrahan, J.P., Lesanovsky, I.: Universal nonequilibrium properties of dissipative Rydberg gases. Phys. Rev. Lett. **113**, 210401 (2014). https://doi.org/10.1103/PhysRevLett.113.210401
133. Marconi, M., Raineri, F., Levenson, A., Yacomotti, A.M., Javaloyes, J., Pan, S.H., El Amili, A., Fainman, Y.: Mesoscopic limit cycles in coupled nanolasers (2019). e-prints arXiv:1911.10830
134. Matus, P., Sacha, K.: Fractional time crystals. Phys. Rev. A **99**, 033626 (2019). https://doi.org/10.1103/PhysRevA.99.033626
135. Medenjak, M., Prosen, T., Zadnik, L.: Rigorous bounds on dynamical response functions and time-translation symmetry breaking (2020). e-prints arXiv:2003.01035
136. Mierzejewski, M., Herbrych, J., Prelovšek, P.: Universal dynamics of density correlations at the transition to many-body localized state. Phys. Rev. B **94**, 224207 (2016). https://doi.org/10.1103/PhysRevB.94.224207. http://arxiv.org/abs/1607.04992
137. Mizuta, K., Takasan, K., Nakagawa, M., Kawakami, N.: Spatial-translation-induced discrete time crystals. Phys. Rev. Lett. **121**, 093001 (2018). https://doi.org/10.1103/PhysRevLett.121.093001
138. Mizuta, K., Takasan, K., Kawakami, N.: High-frequency expansion for floquet prethermal phases with emergent symmetries: application to time crystals and floquet engineering. Phys. Rev. B **100**, 020301 (2019). https://doi.org/10.1103/PhysRevB.100.020301
139. Mondaini, R., Rigol, M.: Many-body localization and thermalization in disordered Hubbard chains. Phys. Rev. A **92**, 041601(R) (2015). https://doi.org/10.1103/PhysRevA.92.041601
140. Monthus, C., Garel, T.: Many-body localization transition in a lattice model of interacting fermions: statistics of renormalized hoppings in configuration space. Phys. Rev. B **81**, 134202 (2010). https://doi.org/10.1103/PhysRevB.81.134202
141. Mori, T., Kuwahara, T., Saito, K.: Rigorous bound on energy absorption and generic relaxation in periodically driven quantum systems. Phys. Rev. Lett. **116**, 120401 (2016). https://doi.org/10.1103/PhysRevLett.116.120401
142. Nalitov, A.V., Sigurdsson, H., Morina, S., Krivosenko, Y.S., Iorsh, I.V., Rubo, Y.G., Kavokin, A.V., Shelykh, I.A.: Optically trapped polariton condensates as semiclassical time crystals. Phys. Rev. A **99**, 033830 (2019). https://doi.org/10.1103/PhysRevA.99.033830
143. Nandkishore, R., Huse, D.A.: Many-body-localization and thermalization in quantum statistical mechanics. Ann. Rev. Cond. Mat. Phys. **6**, 15 (2015)
144. Nicolis, G., Prigogine, I.: Self-Organization in Nonequilibrium Systems: From Dissipative Structures to Order Through Fluctuations, Nicolis, G., Prigogine, I. (eds.) Wiley, New York (1977)

145. Nurwantoro, P., Bomantara, R.W., Gong, J.: Discrete time crystals in many-body quantum chaos. Phys. Rev. B **100**(21), 214311 (2019). https://doi.org/10.1103/PhysRevB.100.214311
146. Oberreiter, L., Seifert, U., Barato, A.C.: Subharmonic oscillations in stochastic systems under periodic driving. Phys. Rev. E **100**, 012135 (2019). https://doi.org/10.1103/PhysRevE.100.012135
147. Oberreiter, L., Seifert, U., Barato, A.C.: Stochastic discrete time crystals: Entropy production and subharmonic synchronization (2020). e-prints arXiv:2002.09078
148. Oganesyan, V., Huse, D.A.: Localization of interacting fermions at high temperature. Phys. Rev. B **75**, 155111 (2007). https://doi.org/10.1103/PhysRevB.75.155111
149. Oleś, B., Ziń, P., Chwedeńczuk, J., Sacha, K., Trippenbach, M.: Bose-Einstein condensate in a double well potential in the vicinity of a critical point. Laser Phys. **20**(3), 671–677 (2010). https://doi.org/10.1134/S1054660X10050130.
150. Opatrný, T., Richterek, L., Opatrný, M.: Analogies of the classical Euler top with a rotor to spin squeezing and quantum phase transitions in a generalized Lipkin-Meshkov-Glick model. Sci. Rep. **8**, 1984 (2018). https://doi.org/10.1038/s41598-018-20486-y
151. O'Sullivan, J., Lunt, O., Zollitsch, C.W., Thewalt, M.L.W., Morton, J.J.L., Pal, A.: Dissipative discrete time crystals (2018). e-prints arXiv:1807.09884
152. Pal, A., Huse, D.A.: Many-body localization phase transition. Phys. Rev. B **82**, 174411 (2010). https://doi.org/10.1103/PhysRevB.82.174411
153. Pal, S., Nishad, N., Mahesh, T.S., Sreejith, G.J.: Temporal order in periodically driven spins in star-shaped clusters. Phys. Rev. Lett. **120**, 180602 (2018). https://doi.org/10.1103/PhysRevLett.120.180602
154. Peng, Y., Refael, G.: Time-quasiperiodic topological superconductors with Majorana multiplexing. Phys. Rev. B **98**, 220509 (2018). https://doi.org/10.1103/PhysRevB.98.220509
155. Pethick, C., Smith, H.: Bose-Einstein Condensation in Dilute Gases. Cambridge University Press, Cambridge (2002)
156. Piazza, F., Ritsch, H.: Self-ordered limit cycles, chaos, and phase slippage with a superfluid inside an optical resonator. Phys. Rev. Lett. **115**, 163601 (2015). https://doi.org/10.1103/PhysRevLett.115.163601
157. Pizzi, A., Knolle, J., Nunnenkamp, A.: Higher-order and fractional discrete time crystals in clean long-range interacting systems (2019). e-prints arXiv:1910.07539
158. Pizzi, A., Knolle, J., Nunnenkamp, A.: Period-n discrete time crystals and quasicrystals with ultracold Bosons. Phys. Rev. Lett. **123**(15), 150601 (2019). https://doi.org/10.1103/PhysRevLett.123.150601
159. Ponte, P., Chandran, A., Papić, Z., Abanin, D.A.: Periodically driven ergodic and many-body localized quantum systems. Ann. Phys. **353**, 196–204 (2015). https://doi.org/10.1016/j.aop.2014.11.008. http://www.sciencedirect.com/science/article/pii/S0003491614003212
160. Prelovšek, P., Herbrych, J.: Self-consistent approach to many-body localization and subdiffusion. Phys. Rev. B **96**(3), 035130 (2017). https://doi.org/10.1103/PhysRevB.96.035130
161. Prigogine, I.: Time, structure, and fluctuations. Science **201**(4358), 777–785 (1978). https://doi.org/10.1126/science.201.4358.777. https://science.sciencemag.org/content/201/4358/777
162. Rademaker, L., Ortuño, M.: Explicit local integrals of motion for the many-body localized state. Phys. Rev. Lett. **116**, 010404 (2016). https://doi.org/10.1103/PhysRevLett.116.010404
163. Raghavan, S., Smerzi, A., Fantoni, S., Shenoy, S.R.: Coherent oscillations between two weakly coupled Bose-Einstein condensates: Josephson effects, π oscillations, and macroscopic quantum self-trapping. Phys. Rev. A **59**, 620–633 (1999). https://doi.org/10.1103/PhysRevA.59.620
164. Reichl, L.E.: The Transition to Chaos: Conservative Classical Systems and Quantum Manifestations. Institute for Nonlinear Science. Springer, Berlin (2004)
165. Ribeiro, P., Vidal, J., Mosseri, R.: Exact spectrum of the Lipkin-Meshkov-Glick model in the thermodynamic limit and finite-size corrections. Phys. Rev. E **78**, 021106 (2008). https://doi.org/10.1103/PhysRevE.78.021106
166. Rigol, M., Dunjko, V., Olshanii, M.: Thermalization and its mechanism for generic isolated quantum systems. Nature **452**, 854–858 (2008). https://doi.org/10.1038/nature06838

167. Ros, V., Müller, M., Scardicchio, A.: Integrals of motion in the many-body localized phase. Nuclear Phys. B **891**, 420 (2015). http://dx.doi.org/10.1016/j.nuclphysb.2014.12.014. http://www.sciencedirect.com/science/article/pii/S0550321314003836
168. Rose, D.C., Macieszczak, K., Lesanovsky, I., Garrahan, J.P.: Metastability in an open quantum ising model. Phys. Rev. E **94**, 052132 (2016). https://doi.org/10.1103/PhysRevE.94.052132
169. Rovny, J., Blum, R.L., Barrett, S.E.: ^{31}P NMR study of discrete time-crystalline signatures in an ordered crystal of ammonium dihydrogen phosphate. Phys. Rev. B **97**, 184301 (2018). https://doi.org/10.1103/PhysRevB.97.184301
170. Rovny, J., Blum, R.L., Barrett, S.E.: Observation of discrete-time-crystal signatures in an ordered dipolar many-body system. Phys. Rev. Lett. **120**, 180603 (2018). https://doi.org/10.1103/PhysRevLett.120.180603
171. Russomanno, A., Iemini, F., Dalmonte, M., Fazio, R.: Floquet time crystal in the Lipkin-Meshkov-Glick model. Phys. Rev. B **95**, 214307 (2017). https://doi.org/10.1103/PhysRevB.95.214307
172. Russomanno, A., Notarnicola, S., Surace, F.M., Fazio, R., Dalmonte, M., Heyl, M.: Homogeneous floquet time crystal protected by gauge invariance. Phys. Rev. Res. **2**, 012003 (2020). https://doi.org/10.1103/PhysRevResearch.2.012003
173. Sacha, K.: Modeling spontaneous breaking of time-translation symmetry. Phys. Rev. A **91**, 033617 (2015). https://doi.org/10.1103/PhysRevA.91.033617
174. Sacha, K., Zakrzewski, J.: Time crystals: a review. Rep. Progr. Phys. **81**(1), 016401 (2017). https://doi.org/10.1088/1361-6633/aa8b38
175. Sacha, K., Müller, C.A., Delande, D., Zakrzewski, J.: Anderson localization of solitons. Phys. Rev. Lett. **103**, 210402 (2009). https://doi.org/10.1103/PhysRevLett.103.210402
176. Schäfer, R., Uhrig, G.S., Stolze, J.: Time-crystalline behavior in an engineered spin chain. Phys. Rev. B **100**, 184301 (2019). https://doi.org/10.1103/PhysRevB.100.184301
177. Schreiber, M., Hodgman, S.S., Bordia, P., Lüschen, H.P., Fischer, M.H., Vosk, R., Altman, E., Schneider, U., Bloch, I.: Observation of many-body localization of interacting fermions in a quasi-random optical lattice. Science **349**, 842 (2015). https://doi.org/10.1126/science.aaa7432
178. Seibold, K., Rota, R., Savona, V.: A dissipative time crystal in an asymmetric non-linear photonic dimer (2019). e-prints arXiv:1910.03499
179. Serbyn, M., Papić, Z., Abanin, D.A.: Local conservation laws and the structure of the many-body localized states. Phys. Rev. Lett. **111**, 127201 (2013). https://doi.org/10.1103/PhysRevLett.111.127201
180. Serbyn, M., Knap, M., Gopalakrishnan, S., Papić, Z., Yao, N.Y., Laumann, C.R., Abanin, D.A., Lukin, M.D., Demler, E.A.: Interferometric probes of many-body localization. Phys. Rev. Lett. **113**, 147204 (2014). https://doi.org/10.1103/PhysRevLett.113.147204
181. Serbyn, M., Papić, Z., Abanin, D.A.: Criterion for many-body localization-delocalization phase transition. Phys. Rev. X **5**, 041047 (2015). https://doi.org/10.1103/PhysRevX.5.041047
182. Shechtman, D., Blech, I., Gratias, D., Cahn, J.W.: Metallic phase with long-range orientational order and no translational symmetry. Phys. Rev. Lett. **53**, 1951–1953 (1984). https://doi.org/10.1103/PhysRevLett.53.1951
183. Shirley, J.H.: Solution of the Schrödinger equation with a Hamiltonian periodic in time. Phys. Rev. **138**, B979–B987 (1965). https://doi.org/10.1103/PhysRev.138.B979
184. Shtanko, O., Movassagh, R.: Unitary Subharmonic Response and Floquet Majorana Modes (2019). e-prints arXiv:1911.05795
185. Sierant, P., Delande, D., Zakrzewski, J.: Many-body localization due to random interactions. Phys. Rev. A **95**, 021601 (2017). https://doi.org/10.1103/PhysRevA.95.021601
186. Smits, J., Liao, L., Stoof, H.T.C., van der Straten, P.: Observation of a space-time crystal in a superfluid quantum gas. Phys. Rev. Lett. **121**, 185301 (2018). https://doi.org/10.1103/PhysRevLett.121.185301
187. Srednicki, M.: Chaos and quantum thermalization. Phys. Rev. E **50**, 888–901 (1994). https://doi.org/10.1103/PhysRevE.50.888

188. Steinigeweg, R., Herbrych, J., Pollmann, F., Brenig, W.: Scaling of the optical conductivity in the transition from thermal to many-body localized phases. Phys. Rev. B **94**, 180401(R) (2016). https://doi.org/10.1103/PhysRevB.94.180401
189. Steurer, W., Sutter-Widmer, D.: Photonic and phononic quasicrystals. J. Phys. D Appl. Phys. **40**(13), R229 (2007). http://stacks.iop.org/0022-3727/40/i=13/a=R01
190. Sun, F.X., He, Q., Gong, Q., Teh, R.Y., Reid, M.D., Drummond, P.D.: Discrete time symmetry breaking in quantum circuits: exact solutions and tunneling. New J. Phys. **21**(9), 093035 (2019). https://doi.org/10.1088/1367-2630/ab3f5e
191. Sun, F.X., He, Q., Gong, Q., Teh, R.Y., Reid, M.D., Drummond, P.D.: Schrödinger cat states and steady states in subharmonic generation with Kerr nonlinearities. Phys. Rev. A **100**, 033827 (2019). https://doi.org/10.1103/PhysRevA.100.033827
192. Surace, F.M., Russomanno, A., Dalmonte, M., Silva, A., Fazio, R., Iemini, F.: Floquet time crystals in clock models. Phys. Rev. B **99**, 104303 (2019). https://doi.org/10.1103/PhysRevB.99.104303
193. Tran, M.C., Ehrenberg, A., Guo, A.Y., Titum, P., Abanin, D.A., Gorshkov, A.V.: Locality and heating in periodically driven, power-law-interacting systems. Phys. Rev. A **100**(5), 052103 (2019). https://doi.org/10.1103/PhysRevA.100.052103
194. Tucker, K., Zhu, B., Lewis-Swan, R.J., Marino, J., Jimenez, F., Restrepo, J.G., Rey, A.M.: Shattered time: can a dissipative time crystal survive many-body correlations? New J. Phys. **20**(12), 123003 (2018). https://doi.org/10.1088/1367-2630/aaf18b
195. Vardeny Z. Valy, Nahata Ajay, Agrawal Amit: Optics of photonic quasicrystals. Nat. Photon. **7**, 177 (2013). http://dx.doi.org/10.1038/nphoton.2012.343
196. Walls, D.F., Drummond, P.D., Hassan, S.S., Carmichael, H.J.: Non-equilibrium phase transitions in cooperative atomic systems. Progr. Theor. Phys. Supp. **64**, 307–320 (1978). https://doi.org/10.1143/PTPS.64.307
197. Wang, Y.K., Hioe, F.T.: Phase transition in the Dicke model of superradiance. Phys. Rev. A **7**, 831–836 (1973). https://doi.org/10.1103/PhysRevA.7.831
198. Wang, R.R.W., Xing, B., Carlo, G.G., Poletti, D.: Period doubling in period-one steady states. Phys. Rev. E **97**, 020202 (2018). https://doi.org/10.1103/PhysRevE.97.020202
199. Watanabe, H., Oshikawa, M.: Absence of quantum time crystals. Phys. Rev. Lett. **114**, 251603 (2015). https://doi.org/10.1103/PhysRevLett.114.251603
200. Watanabe, H., Oshikawa, M., Koma, T.: Proof of the absence of long-range temporal orders in Gibbs states. J. Stat. Phys. **178**(4), 926–935 (2020). https://doi.org/10.1007/s10955-019-02471-5
201. Weimer, H.: Variational principle for steady states of dissipative quantum many-body systems. Phys. Rev. Lett. **114**, 040402 (2015). https://doi.org/10.1103/PhysRevLett.114.040402
202. Weiss, C., Castin, Y.: Creation and detection of a mesoscopic gas in a nonlocal quantum superposition. Phys. Rev. Lett. **102**, 010403 (2009). https://doi.org/10.1103/PhysRevLett.102.010403
203. Wilczek, F.: Quantum time crystals. Phys. Rev. Lett. **109**, 160401 (2012). https://doi.org/10.1103/PhysRevLett.109.160401
204. Yao, N.Y., Potter, A.C., Potirniche, I.D., Vishwanath, A.: Discrete time crystals: rigidity, criticality, and realizations. Phys. Rev. Lett. **118**, 030401 (2017). https://doi.org/10.1103/PhysRevLett.118.030401
205. Yao, N.Y., Nayak, C., Balents, L., Zaletel, M.P.: Classical discrete time crystals. Nat. Phys. **16**, 438–447 (2020). https://doi.org/10.1038/s41567-019-0782-3
206. Yu, W.C., Tangpanitanon, J., Glaetzle, A.W., Jaksch, D., Angelakis, D.G.: Discrete time crystal in globally driven interacting quantum systems without disorder. Phys. Rev. A **99**, 033618 (2019). https://doi.org/10.1103/PhysRevA.99.033618
207. Zaporski, L., Flicker, F.: Superconvergence of topological entropy in the symbolic dynamics of substitution sequences. SciPost Phys. **7**, 18 (2019). https://doi.org/10.21468/SciPostPhys.7.2.018. https://scipost.org/10.21468/SciPostPhys.7.2.018
208. Zel'dovich, Y.B.: Scattering and emission of a quantum system in a strong electromagnetic wave. Sov. Phys. Usp. **16**, 427 (1973)

209. Zhang, J., Hess, P.W., Kyprianidis, A., Becker, P., Lee, A., Smith, J., Pagano, G., Potirniche, I.D., Potter, A.C., Vishwanath, A., Yao, N.Y., Monroe, C.: Observation of a discrete time crystal. Nature **543**(7644), 217–220 (2017). Letter. https://doi.org/10.1038/nature21413
210. Zhao, H., Mintert, F., Knolle, J.: Floquet time spirals and stable discrete-time quasicrystals in quasiperiodically driven quantum many-body systems. Phys. Rev. B **100**, 134302 (2019). https://doi.org/10.1103/PhysRevB.100.134302
211. Zhu, B., Marino, J., Yao, N.Y., Lukin, M.D., Demler, E.A.: Dicke time crystals in driven-dissipative quantum many-body systems. New J. Phys. **21**(7), 073028 (2019). https://doi.org/10.1088/1367-2630/ab2afe
212. Ziń, P., Chwedeńczuk, J., Oleś, B., Sacha, K., Trippenbach, M.: Critical fluctuations of an attractive Bose gas in a double-well potential. Europhys. Lett. **83**(6), 64007 (2008). https://doi.org/10.1209/0295-5075/83/64007
213. Žnidarič, M., Prosen, T., Prelovšek, P.: Many-body localization in the Heisenberg xxz magnet in a random field. Phys. Rev. B **77**, 064426 (2008). https://doi.org/10.1103/PhysRevB.77.064426

Chapter 5
Condensed Matter Physics in the Time Dimension

Abstract Ordinary space crystals can be insulators, conductors, and even superconductors. Number of different solid state phases, including topological phases, is large. They are a subject of basic research which gives a hope for novel practical applications. The present chapter shows that the time domain is also a good *degree of freedom* where non-trivial condensed matter phenomena can be realized. For example, Anderson and many-body localization, Mott insulator phase, and topological phases can all be observed in the time dimension. Time engineering offers also a possibility for invention of completely new objects and an example of such objects is provided in the current chapter.

5.1 Platform for Condensed Matter Physics in the Time Dimension

In the previous chapters we have analyzed systems which can reveal spontaneous breaking of continuous or discrete time translation symmetries. Spontaneous breaking of a time translation symmetry is the demonstration of one of very important properties of solid state systems. However, a question arises if various condensed matter phases can also be observed in the time dimension and it is the subject of the present chapter [38, 51].

Guo et al. [47] were the first who realized that a resonantly driven single-particle system can reveal a quasi-energy band structure similar to energy bands of a particle in a time-independent potential periodic in space [45]. They called this phenomenon a phase space crystal and in subsequent publications the idea of phase space crystals was developed [44, 46, 83]. We describe phase space crystals in the next chapter. In the current chapter we focus on the time domain and show that any system which can reveal nonlinear classical resonances (in the language of the theory of dynamical systems) is suitable for realization of various condensed matter phases in the time domain [38, 96, 98].

K. Sacha, *Time Crystals*, Springer Series on Atomic, Optical, and Plasma Physics 114, https://doi.org/10.1007/978-3-030-52523-1_5

Let us begin with classical mechanics. The key method to describe nonlinear classical resonances is the secular approximation approach that reduces a resonantly driven single-particle system to the problem of a pendulum [64]. Assume that we deal with a particle of the unit mass in the presence of a time-independent potential,

$$H_0 = \frac{p^2}{2} + V(z), \tag{5.1}$$

which can perform periodic motion with a period that depends on energy. This assumption excludes the harmonic oscillator problem where the period of a particle motion is constant independently of its energy. A resonantly driven harmonic oscillator is analyzed in the next chapter where we discuss phase space crystals [44–46, 83]. For periodic motion of a particle we can perform canonical transformation to the angle-action variables θ and I which are described in the background information in Sect. 4.2.1. Then, the Hamiltonian (5.1) depends on the new momentum (the action I) only, $H_0 = H_0(I)$. The action I is a constant of motion and the new position variable (the angle $0 \le \theta < 2\pi$) evolves linearly in time $\theta(t) = \Omega(I)t + \theta_0$ where the frequency of periodic motion of a particle depends on I and thus on energy,

$$\Omega(I) = \frac{dH_0(I)}{dI}. \tag{5.2}$$

Assume that we start driving the system, i.e. we turn on a time-periodic perturbation

$$H_1(z, t) = \lambda f(t) h(z), \tag{5.3}$$

where $f(t) = f(t + T)$ with $T = 2\pi/\omega$, so that the entire Hamiltonian of a particle reads $H = H_0 + H_1$. Time-periodic function $f(t)$ can be expanded in the Fourier series in time and also the spatial part of the perturbation $h(z)$ can be expanded in the Fourier series in the action-angle variables,

$$f(t) = \sum_{l \neq 0} f_l e^{il\omega t}, \qquad h(z) = \sum_n h_n(I) e^{in\theta}. \tag{5.4}$$

We are interested in a resonant driving of a particle. That is, we focus on initial conditions close to an unperturbed periodic orbit corresponding to the frequency $\Omega(I_s)$ which fulfills $s:1$ resonance condition with the driving frequency ω,

$$\Omega(I_s) = \frac{\omega}{s}, \tag{5.5}$$

where s is integer. Let us switch to the moving reference frame,

$$\Theta = \theta - \frac{\omega}{s}t, \qquad I' = I, \tag{5.6}$$

where particle motion close to a resonant periodic orbit is slow. It results in the following form of the Hamiltonian

$$H = H_0(I) - \frac{\omega I}{s} + \lambda \sum_{n,l} f_l h_n(I) e^{in\Theta} e^{i(ls+n)\omega t/s}, \tag{5.7}$$

where we have dropped the prime in I'. If we choose the initial action $I \approx I_s$, then, for weak driving, both I and Θ evolve slowly because $dI/dt = O(\lambda)$ and also $d\Theta/dt \approx O(\lambda)$ due to the resonance condition (5.5). Thus, one can average the Hamiltonian (5.7) over time keeping I and Θ fixed. It leads to

$$H_{\text{eff}} \approx \frac{P^2}{2m_{\text{eff}}} + \lambda \sum_{n} f_n h_{-sn}(I_s) e^{isn\Theta}, \tag{5.8}$$

where $P = I - I_s$ and the effective mass of a particle[1]

$$m_{\text{eff}} = \left(\frac{d^2 H_0(I_s)}{dI_s^2} \right)^{-1}. \tag{5.9}$$

The Hamiltonian (5.8) is an effective Hamiltonian that describes a particle motion close to a resonant orbit. It is the central result of the classical secular approximation method. Its validity can be analyzed either by calculating the second order terms within the secular approximation approach or simply by comparing the phase space portrait generated by the Hamiltonian (5.8) with the exact stroboscopic picture of the phase space obtained by numerical integration of the exact classical equations of motion [37, 38, 96].

The Hamiltonian (5.8) indicates that a resonantly driven particle, in the frame moving basically along a resonant orbit, cf. (5.6), behaves effectively like a particle with a certain effective mass in a time-independent potential. What is the shape of this potential? Well, it depends on time-periodic driving, i.e. on Fourier components f_l in (5.4) and it can be actually shaped at will.

Let us start with simple harmonic driving in time where $f(t) = \cos\omega t$. Then, the effective Hamiltonian (5.8) takes the form

$$H_{\text{eff}} = \frac{P^2}{2m_{\text{eff}}} + V_0 \cos(s\Theta), \tag{5.10}$$

[1]In order to obtain (5.8), we have expanded $H_0(I) - \omega I/s$ in the Taylor series around I_s up to the second order and omitted the constant term $H_0(I_s) - \omega I_s/s$.

where $V_0 = \lambda h_s(I_s)$ and we have assumed that $h_n(I_s)$ are real.[2] If a high principal resonance is chosen ($s \gg 1$), then, in the moving frame, our driven particle is described by the Hamiltonian (5.10) which is similar to the Hamiltonian for an electron in an ordinary space crystal where periodically distributed ions form a space-periodic potential. In order to deal with a solid state problem we have to switch to quantum description which can be done either by quantizing the classical secular Hamiltonian (5.10), i.e. substituting $P \to -i\frac{d}{d\Theta}$, or by applying the quantum secular approach from the very beginning [14].[3] Both methods yield the same resonant quasi-energy spectrum of the system if the resonance condition corresponds to I_s much larger than the Planck constant (here $\hbar = 1$).

For $s \gg 1$ eigenvalues of the quantized version of the effective Hamiltonian (5.10) form energy bands and the corresponding eigenstates are Bloch waves

$$\psi_{m,k}(\Theta) = e^{ik\Theta}u_{m,k}(\Theta), \quad \text{with} \quad u_{m,k}(\Theta + 2\pi/s) = u_{m,k}(\Theta), \tag{5.11}$$

where m is a number of an energy band and k is a quasi-momentum which is an integer number due to the periodic boundary conditions imposed by the Hamiltonian (5.10). The condensed matter-like behavior (5.11) that we observe in the moving frame versus Θ will be reproduced in the time domain when we return to the laboratory. This is the absolutely crucial property which we have to stress. The transformation between the laboratory frame and the moving frame is linear in time, see (5.6). Thus, when we return to the laboratory frame and locate a detector close to a resonant orbit (i.e. we fix I = constant $\approx I_s$ and θ = constant), time evolution of the probability density for detection of a particle at this fixed point will reproduce the condensed matter-like behavior that we have identified in the moving frame [38, 67, 96, 98], i.e. the Bloch waves (5.11) in the laboratory frame read

$$\psi_{m,k}(\theta - \omega t/s) = e^{ik(\theta-\omega t/s)}u_{m,k}(\theta - \omega t/s), \quad \text{where} \quad \theta = \text{constant}. \tag{5.12}$$

The method of the description of condensed matter-like behavior in the time domain that we have just introduced can be applied to any single-particle system which reveals nonlinear classical resonances. In order to illustrate a solid state physics in the time dimension let us consider an example of a particle bouncing resonantly on an oscillating mirror. This system has been already described in details in Sect. 4.2.1 and it will also be used for demonstration of various condensed matter phenomena in the time domain in the following sections. For $s:1$ resonant driving

[2] If the Fourier components are complex, $h_n = h^*_{-n} = |h_n|e^{i\varphi}$, then the phase shift φ is present in the cosine function in (5.10).

[3] The quantum version of the secular approach relies on switching to the rotating frame by means of the unitary transformation $U = e^{i\hat{n}\omega t/s}$ and then averaging the resulting Hamiltonian over time which yields the quantum effective Hamiltonian $\langle n'|H|n\rangle \approx (E_n - n\omega/s)\delta_{n,n'} + (\lambda/2)\langle n'|h|n\rangle(\delta_{n',n+s} + \delta_{n',n-s})$ where $H_0|n\rangle = E_n|n\rangle$ and $\hat{n}|n\rangle = n|n\rangle$, cf. (4.53).

the effective Hamiltonian (5.10) reduces to the Hamiltonian (4.49).[4] The effective mass turns out to be negative, see (4.50), and consequently the first energy band of (5.10) is the highest in energy (cf. Fig. 4.9). Let us analyze how the Bloch wave eigenstates of (5.10) look like but in the laboratory frame and in the Cartesian coordinates. In left panel of Fig. 5.1, a Bloch wave corresponding the first energy band of (5.10) is presented. For clarity we have chosen a small number of the resonance $s = 4$ and consequently we have only four lattice sites in the potential in (5.10) and the Bloch wave in Fig. 5.1 reveals only four localized wavepackets bouncing off an oscillating mirror and evolving along the $4 : 1$ resonant classical trajectory. The Bloch wave is actually a Floquet state obtained within the secular approximation approach. It is plotted in the configuration space but the configuration space is not a domain where we can observe condensed matter physics (or in other words where we can observe a crystalline structure). A crystalline structure can be observed but in the time domain. That is, if we locate a detector close the resonant trajectory, e.g. close to the classical turning point of a particle bouncing on a mirror, and plot how the probability for the detection of a particle changes in time we can see time-periodic behavior because each localized wavepacket passes a given space point in the same way and periodically, see right panel of Fig. 5.1. The probability density corresponding to the Bloch wave evolves with the driving period $T = 2\pi/\omega$ because it is a Floquet state. However, each localized wavepacket appears at a given space point with the period sT. There are s Floquet states $\phi_n(z, t + T) = \phi_n(z, t)$ associated with the first energy band of the Hamiltonian (5.10) which evolve in time like $e^{-iE_nt}\phi_n(z, t)$. Linear combinations of the s Floquet states $\phi_n(z, t)$ corresponding to the first energy band allow us to extract the localized wavepackets $w_j(z, t)$ which fulfill $w_j(z, t + sT) = w_j(z, t)$ and which in the moving frame corresponds to the Wannier states [30] localized in sites of the periodic potential in (5.10).

Let us restrict the description of a particle resonantly bouncing on an oscillating mirror to the first energy band of the Hamiltonian (5.10). In other words we restrict to the resonant Hilbert subspace spanned by s localized wavepackets $w_j(z, t)$ where a particle wavefunction can be expanded as

$$\psi(z, t) = \sum_{j=1}^{s} w_j(z, t)\, a_j, \tag{5.13}$$

with $\sum_{j=1}^{s} |a_j|^2 = 1$. If we do that, the energy of a particle (more precisely quasi-energy E_F) takes a form of a tight-binding model well known in condensed matter physics,

[4]The Hamiltonian (4.49) reads $H_F \approx H_{\text{eff}}(\Theta, P) + p_t$ (where H_{eff} is given in (5.10)) and it is the effective Hamiltonian in the phase space extended by t and the conjugate momentum p_t. The Hamiltonian H_F is the classical counterpart of the effective Floquet Hamiltonian of a particle. Due to the fact that p_t is a constant of motion within the effective Hamiltonian approach, we can set $p_t = 0$ and then (4.49) and (5.10) are identical. Thus, the quantized version of (5.10) is the effective Floquet Hamiltonian corresponding to $p_t = k\omega/s$ with $k = 0$, compare Eq. (4.56).

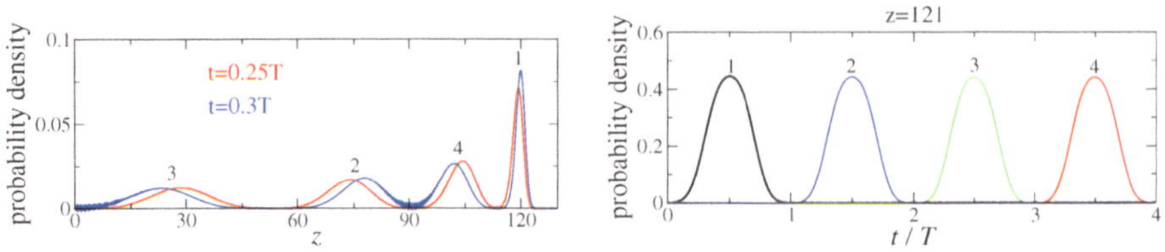

Fig. 5.1 Particle bouncing resonantly on a harmonically oscillating mirror in the presence of the gravitational force in the case of the 4 : 1 resonance. Left panel shows a Bloch wave corresponding to the first energy band of the Hamiltonian (5.10) in the laboratory frame and in the Cartesian coordinates (the energy bands of (5.10) emerge for $s \gg 1$ but for the sake of clarity of the presentation, the $s = 4$ case is chosen). The Bloch wave is actually a Floquet state of the time-periodic system. The probability density is presented at two slightly different moments of time (as indicated in the plot). It consists of four localized wavepackets propagating along the 4 : 1 classical resonant orbit and bouncing off the mirror which is located at $z = 0$. The configuration space is, however, not a domain where a periodic crystalline structure can be observed. The crystalline structure emerges in the time domain. Indeed, in the right panel the probability density for the detection of a particle at $z = 121$ is plotted versus time. Numbers of the wavepackets in both panels indicate that two neighboring wavepackets in the time domain are not always neighbors in the configuration space. Reprinted from [96]

$$E_F \approx -\frac{J}{2}\sum_{j=1}^{s}\left(a^*_{j+1}a_j + \text{c.c.}\right), \tag{5.14}$$

where due to the periodic boundary conditions $a_{s+1} = a_1$ and J is the amplitude that describes tunneling of a particle between wavepackets $w_{j+1}(z, t)$ and $w_j(z, t)$ which are neighbors but in the time domain, see right panel of Fig. 5.1. There is also longer range (or *longer time*) tunneling but it is usually orders of magnitude weaker similarly as it is the case in a typical solid state system that is described by a tight-binding model. The tunneling amplitude can be obtained either with the help of Bloch waves in the moving frame, see (5.11), or in the laboratory frame in the Cartesian coordinates with the help of the Floquet operator $H_F = H - i\partial_t$,[5]

$$J = -\frac{2}{sT}\int_0^{sT} dt \int_0^{\infty} dz\, w^*_{j+1}(z, t)[H - i\partial_t]w_j(z, t), \tag{5.15}$$

where H is the Hamiltonian (4.29) of a particle bouncing on an oscillating mirror (compare also Eqs. 4.62 and 4.96).

[5]It may happen that within the quantum description that goes beyond the effective Hamiltonian (5.10), the tunneling amplitude J is a complex number and, due to the periodic boundary conditions, it is not possible to eliminate its phase by changing global phases of $w_j(z, t)$. Then, one may interpret this situation as the presence of a *magnetic flux* penetrating a closed resonant periodic orbit.

We have shown that a condensed matter-like Hamiltonian turns up in the effective description of a resonantly driven particle. It should be stressed that such crystalline behavior is not a result of the presence of any potential energy periodic in space. An original time-periodic Hamiltonian can have nothing in common with a solid state system, for example, the Hamiltonian (4.29) is the Hamiltonian of a particle bouncing on an oscillating mirror, and still condensed matter behavior can emerge in dynamics of a system due to resonant driving. We can interpret it as solid state physics along a resonant classical orbit. Period of a resonant orbit is sT and it implies that crystalline structures in time that we consider here fulfill periodic boundary conditions in time on a scale which is the longer, the higher number of principal $s:1$ resonance is chosen. Along this time scale we can do condensed matter physics in the time domain.

It should be also stressed that time crystal behavior that we focus on in the present chapter is not related to spontaneous breaking of time translation symmetry [98]. We rather adopt a standard condensed matter approach where the presence of a periodic crystalline structure in space is assumed and properties of a solid state system are analyzed with the help of a Hamiltonian where space-periodic potential is included *by hand*. Similar approach to condensed matter physics is used in ultra-cold atoms where periodic optical lattice potentials in space are created by means of electromagnetic standing wave [84]. Time crystals we analyze in the current chapter can also be considered as time analogues of photonic crystals where periodic structures in space do not form spontaneously but they are engineered [113].

Having sketched the idea of condensed matter physics in the time domain, we can now turn to description of realization of various solid state phases in the time dimension.

5.2 Anderson Localization in the Time Dimension

A classical particle moving in a time-independent disordered potential can perform diffusive motion. However, when we switch from the classical to quantum mechanics, the situation can change completely because interference phenomena, especially destructive interference between different multiple scattering paths, can suppress the diffusion. Indeed, Philip W. Anderson discovered that in the three-dimensional space, the diffusion and consequently transport of non-interacting particles can stop totally if a disordered potential is sufficiently strong [4]. It is the famous Anderson localization phenomenon which can be observed in many different systems described by wave equations ranging from matter waves and electromagnetic waves to acoustic waves [78]. The suppression of transport is accompanied by exponential localization of eigenstates of a particle in the configuration space. That is, eigenstates of a system are localized around different positions $\mathbf{r}_0$ in the space with tails of the probability density decaying like $e^{-|\mathbf{r}-\mathbf{r}_0|/\xi}$ where ξ is the so-called localization length which in general depends on energy of a particle. Anderson localization strongly depends on the dimension of the space. The scaling theory of localization implies that one- and two-dimensional time-reversal invariant and spinless systems reveal Anderson localization regardless how weak a disordered potential is [1]. In three-dimensional space, the scaling theory predicts a second order phase transition. That is, for a fixed disordered potential there is the so-called mobility edge where all eigenstates with eigenenergies below the mobility edge are localized while with eigenenergies greater than the mobility edge are delocalized.

There is also another type of Anderson localization often called dynamical localization which takes place not in the configuration space but in the momentum space and where no disordered external potential is required [19, 34, 35, 43]. A classical particle moving in one-dimensional space and perturbed by a regular periodic force can reveal chaotic behavior when the driving force is sufficiently strong. Then, one can observe diffusive motion of a particle in a chaotic part of the classical phase space. Switching from the classical to quantum mechanics, interference effects of matter waves come on the scene and are able to suppress the diffusion and spreading of the probability density in the momentum space. In other words heating of a particle by a driving force stops in the quantum description and the probability density in the momentum space becomes exponentially localized. The paradigm system that reveals dynamical Anderson localization is the kicked rotor which is a free particle moving on a one-dimensional ring and being kicked periodically with a force depending on its position x on a ring [19, 34],

$$H = \frac{p^2}{2} + K \cos x \sum_n \delta(t - n), \tag{5.16}$$

where K stands for the strength of the kicking and we have assumed that the kicking period $T = 1$. For $K > 5$ the classical motion is fully chaotic, i.e. the stroboscopic picture of the phase space (obtained by plotting x and p after each period of the kicking) shows a fully developed chaotic sea and a particle performs diffusive motion in the phase space. In the quantum description the behavior of the system is totally different. Starting with a quantum state of a particle corresponding to a Gaussian distribution in the momentum space centered around $p = 0$, initially one observes signatures of diffusive motion of a particle. However, the diffusion stops after a characteristic localization time and the distribution in the momentum space becomes exponentially localized, $e^{-|p|/\xi}$ with $\xi \approx K^2/4$ (we have assumed that the Planck constant $\hbar = 1$) [19]. Fishman et al. showed that such localization in the momentum space is indeed an Anderson localization phenomenon [34, 43]. They reduced the eigenvalue problem for the Floquet evolution operator, $U_F|\phi_n\rangle = e^{i\varphi_n}|\phi_n\rangle$ where U_F is the evolution operator over one period, to the Anderson tight-binding model. In the Anderson model [4], energy of an electron in a lattice with random on-site energies ε_j is described by

$$E = -\frac{1}{2} \sum_{jj'} J_{jj'} a_j^* a_{j'} + \sum_i \varepsilon_j |a_j|^2, \tag{5.17}$$

and the corresponding time-independent Schrödinger equation reads

$$\varepsilon_j a_j - \frac{1}{2}\sum_{j'} J_{jj'} a_{j'} = E a_j. \tag{5.18}$$

If, in the one-dimensional Anderson model (5.17), the tunneling amplitudes $J_{jj'}$ drop sufficiently quickly with $|j - j'|$, all eigenstates are exponentially localized regardless how weak disorder is, i.e. how narrow the probability distribution for ε_j is. Fishman and coworkers demonstrated how to rewrite the eigenvalue problem for the Floquet evolution operator for the kicked rotor problem into the form of (5.18) where momentum states of a particle e^{ijx} play a role of lattice sites j. The on-site energies $\varepsilon_j = \tan[(\varphi_n - j^2/2)/2]$ are not random numbers but pseudorandom numbers but it turns out to be sufficient for the observation of the Anderson localization. The tunneling amplitudes $J_{jj'} = -2\int_0^{2\pi} dx e^{i(j-j')x}\tan(K\cos x/2)$ and thus they are controlled by the kicking strength K. In the Anderson model (5.18), the localization length of an eigenstate depends on the corresponding eigenenergy E. In the kicked rotor problem $E = -\int_0^{2\pi} dx \tan(K\cos x/2)$ is a constant and consequently all eigenstates of (5.18) are characterized by the same localization length ξ.

The dynamical Anderson localization described here can be observed in the cold atoms experiment where atoms are periodically kicked by turning on an optical standing wave for very short periods of time. Then, cold atoms are described by the Hamiltonian (5.16) [35].

The kicked rotor is also a suitable system to realize the localized-delocalized Anderson phase transition. Indeed, when we substitute $K \to K[1 + \epsilon\cos(\omega_2 t)\cos(\omega_3 t)]$ and $\{\omega_2, \omega_3, 2\pi\}$ are incommensurate numbers, the system can be mapped on the three-dimensional Anderson model, cf. (5.18), and the localized-delocalized Anderson transition can be observed [20, 63].

To sum up, we have seen that Anderson localization is localization of a particle in the configuration space due to the presence of a time-independent disordered potential. Anderson localization is also possible in the momentum space in periodically driven systems which in the classical description reveal diffusive motion in the phase space. The diffusive motion and thus the randomness in a system result from the chaotic classical dynamics. In the following we will show that yet another version of Anderson localization is possible: Anderson localization in the time domain due to the presence of *disorder in time* [96].

5.2.1 Anderson Localization in Time Crystals

Anderson localization in the time domain was first suggested in the system of a particle bouncing on a mirror which oscillates harmonically in time but with additional random fluctuations [96]. We will, however, start with the introduction of the concept of the Anderson localization in time in a simpler system of a particle moving on a ring [97].

A Particle on a Ring Driven by Fluctuating Force

A particle moving freely on a ring is described by the Hamiltonian $H_0 = p^2/2$. We denote the position of a particle by an angle θ. Actually θ and the conjugate momentum p are the angle-action variables of the system (see the background information in Sect. 4.2.1). The frequency of classical motion of a particle along the ring is $\Omega(p) = p$. Let us assume that we turn on time-periodic perturbation so that the entire Hamiltonian of a particle reads

$$H = \frac{p^2}{2} + \lambda \cos(s\theta)\cos(s\omega t) + Vh(\theta)f(t), \tag{5.19}$$

where λ and V are the amplitudes of the two perturbation terms, s is integer, $f(t + 2\pi/\omega) = f(t) = \sum_{k\neq 0} f_k e^{ik\omega t}$ and $h(\theta) = \theta/\pi$ with $-\pi < \theta \leq \pi$. The function $h(\theta)$ is discontinuous at $\theta = \pi$ but it can be expanded in the Fourier series $h(\theta) = \sum_{n\neq 0} h_n e^{in\theta}$ with $h_n = i(-1)^n/(\pi n)$. Switching to the rotating frame, $\Theta = \theta - \omega t$, and assuming resonant motion of a particle, i.e. $P = p - \omega \approx 0$, we may average the Hamiltonian over t keeping Θ and P fixed because they are slowly varying variables (see Sect. 5.1). It yields the effective classical Hamiltonian

$$H_{\text{eff}} = \frac{P^2}{2} + \frac{\lambda}{2}\cos(s\Theta) + V \sum_{k=-\infty}^{+\infty} h_k f_{-k} e^{ik\Theta}, \tag{5.20}$$

where a constant term has been omitted. This classical Hamiltonian is identical to the quantum effective Hamiltonian which can be obtained within the quantum secular approach [14] because in the classical approach we have used a linear canonical transformation only. It is valid provided the second order terms of the secular approximation are negligible, i.e. when $\lambda^2/\omega^2 \ll 1$ and $V^2/\omega^2 \ll 1$ which can be always fulfilled by choosing sufficiently high frequency of the time-periodic perturbation.

For $V = 0$, the effective Hamiltonian (5.20) describes a particle moving in a spatially periodic potential consisting of s lattice sites. For $s \gg 1$ the eigenenergies of H_{eff} reveal a band structure and the eigenstates are extended Bloch waves, cf. (5.11). Now, let us turn on the second part of the perturbation with the amplitude $V \ll \lambda$ and assume that f_k (which fulfill $f_k = -f^*_{-k}$ otherwise $f(t)$ would not be a real function) are independent random numbers. Specifically let us consider $f_k = |f_k|e^{i\varphi_k}$ so that

$$|h_k f_{-k}| = \frac{1}{\sqrt{k_0}\pi^{1/4}} e^{-k^2/(2k_0^2)}, \tag{5.21}$$

and $0 < \varphi_k \leq 2\pi$ are random numbers chosen from the uniform distribution. Then, the second part of the perturbation in (5.20) is a disordered potential in Θ, with the correlation length $\sigma = \sqrt{2}/k_0$, that fulfills periodic boundary conditions on the ring. When $\sigma < 2\pi/s$ we end up with a particle moving in a crystalline structure where in each potential well the minimal value of the potential energy is randomly distributed with the mean value $-\lambda/2$ and the standard deviation V. Because we assume $V \ll \lambda$, we deal actually with a crystalline structure in the presence of a weak disorder, i.e. a typical situation for realization of Anderson localization. When we restrict to the lowest energy band of the Hamiltonian (5.20), then the only parameters of the system are the disorder strength V and the amplitude J of tunneling of a particle between potential wells (sites). The latter can be estimated as $J = (2^5\lambda^3 s^2/\pi^2)^{1/4} e^{-\sqrt{32\lambda}/s}$ provided the potential wells are deep [115]. If the range of the space was infinite, i.e. $\Theta \in (-\infty, +\infty)$, then for any disorder strength eigenstates belonging to the lowest energy band would be exponentially localized,

$$|\psi(\Theta)|^2 \propto e^{-|\Theta-\Theta_0|/\xi}, \tag{5.22}$$

around different positions Θ_0. The localization length of the eigenstates at the center of the band can be estimated as $\xi \approx 8\pi J^2/(sV^2)$ which is valid for $V \ll J$ [78]. In the present case we have a finite space, $\Theta \in (-\pi, \pi]$, and in order to observe the Anderson localization, the localization length $\xi \ll 2\pi$.

We have shown that in the rotating frame ($\Theta = \theta - \omega t$), the description of a resonantly driven particle can reveal Anderson localization if the driving force fluctuates in time but still fulfills periodic boundary conditions in time, i.e. $f(t + 2\pi/\omega) = f(t)$. Such a *disorder in time* results in the Anderson localization in the time domain when we return to the laboratory frame. Indeed, if in the laboratory frame we fix the position $\theta =$ constant, and ask how the probability for the detection of a particle at this position changes in time [67], we obtain that it is exponentially localized,

$$|\psi(\theta - \omega t)|^2 \propto e^{-|\theta-\Theta_0-\omega t|/\xi}, \quad \text{for} \quad |\theta - \Theta_0 - \omega t| < \pi, \tag{5.23}$$

around a certain moment of time, $t = (\theta - \Theta_0)/\omega$, and such a behavior will be repeated in time with the period $2\pi/\omega$. In Fig. 5.2 we show an example of a Floquet state of the system corresponding to an eigenstate of (5.20) with eigenenergy at the center of the lowest band. In Fig. 5.2, there is also a Floquet state that reveals Anderson localization in time but corresponding to an eigenstate of (5.20) with eigenenergy at the center of the first excited energy band. Both Floquet states reveal exponential localization in the time dimension around a certain moment of time.

A Particle Bouncing on a Fluctuating Mirror

In Sect. 5.1 we have introduced the secular approximation approach which allows us to describe a resonantly driven single-particle system. We have focused on a specific example of a particle bouncing on a harmonically oscillating mirror and on the $s:1$ resonance where a particle behaves like an electron in an ordinary space crystal that consists of s Bravais lattice cells. If we restrict to the first energy band of the effective Hamiltonian (5.10), the system can be described by a tight-binding

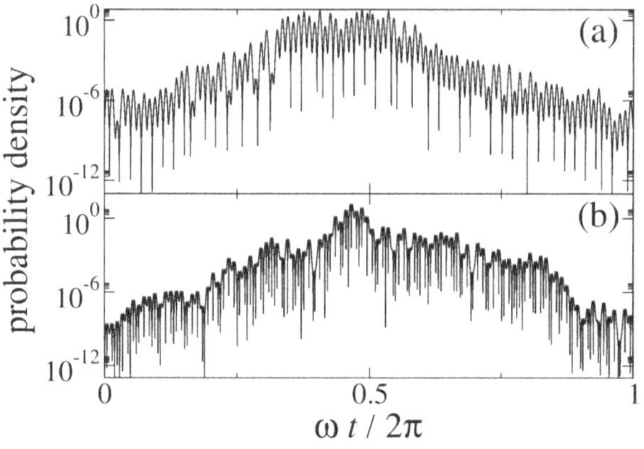

Fig. 5.2 Anderson localization in the time domain of a particle moving along a ring and perturbed resonantly by harmonically oscillating force and additional weak fluctuating force. The 100 : 1 resonance condition is fulfilled between the driving frequency $s\omega$ of the harmonically oscillating force and a particle which is moving with $p \approx \omega$. Such a resonant driving leads to the formation of a crystalline structure in time which in the rotating frame is described by the solid state-like Hamiltonian (5.20) where there is also a disordered term resulting from the presence of the weak fluctuating force. Panel (**a**) shows probability density for the detection of a particle at a fixed space point versus time. The probability density corresponds to an eigenstate of (5.20) with the eigenenergy at the center of the lowest energy band. It is a Floquet state of the original system, described by the Hamitlonian (5.19), that reveals Anderson localization around a certain moment of time. Panel (**b**) presents a similar eigenstate of (5.20) but belonging to the first excited energy band. The parameters of the system are the following: $s = 100$, $\lambda = 2 \times 10^4$, $k_0 = 100$, $V = 10$ (**a**), and $V = 300$ (**b**). The localization length in both panels is of the order of $0.16/\omega$. Reprinted from [97]

model (5.14) where wavefunction of a particle $\psi(z,t) = \sum_{j=1}^{s} a_j w_j$ is expanded in the basis of the localized wavepackets $w_j(z,t)$ which fulfill $w_j(z, t+T) = w_{j+1}(z,t)$ and $w_j(z, t+sT) = w_j(z,t)$.

Let us assume that apart from a simple harmonic motion of a mirror which is characterized by the amplitude λ, we turn on additional modulation of the mirror motion characterized by the amplitude V, so that in the Cartesian coordinates (in the frame oscillating with the mirror, cf. (4.29), and in the gravitational units (4.28)) the Hamiltonian of a particle reads

$$H = \frac{p^2}{2} + z + \lambda z \cos\omega t + V z f(t), \quad \text{where} \quad z \geq 0. \tag{5.24}$$

For $V = 0$, quantum dynamics of a particle corresponding to the $s : 1$ classical resonance can be described by the tight-binding model (5.14). Let us assume that the additional modulation of the mirror motion is described by $f(t)$ that fluctuates randomly in time but still fulfills periodic boundary conditions on a long time scale, i.e. on the scale corresponding to the period of the $s : 1$ resonant orbit which is sT,

$$f(t) = f(t + sT) = \frac{1}{\sqrt{2s}} \sum_{\substack{k=-s \\ k\neq 0}}^{s} f_k e^{ik\omega t/s}, \tag{5.25}$$

where $T = 2\pi/\omega$ and $f_k = -f^*_{-k}$ are random numbers. Then, the tight-binding model (5.14) is supplemented with additional terms, i.e. it reads

$$E_F \approx -\frac{J}{2} \sum_{j=1}^{s} \left(a^*_{j+1} a_j + \text{c.c.} \right) + \sum_{j=1}^{s} \varepsilon_j |a_j|^2, \tag{5.26}$$

where

$$\varepsilon_j = \frac{V}{sT} \int_0^{sT} dt \int_0^{\infty} dz\, z\, f(t)\, |w_j(z,t)|^2. \tag{5.27}$$

If we choose $f_k = (k^2\omega^2/s^2)e^{i\varphi_k}$ where φ_k's are random numbers chosen uniformly in the range $[0, 2\pi)$, then ε_j's are real random numbers corresponding to the Gaussian distribution with zero mean and standard deviation V.[6] Thus, a particle bouncing resonantly on a harmonically oscillating mirror with additional random fluctuations of the mirror oscillations is reduced to the Anderson model (5.17) and Anderson localization occurs provided the localization length is smaller than the length of the lattice in (5.26) [37, 96].

The predicted Anderson localization means that wavefunction of a particle $\psi(z,t) = \sum_{j=1}^{s} a_j w_j(z,t)$ is localized exponentially around a certain localized wavepacket w_{j_0}, i.e. $|a_j|^2 \propto e^{-|j-j_0|/\xi}$. Due to the fact that $w_{j+1}(z,t) = w_1(z, t + jT)$, when we locate a detector close to the mirror, the probability density for the measurement a particle at this position will reveal exponential localization in time around a certain time moment [96]. In Fig. 5.3 we illustrate such Anderson localization in the time domain for experimentally attainable parameters [37].

To conclude this part it is instructive to compare crystalline structures in space and in time. Suppose we have a particle in an ordinary space crystal in one-dimensional space with periodic boundary conditions, i.e. a crystalline structure on a ring, see Fig. 5.4. If a weak disordered potential is added to such a crystalline

[6]In order to show that ε_j's correspond to the normal distribution, we can start with the effective Hamiltonian in the angle-action variables, cf. (5.10), which in the present case and for $f_k = (k^2\omega^2/s^2)e^{i\varphi_k}$ reduces to $H_{\text{eff}} = P^2/2m_{\text{eff}} + V_0\cos(s\Theta) + (V/\sqrt{2s})\sum_k e^{i(k\Theta+\varphi_k)}$. Then, applying the tight-binding approximation with the help of the Wannier wavefunctions $w_j(\Theta)$ [30], which are the localized wavepackets $w_j(z,t)$ moving along the resonant orbit, one obtains $\varepsilon_j = (V/\sqrt{2s})\sum_k \int_0^{2\pi} d\Theta |w_j(\Theta)|^2 e^{i(k\Theta+\varphi_k)} \approx V\sqrt{2/s}\sum_{k=1}^{s}\cos(\Theta_j + \varphi_k)$ where we have assumed that $|w_j(\Theta)|^2$ is strongly localized around $\Theta_j = \pi(2j-1)/s$. Finally, employing the central limit theorem, we obtain that if the angles φ_k's are chosen randomly from the uniform distribution, then ε_j's are random values corresponding to the Gaussian distribution with zero mean and standard deviation V.

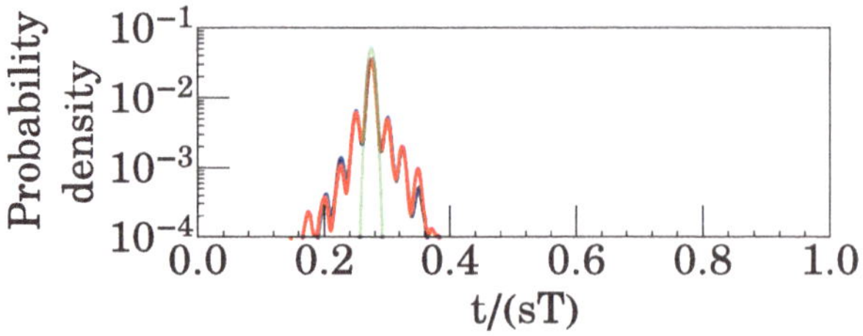

Fig. 5.3 Anderson localization in the time domain of a particle bouncing resonantly on an oscillating mirror. Harmonic oscillations of a mirror, which fulfill the 40 : 1 resonant condition with particle motion, create crystalline structure in time. When there are additionally weak random fluctuations of a mirror around the harmonic oscillations, a particle can reveal Anderson localization in the time domain. The figure shows that when we locate a detector close to the mirror (at $z = 1$), we can observe that the probability density for detection of a particle at this space point is exponentially localized in time. The results presented are obtained by numerical integration of the exact Schrödinger equation with the Hamiltonian (5.24) and for an initial state that is plotted in the figure by means of the green line and that corresponds to $\psi(z, 0) = w_{j_0}(z, 0)$. In the course of time evolution, the probability density $|\psi(z, t)|^2$ first spreads over neighboring localized wavepackets w_j but afterwards it freezes, i.e. $\psi(z, t) = \sum_j a_j w_j(z, t)$ where $|a_j|^2$ is approximately proportional to $e^{-|j-j_0|/\xi}$. Blue and red lines are related to $t = 2210T$ and $t = 5010T$, respectively, where $T = 2\pi/\omega$ (with $\omega = 4.9$) is the period of the harmonic oscillations of a mirror. The standard deviation of the disorder $V = 5.1J$ where the tunneling amplitude $J = 8.6 \times 10^{-4}$, see the tight-binding model (5.26). The presented results are an average over 50 different realizations of the disorder. The chosen parameters are suitable for the ultra-cold atoms experiment [37]. Reprinted from [37]

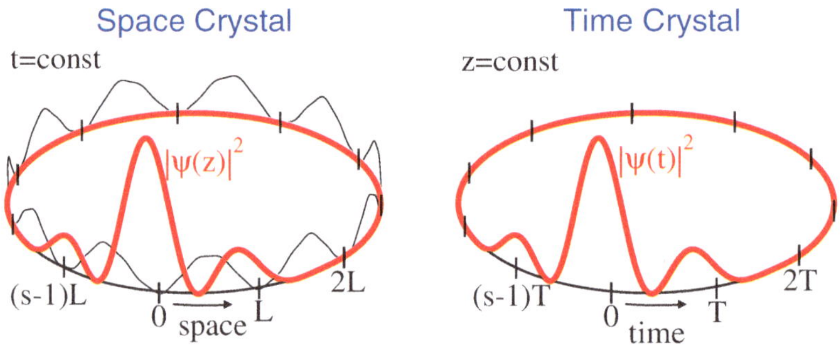

Fig. 5.4 Left panel: one-dimensional space crystal with periodic boundary conditions. If a disordered potential is added to the periodic potential structure in space, Anderson localization can take place. Then, when we travel along the ring we see an exponential localization of the probability density around a certain space point. Right panel: when we switch from space to time crystals, we have to exchange the roles of space and time. That is, we fix position in space and ask how the probability density for the detection of a particle at this space point changes in time. If there is Anderson localization in the time domain, then this probability is exponentially localized around certain moment of time and such behavior is repeated periodically due to the periodic boundary conditions in time. The greater the ring in the left panel, the larger the space crystal. The higher the s : 1 resonance in the right panel, the greater the time crystal. Reprinted from [98]

structure, Anderson localization can occur and when we go around the ring we will see that a particle is exponentially localized around a certain space point. When we switch from space crystals to time crystals we have to exchange the role of space and time. Now, we fix position in space and ask how the probability for detection of a particle at this space point changes versus time [67, 96] and if it is exponentially localized around a certain moment of time we deal with Anderson localization in the time domain, see Fig. 5.4.

5.2.2 Anderson Localization in the Time Dimension Without Crystalline Structure in Time

The fact that a detector, which is located at a certain position, shows large probability of clicking at some moment of time when a particle is passing nearby is not very surprising. However, if such a probability has a universal character of an exponential profile in time and moreover it is the result of the presence of fluctuating force that acts on a particle, the situation becomes more interesting. In Sect. 5.2.1 we have shown that a particle which reveals crystalline behavior in the time dimension can also undergo Anderson localization in the time domain if it experiences additional force that fluctuates in time. In the present section we demonstrate that a time crystalline structure is not needed to observe Anderson localization in the time domain. Indeed, it is sufficient if there is fluctuating force that acts on a periodically moving particle. Thus, the situation is similar like in the standard Anderson localization in the configuration space where the localization can occur in the presence of a disordered potential characterized by a finite correlation length even if there is no crystalline structure in space [78].

Anderson localization in the time domain turns out to be a quite general phenomenon. We have already seen that a particle moving on a ring or a particle bouncing on an oscillating mirror can show up Anderson localization in the time dimension. Now, in order to illustrate Anderson localization in the time domain without a crystalline structure in time, we consider a different system: a Rydberg electron perturbed by a fluctuating microwave field [36].

Let us consider an H atom in the presence of a microwave field polarized along the z axis and a static electric field vector along the same axis. The Hamiltonian of the system, in the atomic units, reads

$$H = H_0 + H_1, \tag{5.28}$$

$$H_0 = \frac{\mathbf{p}^2}{2} - \frac{1}{r}, \tag{5.29}$$

$$H_1 = Fzf(t) - F_s z, \tag{5.30}$$

where F_s is the strength of the electric field, F determines the intensity of the microwave radiation and

$$f(t) = f(t + 2\pi/\omega) = \sum_{\substack{k=-k_0 \\ k\neq 0}}^{k_0} f_k e^{ik\omega t} \tag{5.31}$$

describes temporal behavior of the microwave field. The microwave radiation is assumed to consist of many harmonics with randomly chosen phases and therefore it fluctuates in time within the period $T = 2\pi/\omega$ and such a random behavior is repeated periodically in time.

We begin with a classical analysis of motion of an electron that is resonantly driven by the microwaves. The system is three-dimensional but an unperturbed H atom is an integrable system and we can introduce the angle-actions variables (see background information in Sect. 4.2.1). Even in the presence of the perturbation H_1, the projection of the angular momentum of an electron on the polarization axis of the microwave field is a conserved quantity. We assume that this projection is equal zero. Then, there are only two pairs of canonically conjugate angle-action variables, (θ, J) and (Ψ, L), which determine motion of an electron. The momentum J is the principal action which is the classical analogue of the principal quantum number and which determines the length of the major axis of a Kepler ellipse. The variable θ determines position of an electron on an elliptical Kepler orbit. The total angular momentum of an electron is denoted by L and the conjugate variable Ψ is an angle between the major axis of a Kepler ellipse and the z axis [36]. In the angle-action variables the Hamiltonian of an unperturbed electron in an H atom reads $H_0 = -1/(2J^2)$ and the frequency of electronic motion is $\Omega(J) = dH_0/dJ = 1/J^3$. Assuming that the resonance condition between an electron and the microwave field is fulfilled, i.e. $\Omega(J_0) = \omega$ (where J_0 is the resonant value of the action J), and switching to the moving frame where $\Theta = \theta - \omega t$ and all other variables remain the same, we can obtain an effective secular Hamiltonian of the perturbed system by averaging the full Hamiltonian over time (see description of the secular approximation approach in Sect. 5.1),

$$H_{\text{eff}} = \frac{P^2}{2m_{\text{eff}}} + FJ_0^2 \sum_k U_k(L,\Psi) f_{-k} e^{ik\Theta} + F_s J_0^2 \frac{3\tilde{e}}{2} \cos\Psi, \tag{5.32}$$

where the effective mass $m_{\text{eff}} = -J_0^4/3$,

$$U_k(L,\Psi) = \frac{\mathcal{J}_k'(k\tilde{e})}{k} \cos\Psi + i\frac{\sqrt{1-\tilde{e}^2}}{k\tilde{e}} \mathcal{J}_k(k\tilde{e}) \sin\Psi, \tag{5.33}$$

$$\tilde{e} = \sqrt{1 - \frac{L^2}{J_0^2}}, \tag{5.34}$$

with $\mathcal{J}_k$ and $\mathcal{J}_k'$ the ordinary Bessel functions and their derivatives, respectively, and $\tilde{e}$ is the eccentricity of a Kepler ellipse. The constant term $-3/(2J_0^2)$ has been omitted in (5.32). While averaging over time we have kept Θ, J, L, and Ψ constant because they all are slowly varying quantities. Indeed, $\Theta \approx$ constant because we are interested in motion of an electron close the resonant orbit, i.e. for $P = J - J_0 \approx 0$. Then, in the frame moving along the resonant orbit, the position of an electron Θ changes slowly. The parameters of a Kepler ellipse are determined by J, L, and Ψ and they vary slowly if $FJ_0^4 \ll 1$ and $F_s J_0^4 \ll 1$ [99].

Due to the conservation of the projection of the angular momentum of an electron on the z axis, the system is effectively two-dimensional. Moreover, motion of an electron along a Kepler ellipse approximately decouples from much slower precession of the ellipse and one may apply the Born–Oppenheimer approximation in order to describe quantum states of the system [99]. Thus, we may first quantize P and Θ for frozen L and Ψ and then switch to quantization of L and Ψ. This procedure is especially simple if we are interested in states with low angular momentum L. Indeed, for a sufficiently strong static electric field, analysis of the classical dynamics shows that there exists a stable Kepler orbit which is an ellipse degenerated into a line ($L = 0$) and oriented along the z axis ($\Psi = 0$) [36, 99]. Then, the eccentricity $\tilde{e} = 1$ and the effective Hamiltonian (5.32), for $L = 0$ and $\Psi = 0$ reduces to

$$H_{\text{eff}} = \frac{P^2}{2m_{\text{eff}}} + V_{\text{eff}}(\Theta), \tag{5.35}$$

$$V_{\text{eff}}(\Theta) = FJ_0^2 \sum_k \frac{\mathcal{J}'_k(k)}{k} f_{-k} e^{ik\Theta}, \tag{5.36}$$

where a constant term has been omitted. Let us assume that the Fourier components of the microwave field in (5.31) fulfill

$$\left| \frac{\mathcal{J}'_k(k)}{k} f_{-k} \right| = \frac{1}{\sqrt{2k_0}}, \tag{5.37}$$

for $|k| \leq k_0$ and zero otherwise, and $\varphi_{-k} = -\varphi_k = \text{Arg}(f_{-k})$ are random numbers chosen uniformly from the interval $[0, 2\pi)$.[7] Then, the effective potential (5.36) becomes a disordered potential with uniform superposition of $2k_0$ harmonics that possesses random phases,

$$V_{\text{eff}}(\Theta) = FJ_0^2 \frac{1}{\sqrt{2k_0}} \sum_{\substack{k=-k_0 \\ k \neq 0}}^{k_0} e^{i(k\Theta + \varphi_k)}. \tag{5.38}$$

To sum up, the classical effective Hamiltonian of an electron in an H atom (in the moving frame $\Theta = \theta - \omega t$) is reduced to the system of a particle in a disordered

---

[7]Note that the intensity of the microwave field is given by the formula $I = F^2 \sum_k |f_k|^2/2$.

potential characterized by the correlation length $\sigma = \sqrt{2}/k_0$ with no additional crystalline structure. When we quantize the classical effective Hamiltonian (J_0 corresponds to the principal quantum number n_0 of an H atom), we can observe Anderson localization provided the localization length ξ is smaller than the range of $\Theta \in [0, 2\pi)$. If we choose the limit of weak disorder, i.e. for $F^2 n_0^4 \ll E E_\sigma$ where E is an energy eigenvalue of (5.35) and $E_\sigma = k_0^2/2m_{\text{eff}}$ is the so-called correlation energy, we may apply the Born approximation to predict the localization length ξ which yields [62, 78]

$$\frac{\xi}{\sigma} = \frac{2\sqrt{2}}{\pi} \frac{E E_\sigma}{F^2 n_0^4}, \tag{5.39}$$

for $E \geq E_\sigma/4$.[8] When we return from the moving frame ($\Theta = \theta - \omega t$) to the laboratory frame, the Anderson localization described by the Hamiltonian (5.35) is observed in the time domain because the probability for measurement of an electron close to, e.g. a nucleus is exponentially localized around certain moment of time.

In order to confirm our predictions we switch to the quantum secular approach [14]. Starting with the Hamiltonian (5.28) we apply the time-dependent unitary transformation $\hat{U} = e^{i\hat{n}\omega t}$ (where $\hat{n} = (-2\hat{H}_0)^{-2}$ is the operator of the principal quantum number of an Hydrogen atom) and perform averaging over time which yields

$$\langle n', l'|\hat{H}_{\text{eff}}|n, l\rangle = \left(-\frac{1}{2n^2} - n\omega\right) \delta_{nn'}\delta_{ll'} + \langle n', l'|z|n, l\rangle \left(F f_{n-n'} - F_s \delta_{nn'}\right), \tag{5.40}$$

where $|n, l\rangle$ is a hydrogenic eigenstate with the principal quantum number n, total angular momentum l, and the projection of the angular momentum on the z axis equal zero. Diagonalization of (5.40) results in a bunch of energy levels and in order to identify the desired Anderson localized eigenstates it is very helpful to have the semiclassical prediction based on the classical secular Hamiltonian (5.35). These eigenstates are Floquet states of the system and Anderson localization in the time domain emerges when we analyze them in the laboratory frame, see Figs. 5.5 and 5.6. While the values of the parameters chosen in these figures do not allow for the application of the Born approximation in order to predict the localization length, they correspond to the experimentally attainable parameters. The Anderson localization in the time domain can be detected in a similar way as the detection of non-spreading wavepackets realized in Rydberg atoms [69–71, 111]. Interestingly

[8] It should be mentioned that due to the fact that the effective mass m_{eff} is negative, the Hamiltonian (5.35) is bounded from above not from below. Then, the greatest eigenvalue of (5.35) corresponds to the strongest localization. The localization length increases with a decrease of E. If we want to apply the Born approximation to predict the localization length ξ, the condition $\xi \gg \sigma$ must be fulfilled where σ is the correlation length of a disordered potential [62, 78].

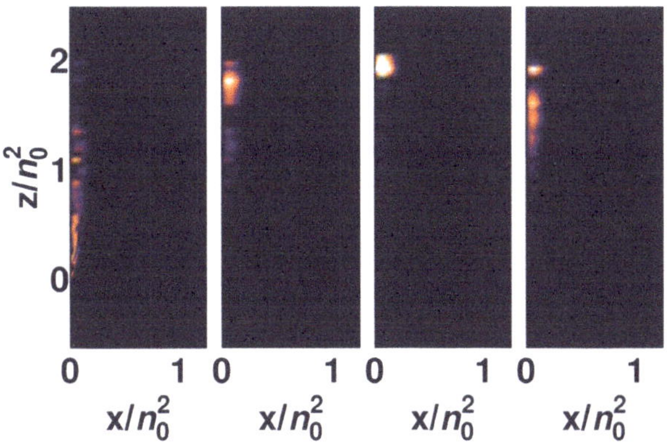

Fig. 5.5 Probability density of an electron in a hydrogen atom corresponding to a Floquet state that reveals Anderson localization in the time domain. The probability density is plotted in the laboratory frame for different moments of time within a period of the fluctuating microwave field, i.e. at $\omega t = \pi$, $3\pi/2$, 2π, and $5\pi/2$ from left to right, respectively. The presented Floquet state is related to an electron moving along the elliptical orbit degenerated into the line along the z axis. The probability density for the measurement of an electron close to a nucleus reveals exponential Anderson localization around a certain moment of time, see Fig. 5.6. The parameters of the system are the following: $\omega = n_0^{-3}$ where $n_0 = 300$, $Fn_0^4 = 0.00016$, $F_s/F = 1.5$, and $k_0 = 5$. Reprinted from [36]

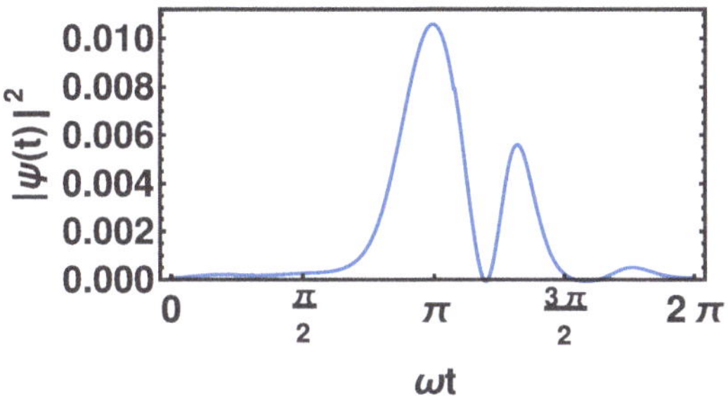

Fig. 5.6 Probability density for the measurement of an electron in a hydrogen atom close to a nucleus versus time. The results correspond to the Floquet state presented in Fig. 5.5. Due to the presence of the fluctuating microwave field, the probability density is Anderson localized around a certain moment of time. Reprinted from [36]

it is also possible to control the shape of a Kepler elliptical orbit along which the Anderson localization takes place by means of the static electric field [36].

It should be stressed that Anderson localization in the time domain without an additional crystalline structure can be realized in any single-particle system which reveals nonlinear classical resonances and which is driven by force fluctuating in time [97].

5.2.3 Anderson Localized-Delocalized Transition in the Time Dimension

The scaling theory of Anderson localization shows that in one- and two-dimensional time-reversal invariant and spinless systems, the localization occurs regardless how weak a spatially disordered potential is [1]. In the three-dimensional case, the situation is more interesting because, the scaling theory predicts a critical strength of a disorder required for Anderson localization. The critical behavior can be also observed versus energy. When one keeps a disordered potential fixed, eigenstates of a three-dimensional system are localized if the corresponding eigenenergies are smaller than a critical energy (the so-called mobility edge) and they are delocalized otherwise.

Time is a single *degree of freedom* and at first glance it seems not possible to observe the Anderson localized-delocalized transition in the time dimension. However, it turns out that if a particle moving in the three-dimensional space is appropriately driven by force fluctuating in time, the transition can occur [28].

Let us consider a classical particle moving freely in the three-dimensional space with periodic boundary conditions, i.e. we deal with a particle moving on a three-dimensional torus. We define three angles θ, ψ, and ϕ which determine position of a particle on a torus and three conjugate momenta p_θ, p_ψ, and p_ϕ, respectively. The Hamiltonian of a particle reads

$$H_0 = \frac{p_\theta^2 + p_\psi^2 + p_\phi^2}{2}. \tag{5.41}$$

The angles and momenta are actually the angle-actions variables of the system (see background information in Sect. 4.2.1). Frequencies of a particle motion along each direction on a torus are determined by the momenta, $\Omega_\theta = dH_0/dp_\theta = p_\theta$ and analogously $\Omega_\psi = p_\psi$ and $\Omega_\phi = p_\phi$. We assume that the following perturbation of a particle is turned on

$$H_1 = V_0\, g(\theta)g(\psi)g(\phi)f_1(t)f_2(t)f_3(t), \tag{5.42}$$

where $g(x)$ is a regular function of $x \in [-\pi, \pi)$ which we choose as[9]

$$g(x) = \frac{x}{\pi} = \sum_{n\neq 0} g_n e^{inx}, \tag{5.43}$$

where the Fourier components $g_n = i(-1)^n/(\pi n)$ while the functions

$$f_j(t) = \sum_{k\neq 0} f_k^{(j)} e^{ik\omega_j t} \tag{5.44}$$

[9]The function $g(x)$, Eq. (5.43), is linear function but due to the periodic boundary conditions on a torus it possesses a discontinuity at $x = \pi$.

are changing periodically in time but with different frequencies ω_j which are incommensurate numbers. It is also assumed that initial conditions of a particle are chosen so that the frequencies of unperturbed motion of a particle fulfill the resonance conditions with the driving frequencies, i.e. $\Omega_\theta = \omega_1$, $\Omega_\psi = \omega_2$, and $\Omega_\phi = \omega_3$. If so, in the moving frame defined by

$$\Theta = \theta - \omega_1 t, \quad \Psi = \psi - \omega_2 t, \quad \Phi = \phi - \omega_3 t, \tag{5.45}$$

all dynamical variables evolve slowly if V_0 is small and if a particle remains close to the resonant trajectory, i.e. $P_\Theta = p_\theta - \omega_1 \approx 0$, $P_\Psi = p_\psi - \omega_2 \approx 0$, and $P_\Phi = p_\phi - \omega_3 \approx 0$. Then, when in the moving frame one averages the Hamiltonian over time, terms with no explicit time dependence survive only and one arrives at the effective Hamiltonian [28]

$$H_{\text{eff}} = \frac{P_\Theta^2 + P_\Psi^2 + P_\Phi^2}{2} + V_{\text{eff}}(\Theta, \Psi, \Phi), \tag{5.46}$$

with

$$V_{\text{eff}}(\Theta, \Psi, \Phi) = V_0 \, h_1(\Theta) \, h_2(\Psi) \, h_3(\Phi), \tag{5.47}$$

where

$$h_j(x) = \sum_{k \neq 0} g_k f_{-k}^{(j)} e^{ikx} = \frac{1}{\sqrt{k_0} \pi^{1/4}} \sum_{k \neq 0} e^{-k^2/(2k_0^2)} e^{i(kx + \varphi_k^{(j)})}. \tag{5.48}$$

The modulus of the Fourier components in (5.44) has been chosen so that $|g_k f_{-k}^{(j)}|$ are the Gaussian function of k and $\varphi_k^{(j)} = \arg(f_k^{(j)}) = -\arg(f_{-k}^{(j)})$ are independent random variables selected uniformly in the interval $[0, 2\pi)$. We have taken such time-periodic functions $f_j(t)$ as an example only. With a proper choice of the Fourier components $f_k^{(j)}$ of $f_j(t)$ it is possible to engineer shape and properties of $h_j(x)$ practically at will.

In order to get (5.46) the fact that the frequencies ω_j's are incommensurate numbers is crucial, otherwise we would deal with the small denominator problem [64]. That is, the second order contribution to the effective Hamiltonian (which we have neglected here because it is expected to be small) is the sum of terms proportional to

$$\left(\frac{V_0}{n_1\omega_1 + n_2\omega_2 + n_3\omega_3} \right)^2 e^{-(n_1^2 + n_2^2 + n_3^2)/4k_0^2}. \tag{5.49}$$

The sum runs over all non-zero integer n_j and would diverge if ω_j's were commensurate numbers. Even if the frequencies ω_j are incommensurate numbers, it can still happen that the denominator of a term (5.49) is very close to zero and then the second order contribution to the effective Hamiltonian must not be neglected. In order to avoid this problem we assume that the frequencies fulfill the additional condition $\omega_1 > 2k_0(\omega_2 + \omega_3)$. The latter requirement ensures that the denominator in (5.49) can be small only when there is also a strong Gaussian suppression and consequently the second order terms are never important when $(V_0/\omega_j)^2 \to 0$ and the effective Hamiltonian (5.46) is valid.

The effective Hamiltonian (5.46) has been derived by means of the classical secular approximation approach [64] but it remains valid also in the quantum description because all applied canonical transformations were linear and they have unambiguous unitary counterparts. Thus, in the moving frame (5.45) a quantum particle is effectively described by the time-independent Hamiltonian with the disordered potential $V_{\rm eff}(\Theta, \Psi, \Phi)$ which is the product of the three one-dimensional disordered potentials $h_j(x)$. The potential $V_{\rm eff}$ is characterized by the two-point correlation function which factorizes into the product of three identical correlation functions along each direction on a torus,

$$\overline{V_{\rm eff}(\Theta', \Psi', \Phi')\, V_{\rm eff}(\Theta' + \Theta, \Psi' + \Psi, \Phi' + \Phi)} = V_0^2 \exp\left(-\frac{\Theta^2 + \Psi^2 + \Phi^2}{2\sigma^2}\right), \tag{5.50}$$

where the bar $\overline{}$ denotes the averaging over disorder realizations, i.e. over different random choices of the phases φ_k in (5.48), and $\sigma = \sqrt{2}/k_0$ is the correlation length of the disordered potential. In order to observe Anderson localization of a particle in the moving frame (5.45), the localization length ξ has to be smaller than 2π because a particle is moving on a finite torus. Suppose for a moment that the three-dimensional space is infinite and let us describe the results of the transfer matrix approach routinely used in order to calculate the localization length. However, before we do it, first let us introduce a reader to the transfer matrix technique.

Transfer Matrix Calculations of Localization Length

Let us start with a particle in the one-dimensional space in the presence of a disordered potential $V(z)$,

$$H = \frac{p^2}{2} + V(z). \tag{5.51}$$

If we discretize the space, i.e. $z \to z_i$ where $dz = z_{i+1} - z_i > 0$ but much smaller than the correlation length of a disordered potential, the time-independent Schrödinger equation, $H\psi = E\psi$, can be written in the form

(continued)

$$\begin{bmatrix}\psi(z_{i+1})\\ \psi(z_i)\end{bmatrix} = T_i \begin{bmatrix}\psi(z_i)\\ \psi(z_{i-1})\end{bmatrix}, \quad \text{with} \quad T_i = \begin{bmatrix}2dz^2[V(z_i)-E]+2 & -1\\ 1 & 0\end{bmatrix}, \tag{5.52}$$

where $d^2\psi(z_i)/dz^2 \approx [\psi(z_{i+1}) - 2\psi(z_i) + \psi(z_{i-1})]/dz^2$. If we start with certain $\psi(z_0)$ and $\psi(z_1)$ and perform multiplication by the transfer matrices T_i,

$$\begin{bmatrix}\psi(z_{N+1})\\ \psi(z_N)\end{bmatrix} = \mathcal{T}_N \begin{bmatrix}\psi(z_1)\\ \psi(z_0)\end{bmatrix}, \quad \text{where} \quad \mathcal{T}_N = T_N T_{N-1} \ldots T_2 T_1, \tag{5.53}$$

we effectively *propagate* a system along the one-dimensional chain towards increasing values of z_i. The matrices T_i are unimodular, i.e. the modulus of their determinant is unity, and they are actually random matrices because $V(z)$ is a disordered potential. Then, Furstenberg theorem [48] implies that

$$\frac{1}{2\xi} = \lim_{N\to\infty} \frac{1}{N} \ln\left(\mathrm{Tr}\left[\mathcal{T}_N\right]\right) > 0. \tag{5.54}$$

Thus, the unimodular matrix $\mathcal{T}_N$ has two eigenvalues $e^{\pm N/2\xi}$. With a special fine-tuning of E it is possible to find $\psi(z_0)$ and $\psi(z_1)$ so that the wavefunction decays like $e^{-N/2\xi}$ in both positive and negative directions of z and represents a normalized and exponentially localized eigenstate of a particle. It means that eigenenergies of a one-dimensional Anderson problem cannot form a continuum and the spectrum is rather discrete [48] (i.e. not all values of eigenenergies E are allowed for normalized eigenstates). If we are interested in the localization length ξ only, we do not have to fine tune energy E and it is sufficient to use the formula (5.54) in order to investigate how $\xi(E)$ depends on energy E.[10] Note that in (5.54), ξ is the localization length in the units of dz, thus the quantity we are actually interested in is ξdz.

Now, let us switch to a higher-dimensional problem, especially to a particle moving in the three-dimensional space in the presence of a disordered potential. In order to calculate the localization length we can also discretize the space and apply a similar approach like in the one-dimensional case but the entire procedure is slightly more involved [27, 68, 85, 106].

(continued)

[10] In practise, in the one-dimensional case, it is more convenient to propagate $R_{i+1} = \psi(z_{i+1})/\psi(z_i)$ instead of (5.53). That is, the propagation of $R_{i+1} = 2dz^2[V(z_i) - E] + 2 - 1/R_i$ does not require a numerical renormalization that is needed in (5.53) in order to tackle exponentially growing numbers.

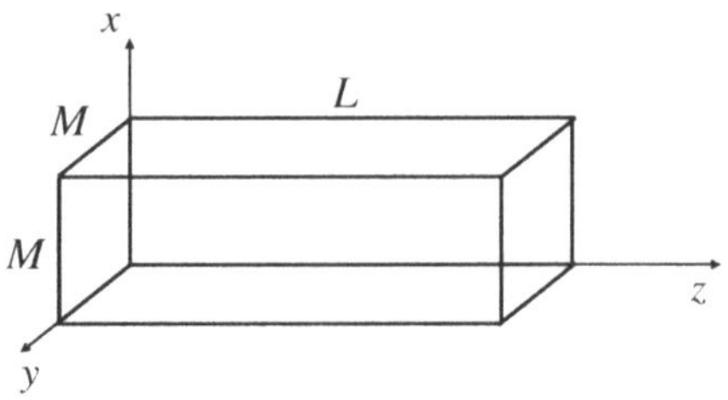

Fig. 5.7 A bar is considered in the discretized three-dimensional space for a particle moving in a disordered potential. For each z_i there are M^2 points of a bar in the xy plane. The transfer matrix approach allows one to calculate how wavefunction in a given slice of a bar, $\psi(x, y, z_i)$, propagates to the next slice, $\psi(x, y, z_{i+1})$. Decay of a wavefunction along the z axis allows one to obtain the localization length ξ_M

Let us consider not a one-dimensional chain of points but an elongated three-dimensional bar of L points along the z axis and the transverse size of $M \times M$ points where $L \gg M$, see Fig. 5.7. For a fixed M and in the presence of a disordered potential $V(x, y, z)$, transport properties along a bar is characterized by a localization length ξ_M because a bar is essentially a one-dimensional system with the finite M^2 number of channels and therefore Anderson localization takes place regardless how weak a potenstial $V(x, y, z)$ is [1]. The localization length ξ_M can be calculated by means of a similar iterative procedure as in (5.53) but the transfer matrices contain now Hamiltonian matrices $\mathbf{H}_i$ defined in the two-dimensional slices of a bar at each z_i,

$$\begin{bmatrix} \mathbf{A}_{i+1} \\ \mathbf{A}_i \end{bmatrix} = T_i \begin{bmatrix} \mathbf{A}_i \\ \mathbf{A}_{i-1} \end{bmatrix}, \quad \text{with} \quad T_i = \begin{bmatrix} E - \mathbf{H}_i & -\mathbf{1} \\ \mathbf{1} & 0 \end{bmatrix}, \tag{5.55}$$

where $\mathbf{H}_i$'s are $M \times M$ matrices and vectors $\mathbf{A}_i$ and $\mathbf{A}_{i+1}$ represent the wavefunction at the i-th and $(i+1)$-th slices of a bar, respectively. The matrices T_i's are $2M \times 2M$ symplectic matrices, i.e. their eigenvalues are pairs whose elements, are the inverse of each other, and they are random matrices due to the presence of a disordered potential $V(x, y, z)$. Then, the Oseledets theorem implies that the limit

$$\Gamma = \lim_{N\to\infty} \left(\mathcal{T}_N^\dagger \mathcal{T}_N\right)^{1/2N} \tag{5.56}$$

exists where $\mathcal{T}_N = T_N T_{N-1} \ldots T_2 T_1$ [68, 85]. Let us denote pairs of the eigenvalues of Γ by $e^{\pm\gamma_r}$. The eigenvalues corresponding to the smallest Lyapunov exponent γ_r are related to the longest length scale on which the

(continued)

system localizes along a bar. Thus the desired localization length $1/(2\xi_M) = \min\{\gamma_r\}$ and it determines transport properties along a bar [85]. There is the alternative Green function method, similar to the transfer matrix approach presented here, that is often used in calculations of the localization length ξ_M along a bar, see [68].

For a fixed disordered potential and for a given choice of particle energy E, the localization length $\xi_M(E)$ can be calculated for different transverse sizes M of a bar. If a particle localizes in the full three-dimensional space and the localization length $\xi(E)$ is much smaller than M, then $\xi_M(E)/M \approx \xi(E)/M$ and this ratio goes to zero when $M \to \infty$. Even if $\xi_M(E)$ is not smaller than M we can expect that in the localized regime, $\xi_M(E)/M$ decreases with M. On the other hand for energy greater than the mobility edge, we are in the diffusive regime and $\xi_M(E)/M$ never stops increasing with M. For E equal to the mobility edge, $\xi_M(E)/M \approx$ constant meaning that the quasi-one-dimensional localization length $\xi_M(E)$ is always comparable to the transverse size of the system [27]. Figure 5.8 illustrates the behavior of $\xi_M(E)/M$ in the localized and diffusive regimes.

Finally we would like to get information not only about the mobility edge but also about values of the localization length $\xi(E)$ in the full three-dimensional space. It can be achieved by employing the finite-size scaling approach in the spirit of the scaling theory of localization [1].

The one-parameter scaling hypothesis assumes that there is a unique single-variable scaling function $\mathcal{F}$ so that [27, 68]

$$\frac{\xi_M(E)}{M} = \mathcal{F}\left(\frac{\xi(E)}{M}\right), \tag{5.57}$$

where $\xi(E)$ is a characteristic length which in the localized regime is proportional to the localization length and can be always chosen to be exactly the localization length. In the three-dimensional case, there are actually two branches of $\mathcal{F}$: one for the localized regime and one for the diffusive regime. In the latter case $\xi(E)$ is proportional to the inverse of the diffusive constant [27]. Numerical calculations of $\xi_M(E)/M$ can be performed for many different values of E and M. For sufficiently small E it is possible to find the exact localization length $\xi(E)$ because it is small and there is no problem to do numerical calculations for $M \gg \xi(E)$. For increasing E, the key point is to always choose the corresponding value of $\xi(E)$ in (5.57) so that new data lie on the same curve as the data for smaller E. In such a way the localization length $\xi(E)$ can be found even if E is close to the mobility edge, see Fig. 5.9.

Having introduced the method for the analysis of the Anderson localization of a particle in the three-dimensional space in the presence of a time-independent disordered potential, we can apply it to the problem of the Anderson localization in the time dimension. In Fig. 5.9, the dependence of the localization length on energy of a particle is shown.[11] The localization length diverges algebraically when the energy of a particle approaches the mobility edge at $E/E_\sigma = 0.32 \pm 0.002$ where the correlation energy $E_\sigma = 1/\sigma^2 = k_0^2/2$. For energies greater than the mobility edge, we enter the diffusive regime where no localization of a particle is observed. Note that in Fig. 5.9 the results are presented in the units of the correlation length σ of the disorder potential (5.50) and the correlation energy E_σ. Also the standard deviation of the disordered potential is chosen to be equal to the correlation energy,

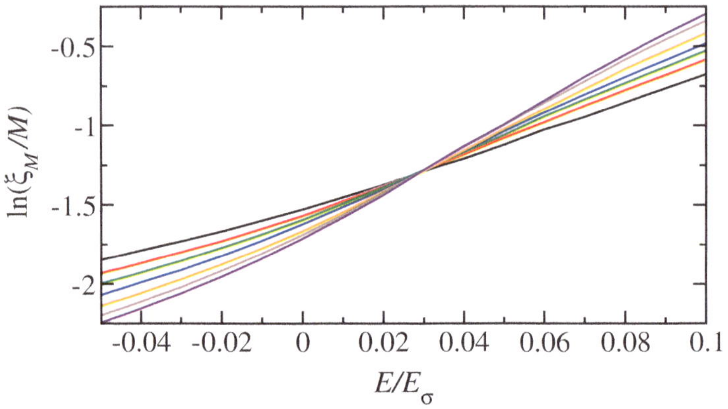

Fig. 5.8 Localization length ξ_M, as a function of energy E, computed numerically for an elongated bar of different transverse sizes M, see Fig. 5.7. Curves with increasing slope correspond to increasing value of M from 25 to 55 with step 5. The results correspond to the Hamiltonian (5.46) with the strength of the disordered potential equaled to the correlation energy, $V_0 = E_\sigma = 1/\sigma^2$. For energies smaller than the mobility edge at $E/E_\sigma = 0.032 \pm 0.002$, the ratio ξ_M/M decreases with M while for energies greater than the mobility edge it increases with M indicating the diffusive regime where the Anderson localization is not present. All results have been obtained assuming that a particle is moving not on a finite torus but in the infinite three-dimensional space. Reprinted from [28]

[11] It should be stressed that the localization length can be defined either as exponential decay of tails of wavefunction ψ or of the probability density $|\psi|^2$. We try to consistently use the latter option. In Ref. [28] the former possibility was chosen and therefore the localization length ξ presented in Fig. 5.9 is twice smaller than originally in Ref. [28].

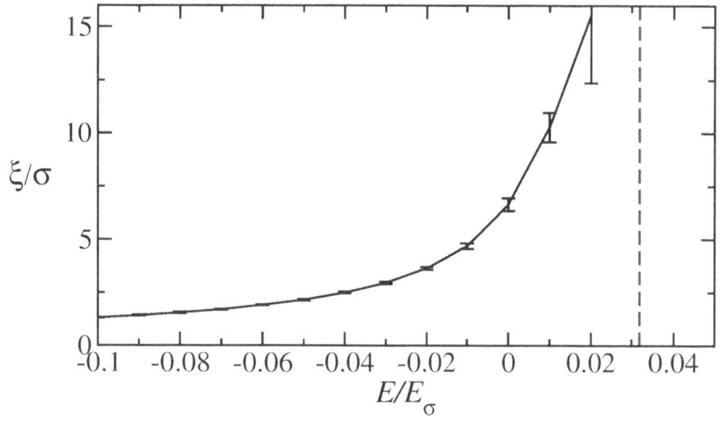

Fig. 5.9 Localization length, in the units of the correlation length σ of the disordered potential (5.50), for a particle described by the Hamiltonian (5.46). The localization length, in the units of σ, diverges algebraically when energy approaches the mobility edge at $E/E_\sigma = 0.032 \pm 0.002$ where the correlation energy $E_\sigma = 1/\sigma^2$. For energies greater than the mobility edge, a particle performs diffusive motion. The standard deviation of the disordered potential $V_0 = E_\sigma$. Reprinted from [28]

i.e. $V_0 = E_\sigma$. The localization length diverges in the units of σ. It allows us to observe the Anderson localized-delocalized transition even if a particle is moving on a finite torus. Indeed, any point in Fig. 5.9 can be realized in the present system because regardless how big ξ/σ is, it is always possible to choose such a small correlation length σ that ξ/σ is big and at the same time the localization length $\xi \ll 2\pi$. The correlation length $\sigma = \sqrt{2}/k_0$ can be easily controlled because it decreases if more harmonics contribute to the driving force, cf. (5.44) and (5.48).

We have demonstrated the Anderson localized-delocalized transition of a particle in the moving frame (5.45). In order to observe the Anderson transition in the time dimension we have to return to the laboratory frame and ask how the probability density for measurement of a particle at a given point on a torus behaves in time. It is shown in Fig. 5.10 where an example of a localized eigenstate of the effective Hamiltonian (5.46) is presented. Probability density for a fixed θ, ψ, or ϕ in the laboratory frame shows exponential localization around a certain moment of time [67]. Increasing the number of harmonics in the drive and consequently decreasing the correlation length of the effective disordered potential, one can follow the behavior presented in Fig. 5.9 and observe the localized-delocalized Anderson transition in the time dimension [28].

We have analyzed the Anderson localized-delocalized transition with the help of a driven particle that moves on a three-dimensional torus. The same phenomenon can be realized for a particle in any separable potential in the three-dimensional space. For example, when one exchanges a torus with a three-dimensional box potential, the predictions presented here need small modification only.

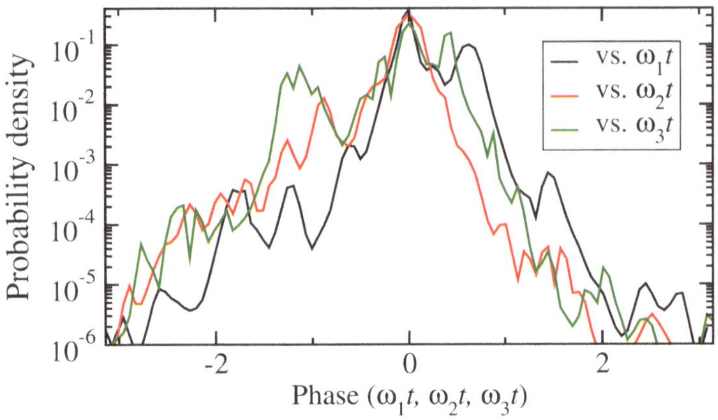

Fig. 5.10 Time evolution of the probability density at a fixed position θ, ψ, or ϕ in the laboratory frame (integrated along two remaining directions[12]) for $\sigma = 0.1$, $V_0 = E_\sigma$, and $E/E_\sigma = -0.049968$. The localization length in the moving frame $\xi \approx 4\sigma$ (see Fig. 5.9) which in the laboratory frame corresponds to the localization in the time dimension over time periods $4\sigma/\omega_{1,2,3}$. Reprinted from [28]

5.3 Single-Particle Condensed Matter Phenomena in the Time Dimension

Anderson localization is a single-particle problem originally investigated in condensed matter systems. In Sect. 5.2 we have demonstrated that it can be also observed in the time dimension. In the present section we show that other single-particle condensed matter phenomena can be realized in the time dimension too. We consider a simple example of a topological time crystal. We also show that an appropriately driven particle can behave in the time dimension like an electron moving in a potential with a quasi-crystalline structure.

5.3.1 Topological Time Crystals

In the current section we show that a topological insulating phase can emerge in dynamics of a periodically and resonantly driven particle [39] (for a photonic topological time crystal [66] see Chap. 7 and for discrete time crystals with topological properties see [17]). Topological insulators are examples of symmetry protected topological phases and attract a lot of interests in condensed matter physics and in quantum simulators [40, 52, 81, 101]. They are insulating systems

[12]For example the black curve in Fig. 5.10 corresponds to $\rho(t) = \int d\psi d\phi |\tilde{\Psi}(\theta, \psi, \phi, t)|^2$ with a fixed value of θ where $\tilde{\Psi}$ is an eigenstate of the Hamiltonian (5.46).

in their interior (bulk). That is, their energy spectra have a band structure and if the Fermi level is chosen in a gap between energy bands, they have insulating properties. However, topological insulators are characterized by a global topological invariant. A perturbation of such a system is not able to change a topological invariant unless it breaks a symmetry of a system or it is so strong that it closes an energy gap between bands.

There is important bulk-edge correspondence in topological insulators. The vacuum is obviously topologically trivial because it is an empty space. A topological insulator is characterized by a global invariant and consequently on the border between an insulator and the vacuum something must happen. Indeed, there appear surface (edge) conducting eigenstates with eigenenergies located in energy gaps of an insulator [40, 52, 81, 101].

We restrict here to topological insulators of non-interacting systems. In condensed matter physics they are considered as non-interacting Fermi systems. If the Fermi level is located in a gap between energy bands, then non-vanishing conductance of a system is related to the presence of edge states. A non-interacting many-body system is actually equivalent to a single-particle system because its spectrum and eigenstates are determined by a single-particle counterpart. We will consider single-particle properties of a topological insulator which can be experimentally investigated in a periodically driven atomic Bose–Einstein condensate. It should be stressed that a topological time crystal that we consider here does not belong to the so-called Floquet topological systems where a crystalline structure in space (e.g. an optical lattice) is periodically driven in order to change effective parameters of a system so that a system becomes topologically non-trivial [15, 18, 72, 89]. Our topological time crystal is also not an example of Floquet–Bloch systems where time is considered as an additional synthetic dimension combined with a crystalline structure in space [33, 59, 86, 94, 95, 110, 112]. We focus on a system that at first glance does not have anything in common with a crystalline structure but due to appropriate resonant driving, dynamics of a system can reveal properties of a topological insulator in the time dimension. More precisely, we will show that dynamics of a periodically and resonantly driven system can reproduce in the time dimension the Su–Schrieffer–Heeger (SSH) model [6, 109]. Before we do it, it should be useful for a reader to describe the SSH model and present its basic topological properties.

Su–Schrieffer–Heeger Model

Let us consider a particle hoping between neighboring sites of a one-dimensional chain. Wavefunction of a particle can be written as $|\psi\rangle = \sum_{i=1}^{s} a_i |i\rangle$ and its energy

$$E = -\frac{1}{2}\sum_{i=1}^{s} J_i \left(a_{i+1}^* a_i + \text{c.c.}\right), \tag{5.58}$$

(continued)

where $|i\rangle$ denotes the state localized in the i-th site of the chain. The staggered real hopping amplitudes are assumed, i.e. $J_{2i-1} = J$ and $J_{2i} = J'$, see Fig. 5.11a. The total number of sites s is an even number and periodic boundary conditions are fulfilled. The introduced system is the Su–Schrieffer–Heeger (SSH) model that describes most important transport properties of the conductive organic polymer trans-polyacetylene. Here we restrict to the static version of the model with no phonon excitation [6].

The chain presented in Fig. 5.11a is actually a Bravais lattice with a two-point basis, i.e. spatial period of the system corresponds to two sites of the chain. Particle wavefunction can be written as $|\psi\rangle = \sum_{j=1}^{s/2}(\alpha_j|2j-1\rangle + \beta_j|2j\rangle)$ where $|2j-1\rangle$ and $|2j\rangle$ form a two-point basis in the j-th site of the Bravais lattice. Then, energy of a particle reads

$$E = -\frac{1}{2}\sum_{j=1}^{s/2}\left(J\beta_j^*\alpha_j + J'\alpha_{j+1}^*\beta_j + \text{c.c.}\right), \tag{5.59}$$

where $\alpha_j = a_{2j-1}$ and $\beta_j = a_{2j}$. It is easy to find eigenenergies of a particle when we switch to the Fourier space,

(continued)

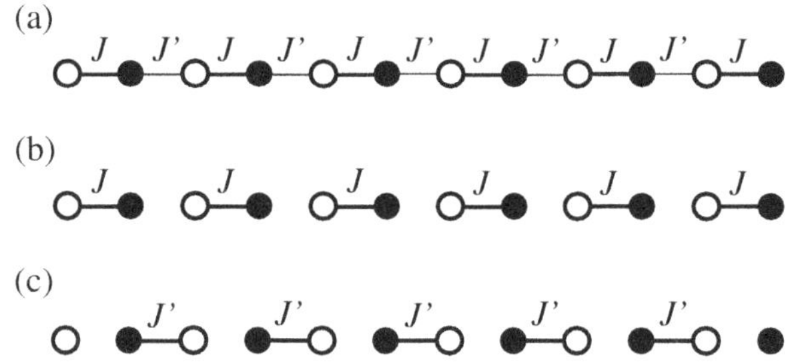

Fig. 5.11 SSH model: one-dimensional chain with staggered hopping amplitudes J and J'. Wavefunction of a particle can be written as $|\psi\rangle = \sum_{i=1}^{s} a_i|i\rangle$ where $|i\rangle$ is the state localized in the i-the site of the chain and s is the total number of sites which is assumed to be an even number. The chain is actually a Bravais lattice with a two-point basis, i.e. particle wavefunction can be written as a sum over the Bravais lattice sites $|\psi\rangle = \sum_{j=1}^{s/2}(\alpha_j|2j-1\rangle + \beta_j|2j\rangle)$ where $|2j-1\rangle$ and $|2j\rangle$ form a two-point basis in the j-th site of the Bravais lattice. Panel (**a**) shows a general case when both tunneling amplitudes J and J' are non-zero. Panels (**b**) and (**c**) present fully dimerized system where two-site states are formed in the interior of the lattice due to the fact that either $J' = 0$ or $J = 0$. In panel (**c**) the end-points of the chain do not have pair and two edge states localized in single sites at the ends of the chain, with zero energy eigenvalues, appear in the system

$$\begin{bmatrix} \alpha_j \\ \beta_j \end{bmatrix} = \sum_k \begin{bmatrix} \tilde{\alpha}(k) \\ \tilde{\beta}(k) \end{bmatrix} \frac{e^{ikj}}{\sqrt{s/2}}. \tag{5.60}$$

Then, the Hamiltonian corresponding to the energy (5.59) has a block-diagonal form with the blocks given by the matrices

$$H(k) = -\frac{1}{2} \begin{bmatrix} 0 & J + J'e^{ik} \\ J + J'e^{-ik} & 0 \end{bmatrix}. \tag{5.61}$$

Diagonalization of each of the 2×2 blocks yields eigenvalues

$$E(k) = \pm\frac{1}{2} \left| J + J'e^{-ik} \right| = \pm\frac{1}{2}\sqrt{J^2 + J'^2 + 2JJ'\cos k}. \tag{5.62}$$

For the infinite system ($s \to \infty$), Eq. (5.62) describes two energy bands with the energy gap $\Delta E = |J - J'|$ which disappears when $J = J'$, see Fig. 5.12.

(continued)

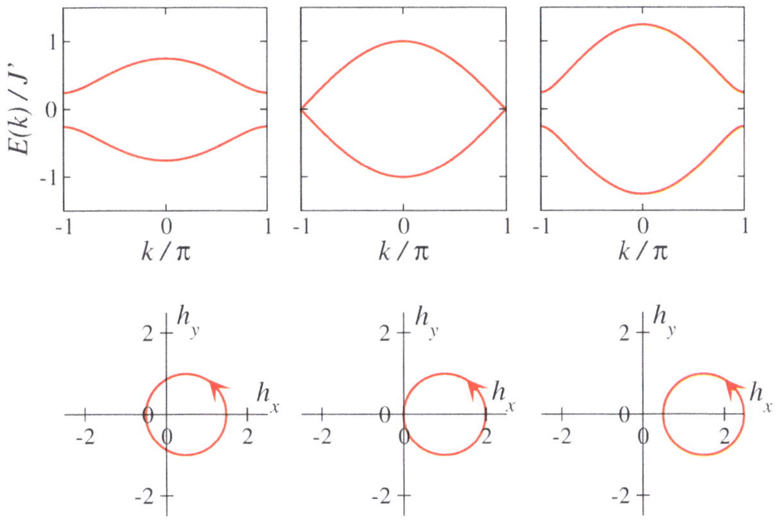

Fig. 5.12 Top row presents energy bands (5.62) for $J'/J = 2$, $J'/J = 1$, and $J'/J = 2/3$ from left to right. Bottom row shows loops encircled by the vector $\mathbf{h}(k)$, Eq. (5.65), when k runs through the Brillouin zone from 0 to 2π for the same values of J'/J as in the corresponding top panels (it does not matter if k runs from 0 to 2π or from $-\pi$ to π because the spectrum $E(k)$ is periodic). When a loop encircles the origin in the $h_x h_y$ plane, the winding number $\nu = 1$ otherwise $\nu = 0$. The middle column corresponds to $J'/J = 1$ when the energy gap between the bands closes and the system is not an insulator. To draw the loops in the bottom row, we have assumed that $J' = 1$ in (5.65)

The SSH Hamiltonian (5.61) possesses the so-called chiral symmetry [6],

$$\Gamma\, H\, \Gamma^{\dagger} = -H, \tag{5.63}$$

where $\Gamma = \sigma_z$ is a unitary (and also Hermitian) operator. Strictly speaking the chiral property is not a symmetry of the Hamiltonian because $[\Gamma, H] \neq 0$ but it has important consequences for the system. It implies that energy levels appear in pairs $\pm E$[13] what agrees with (5.62). Moreover, it allows one to characterize the system by a topological invariant, i.e. by a winding number ν. If we expand the 2×2 matrices in the Hamiltonian (5.61) in terms of the Pauli matrices $\sigma_{x,y,z}$ and the identity matrix, only contributions proportional to σ_x and σ_y are allowed due to the chiral symmetry. Indeed, the Hamiltonian (5.61) takes the form

$$H(k) = -\frac{1}{2}\left(h_x(k)\,\sigma_x + h_y(k)\,\sigma_y\right), \tag{5.64}$$

where

$$h_x(k) = J + J' \cos k, \qquad h_y(k) = J' \sin k. \tag{5.65}$$

Thus, the vector $\mathbf{h}(k) = (h_x(k), h_y(k), 0)$ lies in the $h_x h_y$ plane and for the infinite system (i.e. when the number of lattice sites $s \to \infty$) its endpoint draws a closed, directed loop when k runs through the Brillouin zone from 0 to 2π because $\mathbf{h}(k + 2\pi) = \mathbf{h}(k)$. If $J < J'$, Eqs. (5.65) show that the loop is a circle which encircles the origin of the coordinate frame in the $h_x h_y$ plane while for $J > J'$ the circle does not contain the origin, see Fig. 5.12. In the former case, the origin is encircled once when k goes through the Brillouin zone and therefore the winding number $\nu = 1$. In the latter case the winding number $\nu = 0$. For $J = J'$, the circle touches the origin, i.e. $h_x(\pm\pi) = 0$ and $h_y(\pm\pi) = 0$, and consequently at $k = \pm\pi$ the gap between positive and negative energy bands closes and the system is no longer an insulator. Thus, if we want to change the topological invariant (the winding number ν) by modifying the parameters of the system or by adding perturbation that preserves the chiral symmetry, the gap must be closed before ν switches its value. The winding number can be calculated by integrating the polar angle $\varphi(k)$ of the vector $\mathbf{h}(k)$ in the $h_x h_y$ plane when k runs through the Brillouin zone (BZ),[14]

(continued)

[13]Suppose $H|\psi\rangle = E|\psi\rangle$, then $-H|\psi\rangle = -E|\psi\rangle$ and from (5.63) we get $\Gamma H \Gamma|\psi\rangle = -E|\psi\rangle$ and consequently $H\Gamma|\psi\rangle = -E\Gamma|\psi\rangle$ because Γ is a Hermitian ($\Gamma^{\dagger} = \Gamma$) and unitary ($\Gamma\Gamma^{\dagger} = \Gamma^2 = 1$) operator. Thus, $|\psi\rangle$ and $\Gamma|\psi\rangle$ correspond to eigenenergies $\pm E$, respectively, and $\langle\psi|\Gamma|\psi\rangle \neq 0$ if $E \neq 0$.

[14]The formula (5.66) comes from the relation between the Cartesian and polar coordinate systems. That is, if $x = r\cos\varphi$ and $y = r\sin\varphi$, then $d\varphi = (x dy - y dx)/r^2$.

$$\nu = \frac{1}{2\pi} \oint_{\mathrm{BZ}} d\varphi = \frac{1}{2\pi} \oint_{\mathrm{BZ}} \frac{h_x dh_y - h_y dh_x}{|\mathbf{h}|^2} \tag{5.66}$$

$$= \frac{1}{2\pi} \int_0^{2\pi} \frac{1}{|\mathbf{h}(k)|^2} \left(\mathbf{h}(k) \times \frac{d}{dk} \mathbf{h}(k) \right)_z dk. \tag{5.67}$$

If the endpoint of $\mathbf{h}(k)$ makes a loop around the origin in the $h_x h_y$ plane, then $\varphi(k)$ changes by 2π and consequently $\nu = 1$. If the endpoint does not encircle the origin, the polar angle $\varphi(k)$ first increases but next it decreases and the net change of $\varphi(k)$ is zero and thus also $\nu = 0$.[15]

Let us now consider a finite chain ($s < \infty$) but not with the periodic boundary conditions but with the so-called open boundary conditions, i.e. $a_1 = 0$ and $a_s = 0$ in (5.58). Hence, the system possesses edges at $i = 2$ and $i = s - 1$. For $J < J'$ we have seen that the topology of the infinite system is non-trivial, i.e. the winding number $\nu = 1$. If the lattice is finite, then for $J < J'$ due to the edge-bulk correspondence [6], we may expect edge states which are eigenstates of the system localized close to the end-points of the chain with eigenenergies located in the energy gap of the infinite system. The easiest way to demonstrate the appearance of the edge states is to consider the so-called fully dimerized systems, i.e. when $J \neq 0$ and $J' = 0$ or vice versa, see Fig. 5.11. If $J \neq 0$ and $J' = 0$, each pair of sites in Fig. 5.11b hosts two eigenstates with positive and negative energy $E = \pm J/2$. If, however, $J = 0$ and $J' \neq 0$, then apart from the positive and negative energy eigenstates, $E = \pm J'/2$, hosted by pairs of sites in the interior (bulk) of the system, there are two zero energy eigenstates localized at two single sites on the edge of the system, see Fig. 5.11c. These edge states are present even if $J \neq 0$ provided $J < J'$, i.e. when the topology of the corresponding infinite system is non-trivial. For $J \neq 0$ the edge states are not localized exactly on the single end-points of the chain but penetrate the bulk on a distance given by the localization length $\xi = 1/\ln(J'/J)$.

[15] It should be stressed that changes of the winding number and not the values of ν themselves involve physical consequences [22]. If we change labeling of the lattice sites so that $|\psi\rangle = \sum_j (\alpha_j |2j\rangle + \beta_j |2j+1\rangle)$, instead of the previously used $|\psi\rangle = \sum_j (\alpha_j |2j-1\rangle + \beta_j |2j\rangle)$, and apply (5.60), we actually exchange the role of J and J'. Then, the winding number $\nu = 1$ corresponds to $J > J'$ contrary to the previous case. See also the background information in Sect. 7.1.2 where the Zak phase is introduced which is a quantity related to the winding number that we have introduced in the SSH model.

Now we can come back to topological time crystals. Our aim is to realize a system that reveals topological properties of the SSH model which are observable in the time dimension [39]. While it can be done for many different single-particle systems which are resonantly driven, we will focus on a particle bouncing resonantly on an oscillating mirror which is described in details in Sect. 4.2.1. Here, we assume that the mirror oscillations consist of the harmonic motion with the frequency ω and the subharmonic vibrations with the frequency $\omega/2$. In the frame oscillating with the mirror (see Sect. 4.2.1), the Hamiltonian of a particle reads

$$H = \frac{p^2}{2} + z + z\left[\lambda \cos(\omega t) + \lambda_1 \cos(\omega t/2)\right], \quad z \geq 0, \tag{5.68}$$

where λ and λ_1 are the amplitudes of the harmonic and subharmonic driving, respectively. We consider the $s:1$ resonance, i.e. in the classical description initial conditions of a particle are chosen close to the periodic orbit with the period s times longer than the period $2\pi/\omega$ of the fundamental harmonic in (5.68). Then, applying the approach introduced in Sect. 5.1 one obtains the classical secular effective Hamiltonian (5.8) in the form

$$H_{\text{eff}} = -\left(\frac{P^2}{2|m_{\text{eff}}|} + V_{\text{eff}}(\Theta)\right), \tag{5.69}$$

$$V_{\text{eff}}(\Theta) = V_0 \cos(s\Theta) + V_1 \cos(s\Theta/2), \tag{5.70}$$

with $V_0 = \lambda(-1)^s/\omega^2$, $V_1 = 4\lambda_1(-1)^{s/2}/\omega^2$ and negative effective mass $m_{\text{eff}} = -s^4\pi^2/\omega^4$. The Hamiltonian (5.69) describes a particle in the frame moving along the resonant orbit, cf. (5.6). Figure 5.13 shows an example of the effective potential in (5.70) for $s = 10$. It is a sequence of potential wells but every second barrier between neighboring wells is different. When we switch to the quantum description and follow the procedure that led to Eq. (5.14), i.e. when restrict to the first two energy bands of (5.69),[16] we end up with energy of a particle (more precisely quasi-energy of the driven system) in the tight-bind form,

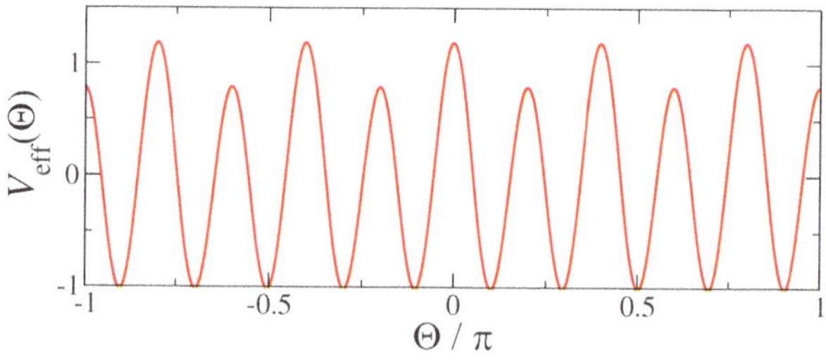

Fig. 5.13 Example of the effective potential in (5.70) for $s = 10$, $V_0 = 1$ and $V_1 = 0.2$

[16] Note that due to the negative effective mass, $m_{\text{eff}} < 0$, the first energy band of (5.69) is actually the highest energy band.

$$E_F = -\frac{1}{2}\sum_{i=1}^{s} J_i \left(a_{i+1}^{*} a_i + \text{c.c.}\right), \tag{5.71}$$

with the staggered hopping amplitudes $J_{2i-1} = J$ and $J_{2i} = J'$ and consequently we arrive at the SSH model (5.58). Wavefunction of a particle corresponding to quasi-energy (5.71) is given by superposition of Wannier states $w_i(\Theta)$ of (5.69) belonging to the first two energy bands [30] which, in the laboratory frame, are localized wavepackets $w_i(z,t)$ moving along the classical resonant orbit (cf. discussion in Sect. 5.1),

$$\psi = \sum_{i=1}^{s} a_i w_i(\Theta) = \sum_{i=1}^{s} a_i w_i(z,t), \tag{5.72}$$

where $w_i(\Theta + 4\pi/s) = w_{i+2}(\Theta)$ but $w_i(\Theta+2\pi/s) \neq w_{i+1}(\Theta)$ if $V_1 \neq 0$. Changing the amplitude λ_1 of the subharmonic modulation of the mirror motion in (5.68) one can change the ratio of the tunneling amplitudes J'/J in (5.71)—for example, for $\lambda_1 = 0$ all potential barriers in Fig. 5.13 are the same and $J'/J = 1$.

The system described by the tight-binding model (5.71) fulfills periodic boundary conditions. In order to demonstrate the edge-bulk correspondence by a particle bouncing resonantly on an oscillating mirror, we need to create an edge in the tight-binding model (5.71). It turns out to be quite easy to realize by turning on an additional modulation of the mirror motion. That is, if every period $s2\pi/\omega$ (which is the period of classical motion along the resonant orbit), the mirror is slightly kicked, an additional barrier in the effective potential (5.70) is created and consequently the open boundary conditions in the tight-binding model (5.71) are realized [39]. Then, wavefunction in (5.72) has $a_1 = a_s = 0$. In practice the edge is realized by adding a perturbation

$$H' = zf(t), \quad \text{with} \quad f(t) = f(t + s2\pi/\omega) = \sum_k f_k e^{ik\omega t/s}, \tag{5.73}$$

to the Hamiltonian (5.68). The Fourier components f_k can be chosen so that the resulting additional potential energy $V_2(\Theta) = \sum_n e^{in\Theta}(-1)^n f_{-ns} s^2/(n^2\omega^2)$ in (5.69) has a shape of a barrier located on two adjacent potential wells of $V_{\text{eff}}(\Theta)$. In the tight-binding approximation (5.71) it means that the open boundary conditions are realized where particle wavefunction has zero amplitudes on two sites, i.e. $a_1 = a_s = 0$.

Figure 5.14 shows the quasi-energy spectrum of a particle bouncing on an oscillating mirror corresponding to the Hilbert subspace described by the tight-binding model (5.71) for $s = 42$[17]—for each value of the parameter λ_1, the

[17] The results presented in Figs. 5.14 and 5.15 are obtained for $f(t)$ with the Fourier components $f_k = \left(10^{-6}\right) k^2(\omega/s)^2 \cos(k\pi/42)\text{sinc}^2(k\pi/121)$.

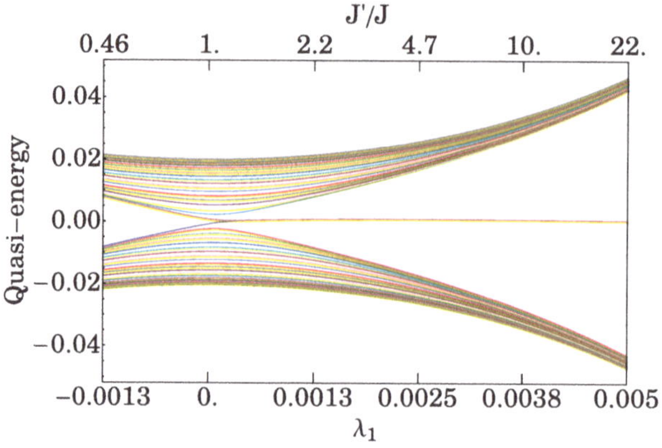

Fig. 5.14 Quasi-energy spectrum of a particle bouncing resonantly on an oscillating mirror. The 42 : 1 resonance condition is fulfilled between particle motion and the fundamental harmonics (described by the amplitude λ) of the mirror oscillations, cf. (5.68). Additional subharmonic oscillations of the mirror, with the amplitude λ_1, allow one to control the ratio J'/J of the tunneling amplitude in the effective SSH model (5.71) of the quantum resonant dynamics of a particle. If $J'/J > 1$, two quasi-energy levels form in the gap between the energy bands—in the plot they are zero quasi-energy levels and correspond to two Floquet states localized on the *edge of time*, see Fig. 5.15. The frequency $\omega = 2.8$, $s = 42$, and the amplitude $\lambda = 0.06$ are chosen in (5.68). Values of quasi-energies are given in the gravitational units (4.28) but multiplied by 10^5. Reprinted from [39]

spectrum has been shifted so that its center is always at zero energy. For $J'/J < 1$ ($\lambda_1 < 0$), the winding number ν of the corresponding bulk system is zero, the topology is trivial and there are no edge states with quasi-energies located in the gap between the bands. For $J'/J > 1$ (i.e. for $\lambda_1 > 0$), the corresponding winding number $\nu = 1$ and in the case of the finite system with the edge, there are two eigenstates (more precisely Floquet states) with zero quasi-energy localized on the edge of the chain described by the SSH model (5.71).

Switching from the moving frame (5.6) to the laboratory frame, we can observe the bulk-edge correspondence but in the time dimension [39]. That is, locating a detector close to the mirror (at $z \approx 0$), we can investigate how the probability density for the measurement of a particle at this point changes in time. If a particle is prepared in one of the edge Floquet states and if the *edge in time* is passing close to the detector, we can observe a particle localized on the edge, see Fig. 5.15. In contrast, if a particle is prepared in one of a bulk Floquet state, it will be delocalized along the entire classical resonant orbit which will be visible as an extended train of probability maxima in the time dimension, see Fig. 5.15. Figure 5.15 presents also how the edge and bulk Floquet states look like in the laboratory frame but in the configuration space for a fixed moment of time.

Experimental demonstration of the edge and bulk Floquet states with the help of atomic Bose–Einstein condensate is quite straightforward. Suppose that a Bose–Einstein condensate is initially prepared in a trap above an atom mirror at the turning

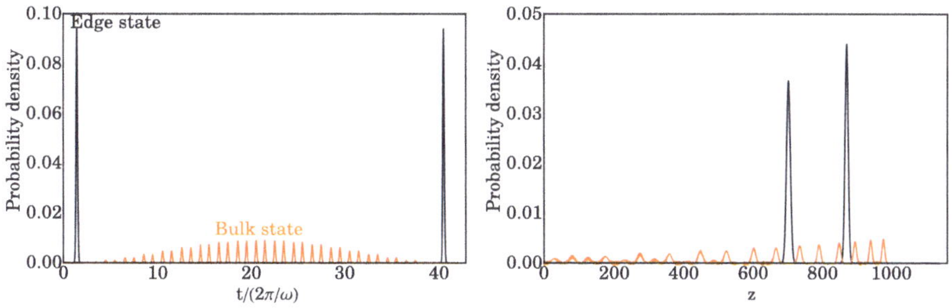

Fig. 5.15 Edge (black) and bulk (orange) Floquet states of a particle bouncing resonantly on an oscillating mirror. A particle is effectively described by the SSH model (5.71) and for $J'/J > 1$ edge states emerge in the system and they are localized close to the *edge in time* (left panel). That is, if a detector is located in the laboratory frame close to the mirror and an edge created by kicking the mirror (5.73) is passing near the detector, two Floquet states localized close to the edge can be observed. In contrast, Floquet states corresponding to bulk states of the SSH model are delocalized along the entire system, i.e. along the whole resonant orbit. The edge Floquet state presented in the figure corresponds to a linear combination of two states localized on both ends of the SSH model. In the right panel the edge and bulk Floquet states are presented but in the configuration space for a fixed moment of time ($\omega t = 39\pi/2$). The mirror is located at $z = 0$ and the turning point of the classical orbit at $z \approx 1110$. Comparing left and right panels it is apparent that signatures of the SSH model are observable in the time dimension only. The amplitude $\lambda_1 = 0.005$ ($J'/J = 22$) and the other parameters of the system are the same as in Fig. 5.14. Reprinted from [39]

point of a classical particle moving along an $s : 1$ resonant orbit. If the condensate is released from the trap at the moment when one of the *edges* is passing the classical turning point, the atomic cloud is loaded to the edge state. That is, atoms occupy $w_2(z, t)$ or $w_{s-1}(z, t)$ wavepacket and remain localized there in the course of time evolution. On the other hand if atoms are loaded to one of the wavepackets in the bulk, i.e. $w_i(z, t)$ with $3 \leq i \leq s - 2$, they start tunneling to neighboring wavepackets. If populations $|a_i(t)|^2$ of wavepackets $w_i(z, t)$ are measured in time, the winding number can be determined by means of the mean chiral displacement, i.e. $\nu \approx 2\sum_j (j - j_0)(|a_{2j+1}(t)|^2 - |a_{2j}(t)|^2)$ where j_0 is a number of a cell of the Bravais lattice where the atomic cloud is initially loaded [18, 72, 73]. This relation between ν and the mean chiral displacement is valid after a long time evolution when time averaged [73].

We have shown that the SSH model which is a model of a topological insulator can be realized in the system of a periodically driven particle and the edge-bulk correspondence is observed in the time dimension. In Chap. 7 similar behavior in photonic time crystals is analyzed [66]. In Sect. 5.4.3 realization of the topological Haldane insulator phase [16, 24, 32, 93] (i.e. a bosonic analog of the Haldane insulator in spin-1 chain [49, 50]) in a periodically driven many-body system is presented [39].

5.3.2 Engineering of Effective Potentials: Quasi-Crystalline Structure in the Time Dimension

Let us consider a periodically driven particle described by the Hamiltonian

$$H = \frac{p^2}{2} + V(z) + \lambda f(t)h(z), \quad \text{where} \quad f(t + 2\pi/\omega) = f(t). \tag{5.74}$$

In Sect. 5.1 we have shown that in the vicinity of an $s:1$ resonant classical orbit and in the moving frame (5.6), a particle can be described by the effective Hamiltonian (5.8) where it acquires a certain effective mass m_{eff} and experiences an effective potential

$$V_{\text{eff}}(\Theta) = \lambda \sum_n f_n h_{-sn}(I_s) e^{isn\Theta}, \tag{5.75}$$

where Θ is the position variable (the angle of the angle-action (Θ, I) variables of the unperturbed particle described by the Hamiltonian $H_0 = p^2/2 + V(z)$, see background information in Sect. 4.2.1), f_n and $h_n(I)$ are Fourier components of $f(t)$ and $h(z)$, Eqs. (5.4), and I_s is the resonant value of the action. While the components $h_n(I)$ depend on a specific system under consideration, values of f_n can be chosen as we wish because they depend on a way we drive a system in time, i.e. they depend on $f(t)$.

Suppose that we want to create an effective potential of a certain shape which fulfills periodic boundary conditions, $V_{\text{eff}}(\Theta + 2\pi) = V_{\text{eff}}(\Theta)$. A desired effective potential can be expanded in the Fourier series,

$$V_{\text{eff}}(\Theta) = \lambda \sum_n v_n e^{in\Theta}, \tag{5.76}$$

where v_n's depend on a shape of $V_{\text{eff}}(\Theta)$. Comparing (5.75) and (5.76) we get a prescription how to drive a particle in order to realize an effective potential we need [38]. That is, we have to choose $f(t)$ so that the fundamental $1:1$ resonance ($s = 1$) is fulfilled and whose Fourier components $f_n = v_n/h_{-n}(I_1)$. If $s > 1$, a potential structure of any shape we want will be duplicated s times in the interval $0 \le \Theta < 2\pi$. For $s \gg 1$, $V_{\text{eff}}(\Theta)$ allows us to reproduce condensed matter problems where a particle moves in a potential with s identical wells with periodic boundary conditions and the wells can have any shape we want.

As an example of engineering of effective potentials by means of periodic driving of a single-particle system, let us show how to create $V_{\text{eff}}(\Theta)$ with the shape of the Fibonacci quasi-crystal structure. In Sect. 4.5 spontaneous emergence of the Fibonacci sequence in the time dimension is investigated. Here, we will not deal with spontaneous formation of a quasi-crystal but we will create a potential $V_{\text{eff}}(\Theta)$ with a quasi-crystal structure intentionally by a proper driving of a particle. Note that $V_{\text{eff}}(\Theta)$ describes potential energy of a particle in the moving frame (5.6). In the quantum description when we return to the laboratory frame, quasi-crystal behavior of a particle will be observed in the time dimension. That is, if we locate a detector close to the resonant orbit and investigate how the probability of clicking of the detector behaves in time [67], it will reproduce what we see in the moving frame versus Θ.

The one-dimensional Fibonacci sequence can be generated by the inflation rule [55] sometimes called the rabbit sequence. That is, in each new generation a small rabbit will grow and become a big rabbit (i.e. $S \rightarrow B$) and the big rabbit reproduces to make a small one ($B \rightarrow BS$). Successive application of the rabbit reproduction cycle generates a sequence of letters ($BSBB\dots$) with no periodicity but with discrete peaks in the Fourier spectrum (cf. Sect. 4.5).

Let us show how to create an effective potential $V_{\text{eff}}(\Theta)$ which reveals a series of small (S) and big (B) potential wells identical to the sequence of S's and B's generated by the six applications of the rabbit reproduction cycle, i.e. the sequence of the length equal to the seventh Fibonacci number. We demonstrate it with the help of a particle bouncing resonantly on an oscillating mirror, i.e. $V(z) = h(z) = z$, with the restriction $z \geq 0$, in the Hamiltonian (5.74), see Sect. 4.2.1 for the detailed description of a particle bouncing on a mirror. The $1:1$ resonance condition between the motion of a particle and the mirror motion is assumed which results in the effective potential (5.75) where $s = 1$ and the Fourier components $h_n(I_1)$ are given by Eq. (4.40) with $I_1 = \pi^2/(3\omega^3)$. In Fig. 5.16 we present the Fourier components f_k^c and f_k^s of

$$f(t) = \omega^2 \sum_k \left[f_k^c \cos(k\omega t) + f_k^s \sin(k\omega t) \right], \tag{5.77}$$

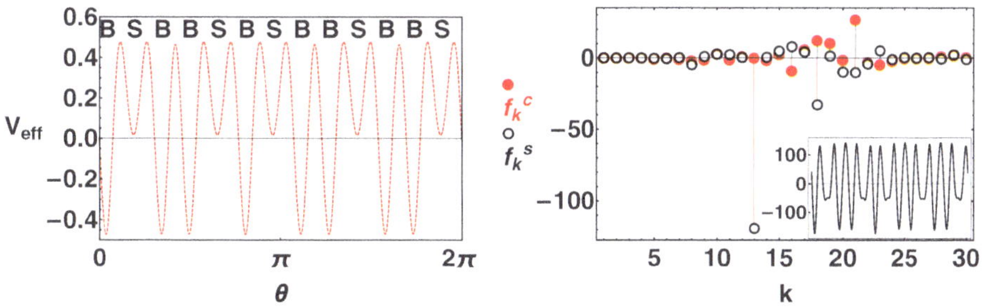

Fig. 5.16 Left panel shows the effective potential $V_{\text{eff}}(\Theta)$ where a series of small (S) and big (B) potential wells reproduces the Fibonacci sequence $BSBBSBSBBSBBS$. The potential $V_{\text{eff}}(\Theta)$ describes a particle bouncing on an oscillating mirror in the vicinity of the $1:1$ resonant classical trajectory—$s = 1$ in (5.75) and we have also set $\lambda = 1$. Right panel shows the Fourier components f_k^c and f_k^s of the time-periodic oscillations of the mirror, cf. (5.77), that lead to the effective potential presented in the left panel. How $f(t)/\omega^2$ changes in time over one period $2\pi/\omega$ of the mirror oscillations is presented in the inset of the right panel. Reprinted from [38]

which lead to the effective potential $V_{\rm eff}(\Theta)$ with the desired Fibonacci quasi-crystal structure. Switching to the quantum description we end up with the system similar to a condensed matter problem where an electron moves in a quasi-crystal potential created by ions. When we return from the moving frame to the laboratory, behavior of the probability density $|\psi(\Theta)|^2$ of a particle versus Θ will be reproduced versus time $|\psi(\theta-\omega t)|^2$ when we fix position $\theta=$ constant in the laboratory frame.

The presented quasi-crystal potential is an example of effective potentials which can be engineered by means of periodic driving of single-particle systems.

5.4 Many-Body Condensed Matter Physics in the Time Dimension

So far in the present chapter we have considered single-particle condensed matter phenomena in the time dimension. Is many-body condensed matter physics also available in the time domain? The answer is yes and it requires periodic and resonant driving of quantum many-body systems. We will often illustrate these phenomena with the help of atoms bouncing resonantly on an oscillating atom mirror but one may implement them in any many-body system if it is possible to drive it resonantly. We start with driven systems that are able to reproduce one-dimensional condensed matter physics and latter show that two- and three-dimensional condensed matter phases can be realized in periodically driven systems too.

5.4.1 Mott Insulator Phase in the Time Dimension

In Sect. 5.1 it is shown that quantum description of a single-particle system in the presence of a perturbation that evolves harmonically in time,

$$H=\frac{p^2}{2}+V(z)+\lambda h(z)\cos\omega t, \tag{5.78}$$

can be reduced to the tight-binding model (5.14) if a particle is resonantly driven.[18] The total number s of the lattice sites of the model is determined by the $s:1$ resonance between periodic motion of a particle and the periodic perturbation—

[18] Actually, in Sect. 5.1 it is shown that resonant motion of a particle described by the Hamiltonian of the form (5.78) can be reduced to the effective Hamiltonian (5.10). The subsequent analysis in Sect. 5.1 is carried out for a specific problem of a particle bouncing resonantly on an oscillating mirror in order to provide a concrete example how Bloch waves of the Hamiltonian (5.10) can look like in the laboratory frame. However, the obtained tight-binding model (5.14) applies not only for a particle bouncing on a mirror but for any periodically driven single-particle system with the Hamiltonian like in (5.78).

the period of unperturbed particle motion is s times longer than the driving period $T = 2\pi/\omega$ where s is an integer number.

In the present section we switch to the N-body counterpart of the single-particle system (5.78). We focus on bosons but similar approach can be applied to fermions. In order to describe the resonantly driven many-body system we restrict to the resonant Hilbert subspace spanned by Fock states $|n_1, \ldots, n_s\rangle$ where n_j is a number of bosons occupying the j-th lattice site of the tight-binding model (5.14). The j-th lattice site corresponds to the $w_j(z, t)$ localized wavepacket evolving along the resonant orbit, cf. (5.13). It is straightforward to derive the many-body Floquet Hamiltonian in the resonant Hilbert subspace. Indeed, in the second quantization formalism (see the background information in Sect. 2.1.2 and the background information on Many-body Floquet Hamiltonian in Sect. 4.1.1) truncating the bosonic field operator,

$$\hat{\psi}(z,t) \approx \sum_{j=1}^{s} w_j(z,t)\, \hat{a}_j, \tag{5.79}$$

where $[\hat{a}_i, \hat{a}_j^\dagger] = \delta_{ij}$, we obtain the Floquet Hamiltonian in the form of the Bose–Hubbard model [96], cf. Eq. (4.25),

$$\hat{H}_F = \frac{1}{sT}\int_0^{sT} dt \int_{-\infty}^{\infty} dz\, \hat{\psi}^\dagger \left(H - i\partial_t + \frac{g_0}{2}\hat{\psi}^\dagger\hat{\psi}\right)\hat{\psi}, \tag{5.80}$$

$$\approx -\frac{J}{2}\sum_{i=1}^{s}(\hat{a}_{i+1}^\dagger \hat{a}_i + \text{h.c.}) + \frac{1}{2}\sum_{i,j=1}^{s} U_{ij}\hat{a}_i^\dagger \hat{a}_j^\dagger \hat{a}_j \hat{a}_i, \tag{5.81}$$

where a constant term is omitted, H is given in (5.78), J in (5.15) and

$$U_{ij} = \frac{(2-\delta_{ij})g_0}{sT}\int_0^{sT} dt \int_{-\infty}^{\infty} dz\, |w_i(z,t)|^2\, |w_j(z,t)|^2. \tag{5.82}$$

In (5.80) we have assumed that bosons interact via the zero-range Dirac-delta potential $g_0\delta(z)$. In ultra-cold atomic gases the strength of the potential g_0 is proportional to the atomic s-wave scattering length [84]. In (5.80) the integration over the period sT is performed because the single-particle basis in (5.79) changes in time with the period of a classical motion of a particle, i.e. $w_j(z, t + sT) = w_j(z, t)$ but $w_j(z, t+T) = w_{j+1}(z, t)$. It means we deal with the periodic boundary conditions in time with the period sT. The product $\hat{\psi}^\dagger\hat{\psi}^\dagger\hat{\psi}\hat{\psi}$ is a sum of the terms $\hat{a}_i^\dagger\hat{a}_j^\dagger\hat{a}_k\hat{a}_l$ with any values of the indices. However, in the derived Bose–Hubbard model only the density–density interactions are present, cf. (5.82), because other terms are negligible. The validity of the derived Bose–Hubbard Hamiltonian requires that the interaction energy per particle must be much smaller than the gap between the single-particle energy bands of the Hamiltonian (5.10) otherwise we would not be allowed to restrict to the first energy band of (5.10). In practice there is no problem to fulfill this condition because in order to investigate interesting phases of the Bose–Hubbard model (5.81) we need the interaction energy per particle of the order of the tunneling amplitude J. The latter can be a tiny energy as compared to the energy gap between the bands, see also a similar discussion in Sect. 4.2.2 and Fig. 4.9 where such a gap is illustrated in the case of the $40:1$ resonant driving.

To sum up, a resonantly driven many-body system can be effectively described by the Bose–Hubbard model (5.81) [96]. If the interactions between bosons are negligible (i.e. $g_0 N \to 0$, where N is the total number of bosons) the ground state of the Bose–Hubbard Hamiltonian (i.e. the lowest quasi-energy Floquet state in the resonant Hilbert subspace of the driven system) is a perfect Bose–Einstein condensate,

$$|\psi_0\rangle = \left(\frac{1}{\sqrt{s}} \sum_{j=1}^{s} \hat{a}_j^\dagger \right)^N |0\rangle, \tag{5.83}$$

and possesses properties of a superfluid state [30, 115]. However, when the interactions are repulsive (i.e. $g_0 > 0$) and the on-site interactions in the Bose–Hubbard model dominate (i.e. $U_{ii} \gg U_{ij\neq i}$), then for $U_{ii}N/J \gg 1$, the ground state of the model is a single Fock state,

$$|\psi_0\rangle = |N/s, N/s, \ldots, N/s\rangle, \tag{5.84}$$

and the system possesses properties of the Mott insulator phase.[19] The coherence between bosons occupying different localized wavepackets $w_j(z,t)$ is lost and if we fix the position in the laboratory frame close to the resonant orbit, we will observe that well defined numbers of particles arrive periodically at this fixed space point [96], see Fig. 5.17. In contrast in the superfluid regime, each boson is in superposition of all localized wavepackets, cf. (5.83), and there are fluctuations of the occupations of the wavepackets but time coherence between arriving particles is not lost.

Such a quantum phase transition between the superfluid and Mott insulator phases can be observed in various resonantly driven Bose systems. In Section "A Particle on a Ring Driven by Fluctuating Force" we have analyzed a single particle that moves on a ring and is resonantly driven by the force harmonically changing in time, see the Hamiltonian (5.19) for $V = 0$. When we switch to the N-body counterpart of such a single-particle system, the effective Bose–Hubbard model (5.81) will possess the on-site interaction terms only, i.e. $U_{ij\neq i} = 0$. It is due

[19] The single Fock state (5.84) is the ground state provided the total number of bosons N is an integer multiple of the number of the lattice sites s of the Bose–Hubbard model. For $N/s = 1$, the critical interaction strength for the transition between the superfluid phase and the Mott insulator phase corresponds to $U_{ii}/J \approx 3$ [84].

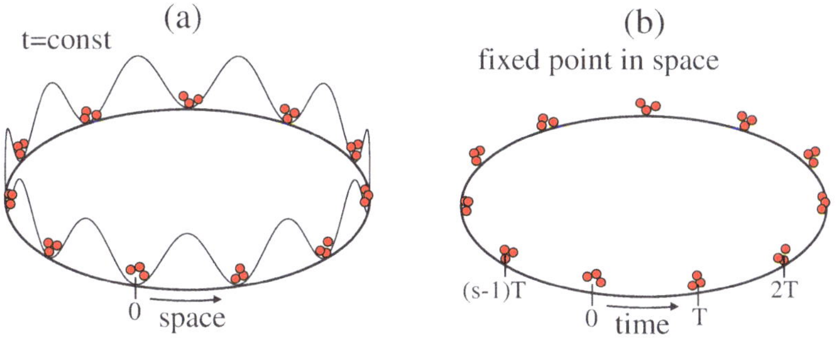

Fig. 5.17 Comparison of the Mott insulator phase in an ordinary one-dimensional crystalline structure in space and in the time dimension. Panel (**a**) illustrates well defined numbers of bosons that occupy each well of a one-dimensional spatially periodic potential on a ring. When we want to switch from space crystals (**a**) to time crystals (**b**) we have to exchange the role of space and time. That is, we fix position in space and ask how probability for the detection of particles at this fixed position changes in time. In the Mott insulating phase we observe that well defined numbers of bosons are arriving at the chosen space point like on a conveyor belt or in a machine gun. The larger the ring in (**a**), the greater the space crystal. The greater number of the $s:1$ resonance, the larger crystalline structure in the time dimension. Reprinted from [96]

to the fact that the localized wavepackets w_j will evolve along the ring one by one and never overlap.

However, if we consider a system of N atoms bouncing resonantly on a harmonically oscillating mirror, then the wavepackets $w_j(z,t)$ are passing each other in the course of time evolution along a resonant orbit because the orbit is a one-dimensional elongated trajectory, see, for example, Fig. 4.2 or Fig. 5.1. The long-range interactions in the Bose–Hubbard model (5.81) are present but for typical parameters the on-site interaction strength U_{ii} is at least an order of magnitude greater than $U_{ij\neq i}$ [96] (in Sect. 5.4.3 we discuss how to engineer the long range interactions).

In order to observe the superfluid Mott insulator transition one has to either change the strength of the interactions U_{ij} or the tunneling amplitude J. The latter can be controlled by changing localization properties of the wavepackets w_j by means of the amplitude λ of the time-periodic driving in (5.78). That is, λ controls the strength of the effective potential in (5.10) and determines the hopping amplitude between neighboring wells of the potential. The interaction strength U_{ij} can be controlled in ultra-cold atomic gases by employing Feshbach resonances which occurs when two colliding atoms are resonantly coupled to a molecular bound state [84]. In the vicinity of a Feshbach resonance, the s-wave scattering length and consequently the parameter g_0 can be precisely controlled.

5.4.2 Many-Body Localization with Temporal Disorder

A single-particle system in the presence of a time-independent disordered potential can reveal Anderson localization. In Sect. 5.2 we have seen that a particle driven by force that fluctuates in time also shows up Anderson localization but in the time dimension. A few different driven systems that reveal Anderson localization in the time domain have been presented in Sect. 5.2. Especially we have seen that description of a particle bouncing resonantly on a harmonically oscillating mirror, e.g. (5.24), reduces to the Anderson model, Eq. (5.26), if there are additional random fluctuations in the mirror motion. That is, a particle is moving along a resonant orbit but in the presence of the temporal disorder it is not delocalized along the orbit but appears at a given point of the resonant trajectory with the probability which is exponentially localized around a certain moment of time.

We have also seen in the previous section that resonantly driven many-body systems of interacting bosons can be described by the Bose–Hubbard Hamiltonian (5.81).[20] Let us turn on an additional force that fluctuates randomly in time. That is, the single-particle Hamiltonian in (5.80) reads

$$H = \frac{p^2}{2} + V(z) + \lambda h(z)\cos\omega t + Vh(z)f(t), \tag{5.85}$$

where $f(t)$ changes randomly but fulfills periodic boundary conditions in time with the period of resonant orbit, i.e. $f(t+sT) = f(t)$, similarly like in the case of the Anderson localization described in Section "A Particle Bouncing on a Fluctuating Mirror". Then, the analogues tight-binding approximation as we have applied in the previous section leads to the many-body Hamiltonian

$$\hat{H}_F \approx -\frac{J}{2}\sum_{i=1}^{s}(\hat{a}_{i+1}^\dagger \hat{a}_i + \text{h.c.}) + \frac{1}{2}\sum_{i,j=1}^{s} U_{ij}\hat{a}_i^\dagger \hat{a}_j^\dagger \hat{a}_j \hat{a}_i + \sum_{i=1}^{s}\varepsilon_i \hat{a}_i^\dagger \hat{a}_i, \tag{5.86}$$

where U_{ij} are given in (5.82) and

$$\varepsilon_j = \frac{V}{sT}\int_0^{sT} dt \int_{-\infty}^{\infty} dz\, h(z)\, f(t)\, |w_j(z,t)|^2 \tag{5.87}$$

are random numbers corresponding to the Gaussian distribution with zero mean and standard deviation V if

[20] For fermions similar tight-binding approximation can be applied. For contact interactions, a mixture of two kinds of fermions has to be periodically driven otherwise identical fermions cannot interact via the zero-range potential and the system consists of non-interacting particles. The resulting tight-binding approximation for a mixture of two kinds of fermions leads to the Hubbard Hamiltonian.

$$f(t) = \frac{1}{\sqrt{2s}} \sum_{\substack{k=-s \\ k\neq 0}}^{s} f_k e^{ik\omega t/s}, \tag{5.88}$$

with $f_k = e^{i\varphi_k}/h_{-k}(I_s)$ where $\varphi_k = -\varphi_{-k}$ are random phases chosen uniformly in the interval $[0, 2\pi)$ and I_s is the resonant value of the action.[21] For details of the derivation of ε_j's and the analysis of their properties see Section "A Particle Bouncing on a Fluctuating Mirror". The validity of the Hamiltonian (5.86) requires the interaction energy per particle and the on-site energies ε_j to be much smaller than the gap between the single-particle energy bands of the effective secular Hamiltonian (5.10)—see the discussion in the previous section and Fig. 4.9 where such a gap is illustrated in the case of the 40 : 1 resonant driving.

In Sect. 4.3.1 we have already described shortly the many-body localization phenomenon [12, 54, 79, 80, 114]. Despite the fact that a generic interacting many-body system is expected to follow the eigenstate thermalization hypothesis [23, 29, 91, 107], a translationally invariant system in the presence of strong spatial disorder does not obey the hypothesis. That is, local properties of a strongly disordered system prepared in an eigenstate with an energy E cannot be modelled by the canonical ensemble whose average energy corresponds to E. There are quasi-local integrals of motion which prevent the thermalization [3, 7, 21, 25, 26, 31, 41, 42, 53, 57, 65, 79, 82, 90, 92, 100, 102–105]. The many-body localization is characterized by the lack of dc transport [2, 8, 10, 11, 13, 61, 108], extremely slow dynamics of various correlation functions [65, 74, 76, 77, 82, 87], and the logarithmic growth of the entanglement entropy [9, 60, 65, 104, 114].

The effective Floquet Hamiltonian (5.86) is derived in the time-periodic basis (5.79) but it has the form of a typical Bose Hamiltonian investigated in the field of the many-body localization [105]. Thus, if the disordered terms are sufficiently strong, all eigenstates of (5.86) reveal many-body localization properties. The key difference between driven systems considered here and standard time-independent many-body localized systems relies on the fact that we deal with temporal disorder and signatures of the many-body localization are observed in the time dimension not in the configuration space if we perform measurements in the laboratory frame [75].

An example of an experimental setup for the realization of many-body localization induced by temporal disorder is shown in Fig. 5.18 for ultra-cold atoms moving on a ring where the position variable z in (5.85) is substituted by the angle θ. We assume that the Hamiltonian (5.85) takes the form of the Hamiltonian (5.19) used to describe Anderson localization in time crystals. Also similar form of the perturbation $h(\theta)$ as in Section "A Particle on a Ring Driven by Fluctuating Force" is assumed. In the case of ultra-cold atoms on a ring, there are only on-site

[21]The resonant action I_s is determined by $\Omega(I_s) = \omega/s$ where $\Omega(I_s)$ is the frequency of motion of the unperturbed particle, i.e. for $\lambda = 0$ and $V = 0$ in (5.85). An integer s denotes the s : 1 resonance.

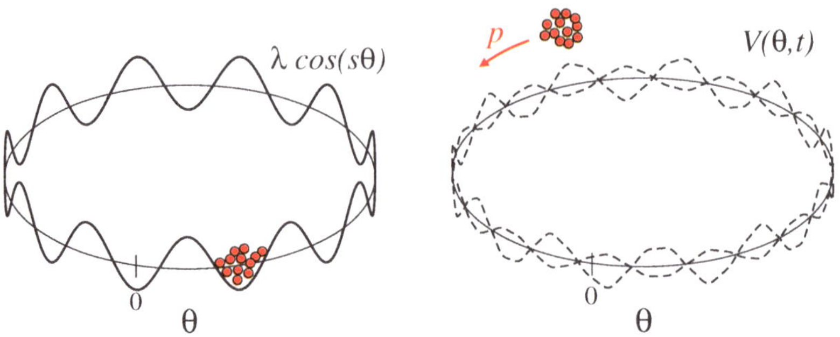

Fig. 5.18 Example of an experimental setup for the realization of the many-body localization induced by temporal disorder. Left panel presents an initial stage of the experiment where ultra-cold atoms are prepared in a local well of the spatially periodic potential along a ring which is a toroidal trap with strong transverse confinement. Then, the temporal harmonic modulation of the potential is turned on together with the term $Vh(\theta)f(t)$ that introduces force randomly fluctuating in time (see the right panel)—the entire potential reads $V(\theta,t) = \lambda\cos(s\theta)\cos(\omega t) + Vh(\theta)f(t)$. At the moment when the time-dependent potential $V(\theta,t)$ is turned on, atoms are kicked so that their momenta $p \approx \omega/s$ fulfill the $s:1$ resonance condition with the harmonically changing force. In the many-body localization regime, the cloud of atoms will evolve along the ring and do not spread even if the interaction between atoms are repulsive. Reprinted from [75]

interactions U_{ii} in the effective Bose–Hubbard model (5.86), i.e. $U_{ij\neq i} = 0$, which are chosen to be repulsive $U_{ii} > 0$. Ultra-cold atoms are initially prepared in a local minimum of the potential $\lambda\cos(s\theta)$ on a ring and then the driving harmonic in time, $\lambda\cos(s\theta)\cos(\omega t)$, and disorder in time, $Vh(\theta)f(t)$, are turned on. Simultaneously, atoms are kicked so that their momenta $p \approx \omega/s$ fulfill the $s:1$ resonance condition with the harmonically changing force. In the presence of sufficiently strong disorder the cloud of atoms will be moving along the ring without much spreading even if the interactions between bosons are assumed to repulsive [75]. In the laboratory frame and with a detector placed close to a ring, the lack of the spreading of the cloud and thus the signatures of the many-body localization will be observed versus time.

It is also quite straightforward to realize similar behavior in ultra-cold atoms bouncing on an oscillating mirror. That is, the Bose–Hubbard model (5.81) for such a system (which is described in the previous section) needs only to be supplemented with disordered terms which can be introduced in the way described in Section "A Particle Bouncing on a Fluctuating Mirror". Then, long-range interactions in the Bose–Hubbard model (5.86) are also present, i.e. $U_{ij\neq i} \neq 0$, but usually they are an order of magnitude weaker than the on-site interaction coefficients U_{ii}. However, as we will see in the next section, the strength of the long-range interactions can be controlled and manipulated and consequently the influence of long-range interactions on the many-body localization phenomena can be investigated in the laboratory [75].

5.4.3 Engineering of Long-Range Interactions in Time Crystals

In the present section we show that the effective interactions in the Bose–Hubbard model (5.81) that describe resonant dynamics of periodically driven many-body Bose systems can be controlled and engineered [38]. We illustrate the idea on ultra-cold atoms bouncing on a harmonically oscillating mirror. The same strategy can be applied for any one-dimensional system provided particles moving on a resonant trajectory retrace the same path after reaching a turning point. It excludes particles on a ring if resonant trajectories correspond to particles circulating along a ring in the same direction.

We focus on the $s:1$ resonant dynamics of ultra-cold atoms bouncing on a harmonically oscillating atom mirror which is described by the Bose–Hubbard model (5.81)—see Sect. 5.4.1 for more details. Although the original interactions between ultra-cold atoms are contact, the effective interactions in the Bose–Hubbard model are long-range (for realization of long-range interactions in ultra-cold atomic gases see also [5, 44, 83]). It is due to the fact that atoms occupying different wavepackets $w_j(z,t)$, that evolve along the resonant orbit, meet together when the wavepackets are passing each other, see an example of the $2:1$ resonance case in Fig. 4.2. In a typical situation the coefficients for the long-range interactions $U_{ij\neq i}$, Eq. (5.82), are an order of magnitude smaller than U_{ii}. However, in ultra-cold atoms one can use Feshbach resonances [84] to enhance or diminish the contact atom–atom interactions at the moment when different wavepackets $w_j(z,t)$ overlap. It opens a possibility for control and even engineering the effective long-range interactions in the Bose–Hubbard model (5.81) [38]. Let us assume that the atomic s-wave scattering length is modulated periodically in time by means of the Feshbach resonance, i.e. $g_0(t+T) = g_0(t)$ where $T = 2\pi/\omega$ is the period of the mirror oscillations. Then, the interaction coefficients in (5.81)

$$U_{ij} = \frac{2-\delta_{ij}}{sT}\int_0^{sT} dt \int_0^{\infty} dz\, g_0(t)\, |w_i(z,t)|^2\, |w_j(z,t)|^2. \tag{5.89}$$

In order to analyze what kinds of the effective interactions in the resonant dynamics one can achieve and how to choose $g_0(t)$ that leads to desired U_{ij}'s, let us consider the coefficients U_{ij} as components of the vector $\mathrm{vec}(U)$ and values of $g_0(t)$ for $0 \le t < sT$ as components of the vector $\mathrm{vec}(g_0)$. It allows us to write Eq. (5.89) in the form

$$U_{ij} = \int_0^{sT} dt\, u_{ij}(t)\, g_0(t), \quad \text{or} \quad \mathrm{vec}(U) = \mathbf{u}\,\mathrm{vec}(g_0), \tag{5.90}$$

where the elements of the matrix $\mathbf{u}$ read

$$u_{ij}(t) = \frac{2-\delta_{ij}}{sT} \int_0^\infty dz\, |w_i(z,t)|^2\, |w_j(z,t)|^2, \tag{5.91}$$

with (i, j) and t treated as indices of rows and columns, respectively. When we apply the singular value decomposition for the matrix $\mathbf{u}$ we obtain the left and right singular vectors[22] [88]. The left singular vectors tell us which sets of the interaction coefficients U_{ij} can be realized, while the corresponding right singular vectors give the recipes for $g_0(t)$ to get them. Suppose we would like to realize a certain set of the coefficients U_{ij} represented by a vector vec(U). We have to expand vec(U) in the basis of the left singular vectors of $\mathbf{u}$ corresponding to the non-zero singular values. Such expansion tells us how well a desired vec(U) can be reproduced and also allows us find which combination of the right singular vectors of $\mathbf{u}$ leads to its realization. The latter provides the prescription for $g_0(t)$ that we are looking for.

In Fig. 5.19 we show an example of the long-range effective interactions in the Bose–Hubbard model that can be realized in ultra-cold atoms bouncing resonantly

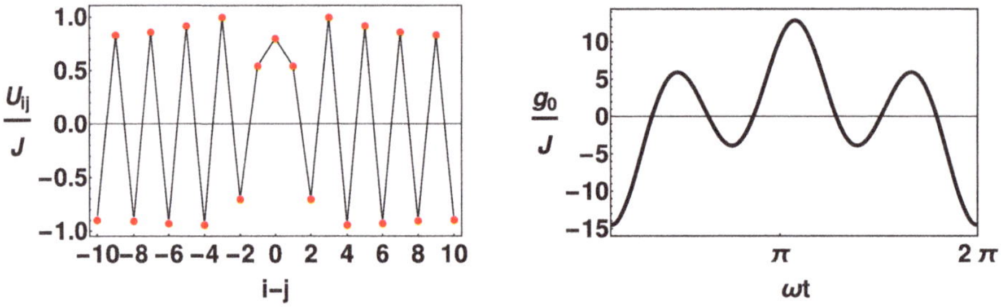

Fig. 5.19 Ultra-cold atoms bouncing resonantly on a harmonically oscillating atom mirror. The 20 : 1 resonance condition is fulfilled which means that there are $s = 20$ lattice sites in the effective Bose–Hubbard model (5.81) that describes the resonant dynamics of atoms. Modulation of the atomic s-wave scattering length in time allows one to engineer exotic long-range interactions in the Bose–Hubbard model. Left panel shows an example where the interactions change between the repulsive and attractive character in an alternate way versus the distance between lattice sites. These kind of long-range interactions are obtained if the scattering length, and thus the coefficient $g_0(t)$ in (5.89) is modulated in time in a way presented in the right panel. The frequency $\omega = 2.8$ of the mirror oscillations and the amplitude parameter $\lambda = 0.1$, cf. (4.29), which result in $J = 3.7 \times 10^{-5}$ in (5.81). The gap between the first and second energy bands of the single-particle secular Hamiltonian (5.10) equals $302J$. Temporary interaction coefficients $(s2\pi/\omega)|g_0(t)|u_{ij}(t) \leq 85J$ and thus they are always smaller than the energy gap indicating that the Bose–Hubbard model (5.81) is the valid description of the many-body system. Reprinted from [38]

[22]The singular value decomposition allows one to write a real rectangular $m \times n$ matrix $\mathbf{M}$ as a product $\mathbf{M} = \mathbf{U}\Sigma\mathbf{V}^{\mathrm{T}}$ where $\mathbf{U}$ is an orthogonal $m \times m$ matrix whose columns are the left singular vectors and $\mathbf{V}$ is an orthogonal $n \times n$ matrix with columns corresponding to the right singular vectors. The rectangular $m \times n$ matrix Σ is diagonal with non-negative values (the so-called singular values) on the diagonal which are roots of non-negative eigenvalues of $\mathbf{M}\mathbf{M}^{\mathrm{T}}$ or $\mathbf{M}^{\mathrm{T}}\mathbf{M}$. For details see, e.g., [88].

on an oscillating atom mirror if the s-wave scattering length is appropriately modulated in time [38]. The interactions presented are quite exotic because with increasing distance between lattice sites of the Bose–Hubbard model (5.81), their character changes between repulsive and attractive in an alternate way.

On the one hand engineering of the effective interactions in resonantly driven ultra-cold atoms by modulating the atomic s-wave scattering length in time seems very convenient experimentally because it requires only to change magnetic field in time around the value corresponding to the Feshbach resonance. On the other hand number of non-zero singular values of the matrix $\mathbf{u}$ we obtain in this way is not very large and consequently not all possible sets of the coefficients U_{ij} can be realized with good accuracy. We can, however, broaden a range of possible U_{ij}'s if we allow for modulation of the scattering length not only in time but also in space. In ultra-cold atoms it can be realized by means of non-homogeneous magnetic field. Suppose that the scattering length, and thus the coefficient g_0, is modulated in time and space in the following way

$$g_0(z,t) = \sum_{m=0}^{M} \alpha(t,m)\, z^m, \tag{5.92}$$

where we have $M+1$ functions $\alpha(t,m)$ which can be independently modulated in time. Then, the interaction coefficients in the Bose–Hubbard model can be written as

$$U_{ij} = \sum_{m=0}^{M} \int_0^{sT} dt\; u_{ij}(t,m)\, \alpha(t,m), \quad \text{or} \quad \mathrm{vec}(U) = \mathbf{u}\, \mathrm{vec}(\alpha), \tag{5.93}$$

where the elements of the matrix $\mathbf{u}$ read

$$u_{ij}(t,m) = \frac{2-\delta_{ij}}{sT} \int_0^{\infty} dz\; |w_i(z,t)|^2\, |w_j(z,t)|^2\, z^m, \tag{5.94}$$

with (i,j) and (t,m) treated as indices of rows and columns, respectively. Similarly as in the case of (5.90), the singular value decomposition of the matrix $\mathbf{u}$ can be performed in order to find how to control and engineer the effective interactions but the number of non-zero singular values of $\mathbf{u}$ is now much greater. Hence, we are much more flexible in engineering different exotic effective interactions in the Bose–Hubbard model (5.81) of the driven system.

In Fig. 5.20 we present how to modulate $g_0(z,t)$ in time and space for atoms bouncing resonantly on a harmonically oscillating atom mirror in order to deal with the dynamics effectively described by the Bose–Hubbard model of the form [39]

$$\hat{H} = -\frac{J}{2}\sum_{i=1}^{s}\left(\hat{a}_{i+1}^{\dagger}\hat{a}_i + \mathrm{h.c.}\right) + \frac{U}{2}\sum_{i=1}^{s}\hat{n}_i(\hat{n}_i - 1) + V\sum_{i=1}^{s}\hat{n}_{i+1}\hat{n}_i, \tag{5.95}$$

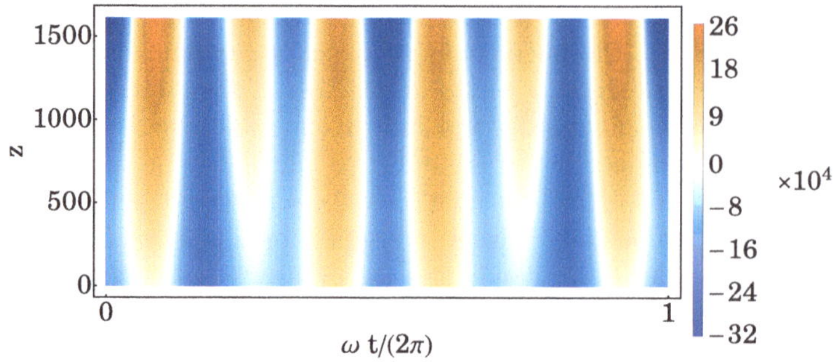

Fig. 5.20 Figure shows how to modulate in space and time the atomic s-wave scattering length $g_0(z,t)/J$, cf. (5.92), of ultra-cold atoms bouncing resonantly on a harmonically oscillating mirror in order to realize the Bose–Hubbard model (5.95) with the parameters corresponding to the topological Haldane insulator—$U/J = 3$, $V/J = 2.5$ and the unit mean boson filling factor of the lattice sites. The other parameters of the system are the following: $\omega = 3.58$, $s = 64$ (i.e. the 64 : 1 resonance), $\lambda = 0.14$ and $M = 2$, cf. (4.29) and (5.92). The mirror is located at $z = 0$ and the turning point of the classical resonant orbit at $z \approx 1600$. The maximal temporal interaction energy per particle is about $1.6 \times 10^4 J$ and thus much smaller than the gap, $1.1 \times 10^5 J$, between the first and second energy bands of the single-particle secular Hamiltonian (5.10), where $J = 5.5 \times 10^{-8}$. Reprinted from [39]

where $\hat{n}_i = \hat{a}_i^\dagger \hat{a}_i$, U determines the on-site interactions and V the nearest neighbor interactions. The interactions are assumed to be repulsive (i.e. $U \geq 0$ and $V \geq 0$). If both on-site and nearest neighbor interactions are negligible, the ground state of the model (5.95), i.e. the lowest quasi-energy Floquet state of the driven system in the resonant Hilbert subspace, describes superfluid phase of bosons. For the dominant on-site interactions, the Mott insulator phase emerges. When the nearest neighbor repulsion dominates, the density-wave phase appears where the translation symmetry of the Hamiltonian (5.95) is spontaneously broken because in the ground state it is not energetically favorable to have bosons uniformly distributed in all lattice sites but to modulate the density with the spatial period twice longer than the lattice period. However, between the Mott insulator phase and the density-wave phase there is the topological Haldane insulator phase [16, 24, 32, 93]. It is a bosonic analog of the Haldane insulator in spin-1 chain [49, 50] related to a highly nonlocal string order parameter [24, 56]. All these phases can be realized in periodically driven many-body systems, e.g. in the considered here ultra-cold atoms bouncing on an oscillating atom mirror [39].

We should remember that the validity of the Bose–Hubbard model, that we use for description of a resonantly driven many-body system, requires that the interaction energy per particle must be smaller than the gap between energy bands of the single-particle secular Hamiltonian (5.10) otherwise we are not allowed to restrict to the Hilbert subspace corresponding to the first energy band of the Hamiltonian (5.10) [39].

5.4.4 *Time Lattices with Properties of Multi-Dimensional Space Crystals*

Dynamics of resonantly driven single-particle or many-body systems can reveal condensed matter behavior. That is, they can be prepared in resonant Hilbert subspaces where their description reduces to solid state-like Hamiltonians. If we observe these systems in the laboratory frame, signatures of condensed matter behavior emerge in the time dimension meaning that they behave in time like, for example, an electron does in space in an ordinary space crystal.

Time is a single degree of freedom and it is hard to imagine multi-dimensional time crystals. However, it turns out that crystalline structures in time with properties of two- or three-dimensional condensed matter systems can be realized [38]. Actually the strategy for the realization of a periodically driven system that is effectively described by the two-dimensional Bose–Hubbard model has been already outlined in Sect. 4.5 where we describe spontaneous formation of time quasi-crystals. Let us remind a reader only the basic elements needed to obtain the two-dimensional Bose–Hubbard model (4.164) as an effective description of ultra-cold atoms bouncing resonantly between two oscillating atom mirrors.

Consider two orthogonal mirrors that oscillate with the same frequency ω and are inclined at an angle of 45° to the gravitational force vector, see Fig. 5.21. A particle bouncing between the mirrors, in the reference frame oscillating with the mirrors, is described by the Hamiltonian (4.162). Motion of a particle separates into two independent one-dimensional motions along the directions orthogonal to the mirrors and consequently the description of a single particle is very simple. Assume that a particle moves in the vicinity of a periodic orbit that fulfills the s : 1 resonance condition with the mirror's oscillations. Description of the resonant dynamics of a particle is convenient when we perform the canonical transformation from the Cartesian coordinates (x, p_x) and (y, p_y), where x and y correspond to the

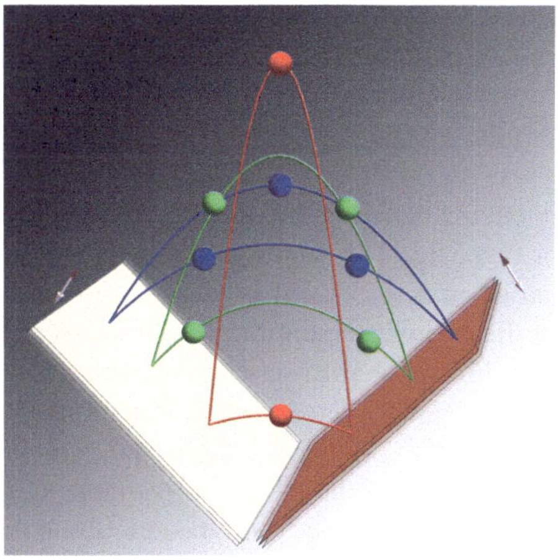

Fig. 5.21 Schematic plot of ultra-cold atoms bouncing resonantly between two orthogonal oscillating mirrors. In the single-particle case motion of a particle separates into two independent degrees of freedom corresponding to two orthogonal directions perpendicular to the mirrors. In the many-body case, interactions between particles introduce coupling between these two degrees of freedom (graphic courtesy: Artur Miroszewski)

axes orthogonal to the mirrors, to the angle-action variables (θ_x, I_x) and (θ_y, I_y), respectively. Then, switching to the moving frame,

$$\Theta_x = \theta_x - \frac{\omega}{s}t, \qquad \Theta_y = \theta_y - \frac{\omega}{s}t, \tag{5.96}$$

and averaging the single-particle Hamiltonian (4.162) over time, we obtain the classical secular effective Hamiltonian,

$$H_{\text{eff}} = \frac{P_x^2}{2m_{\text{eff}}} + \frac{P_y^2}{2m_{\text{eff}}} + V_0\left[\cos(s\Theta_x) + \cos(s\Theta_y)\right], \tag{5.97}$$

which is a two-dimensional counterpart of the one-dimensional secular Hamiltonian (5.10). The parameters in (5.97) are the same as in (5.10) and we have assumed that the amplitudes of the mirror's oscillations in (4.162) are identical, i.e. $\lambda = \lambda_x = \lambda_y$. The Hamiltonian (5.97) indicates that a resonantly driven particle behaves like an electron moving in a potential with a two-dimensional crystalline structure in space. In the quantum description there are energy bands of the quantized version of the Hamiltonian (5.97) and when we restrict to the first band we arrive at the tight-binding approximation. In the Hilbert subspace corresponding to the first energy band we can define the basis of the Wannier states $W_{\mathbf{i}}(\Theta_x, \Theta_y)$ localized in local wells of the potential in (5.97) [30]. In the laboratory frame, these Wannier states appear to be localized wavepackets moving along the classical orbit,

$$W_{\mathbf{i}=(i_x,i_y)}(\Theta_x, \Theta_y) = w_{i_x}(\Theta_x)\, w_{i_y}(\Theta_y) = w_{i_x}(x,t)\, w_{i_y}(y,t), \tag{5.98}$$

where w_{i_x} and w_{i_y} are localized wavepackets that we have defined in the one-dimensional case, see Sect. 5.1 especially the discussion leading to Eq. (5.13).

Let us switch to N ultra-cold atoms which are bosons and which are bouncing between the mirrors and fulfill the $s:1$ resonance condition with the mirror's oscillations. When we restrict to the resonant Hilbert subspace, we end up with the two-dimensional Bose–Hubbard model (4.164), similar to the one-dimensional counterpart (5.81), where the bosonic operators $\hat{a}_{\mathbf{i}}$ annihilate a boson in the states corresponding to the localized wavepackets (5.98), see Sects. 4.5 and 5.4.1 for details.

As an example of the *two-dimensional time lattice* created by ultra-cold atoms bouncing between the oscillating mirrors we present the case of the $5:1$ resonance in Fig. 5.22. The single-particle effective potential in (5.97) describes the squared lattice of 5×5 local potential wells and with the periodic boundary conditions. In the N-body quantum case, the description of the system can be reduced to the two-dimensional Bose–Hubbard model (4.164) corresponding to the squared lattice of 5×5 sites. The on-site interactions in the model dominate over the long-range interactions, i.e. $U_{\mathbf{ii}} \gg U_{\mathbf{ij}\neq\mathbf{i}}$, and if they are repulsive and sufficiently strong, the ground state of the model corresponds to the Mott insulating phase, see Sect. 5.4.1. Then, assuming the unit boson filling factor of the lattice, the ground state is

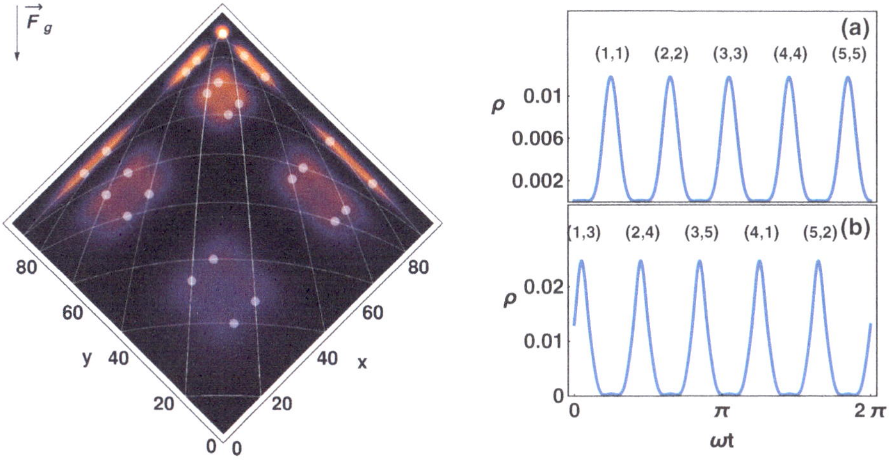

Fig. 5.22 Time lattice created by ultra-cold atoms bouncing resonantly between two orthogonal oscillating atom mirrors, see Fig. 5.21. Both mirrors oscillate with the same frequency $\omega = 1.05$ and with the same amplitude $\lambda = \lambda_x = \lambda_y = 0.02$, and with the phase difference $\Delta\phi = -\pi/2$, see the Hamiltonian (4.162). The 5 : 1 resonance condition is fulfilled between motion of atoms and the mirrors' oscillations. Values of the interaction coefficients in the effective Bose–Hubbard description, Eq. (4.164), are in the ranges: $U_{\mathbf{ii}}/g_0 \in [0.2, 0.3]J$ and $U_{\mathbf{ij}\neq\mathbf{i}}/g_0 < 0.07J$ and consequently the on-site interactions dominate over the long-range interactions. If the interactions are repulsive and sufficiently strong, the ground state of the effective Bose–Hubbard model corresponds to the Mott insulating phase where, for the unit boson filling factor, the average density is the uniform sum of the wavepackets densities $\rho(x, y, t) = \sum_{\mathbf{i}} |W_{\mathbf{i}}(x, y, t)|^2$. The density $\rho(x, y, t)$ is plotted in the left panel for $\omega t = 4$. Right panels present $\rho(x, y, t)$ at $(x, y) = (90, 90)$ (**a**) and $(82, 82)$ (**b**) as a function of time t. The plots reflect cuts of the 5×5 square lattice described by the Bose–Hubbard model. Numbers in parentheses indicate which lattice sites $\mathbf{i} = (i_x, i_y)$ are located along the cuts. Reprinted from [38]

given by the Fock state $|1, 1, \ldots, 1\rangle$ where each localized wavepacket $W_{\mathbf{i}}(x, y, t)$ is occupied by one boson. The ground state corresponds to the lowest quasi-energy Floquet state in the resonant Hilbert subspace of the many-body driven system. The average density of atoms, $\rho(x, y, t) = \sum_{\mathbf{i}} |W_{\mathbf{i}}(x, y, t)|^2$, is plotted in Fig. 5.22 at a certain moment of time. In the course of time evolution, the wavepackets $W_{\mathbf{i}}$ are traveling along the resonant paths. In the moving frame (5.96) we deal with a squared lattice structure that in the many-body case is described by the two-dimensional Bose–Hubbard model. Such a two-dimensional crystalline structure can be detected in the time dimension when we locate a detector at different points in the laboratory frame and measure how the probability density for clicking of the detector changes in time. That is, when we switch from the moving frame (5.96) to the laboratory frame and locate a detector close to the resonant orbit, we observe that the probability for the detection a single boson,

$$\rho(\Theta_x, \Theta_y) = \sum_{\mathbf{i}} |W_{\mathbf{i}}(\Theta_x, \Theta_y)|^2 = \sum_{\mathbf{i}} \left| W_{\mathbf{i}} \left(\theta_x - \frac{\omega}{s} t,\ \theta_y - \frac{\omega}{s} t \right) \right|^2, \qquad (5.99)$$

where $\theta_x = \text{const}$ and $\theta_y = \text{const}$, reflects in time a cut of the squared lattice in the $\Theta_x\Theta_y$ space along a line $\Theta_y = \Theta_x + \theta_y - \theta_x$.[23] Different choice of the detector location, i.e. different choice of constant values of θ_x and θ_y, allows one to scan the entire squared lattice in the $\Theta_x\Theta_y$ space. In Fig. 5.22 examples of two cuts are presented.

Generalization of the described phenomenon to a time lattice corresponding to a three-dimensional cubic lattice in space is straightforward. One has to exchange only two orthogonal oscillating mirrors by three mutually orthogonal and oscillating mirrors.

It is also worth mentioning that time lattices corresponding to two- or three-dimensional space lattices, which we have described here, can be created not only in ultra-cold atoms bouncing between atom mirrors. Any two- or three-dimensional separable single-particle problem is a suitable base for engineering such time lattices. Moreover, resonantly driven two- or three-dimensional systems which are not separable are promising platform for realization of non-separable time lattices.

5.5 Time Engineering: Anderson Molecules

It is possible to observe spontaneous breaking of discrete time translation symmetry in isolated periodically driven systems [58, 98]. It is also possible to observe various condensed matter phases in the time dimension in the driven systems [47, 98]. Time is a degree of freedom which provides a knob that we can use to create interesting phenomena. Does time engineering allow one to realize completely novel objects that have not been attainable in the laboratory so far? It is hard to predict what can come out in the future. Here, we would like to present an example of objects that we call Anderson molecules which can be realized with the help of engineering of a physical system in time [38].

Two or more atoms can form a molecule when they attract each other. Interaction potentials between atoms can support bound states which correspond to various molecular states. We would like to pose a question if two or more atoms can form a bound state not due to attractive interactions between them but because of destructive quantum interference effects. If a pair of atoms interacted via a potential that changes in a disorder way versus the relative distance between the atoms, then one would expect that Anderson localization phenomenon could be responsible for the formation of an exponentially localized bound state of the atoms. However, such a potential does not exist in nature. We are going to show that it can be created by means of time engineering [38].

[23]The canonical transformation between the Cartesian coordinates and the angle-action variables introduces the relations $x = x(I_x, \theta_x)$ and $y = y(I_y, \theta_y)$. Location of a detector close to the resonant orbit means $I_x \approx I_s$ and $I_y \approx I_s$ (where I_s is the resonant value of the actions, see Sect. 5.1) and $\theta_x = \text{const}$ and $\theta_y = \text{const}$. In the Cartesian coordinates it means that $x \approx x(I_s, \theta_x = \text{const})$ and $y \approx y(I_s, \theta_y = \text{const})$.

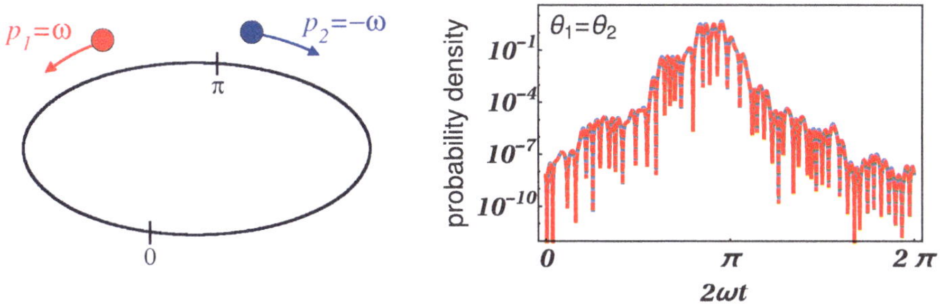

Fig. 5.23 Left panel: two distinguishable atoms moving on a ring in the opposite directions with the momenta $p_{1,2} = \pm\omega$ that fulfill the $1:1$ resonance condition with the time-periodic driving. The frequency ω is the frequency of the repetition of disordered modulations of the interaction potential between atoms, cf. (5.102). Right panel presents an example of an Anderson localized state of the Hamiltonian (5.105) in the laboratory frame. In the laboratory frame a signature of the formation of an Anderson molecule can be visible, e.g. in the time evolution of the probability for detection of the atoms at the same position ($\theta_1 = \theta_2$) which is exponentially localized around a certain moment of time. In the right panel there are two hardly distinguishable curves: one is related to the eigenstate of (5.105) with the energy $E = 616$ and the other to the corresponding exact Floquet state of the full Hamiltonian (5.102). The following parameters are chosen: $k_0 = 100$, $\lambda = 600$, and $\omega = 3 \times 10^5$. Reprinted from [38]

Let us consider two different atoms which are moving on a one-dimensional ring and interact via the zero-range Dirac-delta potential $g_0\delta(\theta_1 - \theta_2)$ where $\theta_{1,2}$ are angles which describe positions of atoms on the ring, see Fig. 5.23. Such a potential models perfectly the interactions between ultra-cold atoms. The strength of the potential is proportional to the atomic s-wave scattering length and it can be changed and controlled by means of the Feshbach resonance which is a standard technique in the ultra-cold atomic gases [84]. Assume that the scattering length is modulated in time, $g_0 = 2\pi\lambda f(t)$, in a disordered way. That is, $f(t)$ fluctuates but the random modulations repeat periodically in time with the period $2\pi/\omega$, i.e.

$$f(t + 2\pi/\omega) = f(t) = \sum_{k=-k_0}^{k_0} f_k e^{ik\omega t}, \tag{5.100}$$

with, for example,

$$f_k = \frac{1}{\sqrt{k_0}} e^{i\varphi_k}, \tag{5.101}$$

where $\varphi_k = -\varphi_{-k}$ are random phases chosen from the uniform distribution in $[0, 2\pi)$. The Hamiltonian of such a pair of atoms reads

$$H = \frac{p_1^2 + p_2^2}{2} + 2\pi\lambda f(t)\,\delta(\theta_1 - \theta_2). \tag{5.102}$$

We begin with the classical description of the system. Suppose that the atoms are moving along the ring in the opposite directions with the momenta that fulfill the 1 : 1 resonance condition with the frequency ω of the repetition of the random behavior of $f(t)$, i.e. $p_1 = \omega$ and $p_2 = -\omega$. Applying the secular approximation approach we are able to derive the effective Hamiltonian of the system valid in the vicinity of the resonant trajectory. To this end we switch to the frame moving with the atoms,

$$\begin{aligned} \Theta_1 &= \theta_1 - \omega t, \qquad & P_1 &= p_1 - \omega, \\ \Theta_2 &= \theta_2 + \omega t, \qquad & P_2 &= p_2 + \omega, \end{aligned} \tag{5.103}$$

where the Hamiltonian takes the form

$$H = \frac{P_1^2 + P_2^2}{2} + \lambda \left(\sum_k f_k e^{ik\omega t} \right) \sum_n e^{in(\Theta_1 - \Theta_2 + 2\omega t)}, \tag{5.104}$$

with a constant term omitted and where we have expanded the interaction contact potential in the series $\delta(\theta_1 - \theta_2) = \sum_n e^{in(\theta_1 - \theta_2)}/(2\pi)$. Assuming that all dynamical variables vary slowly, which is true if we choose initial conditions close to the resonant trajectory ($P_{1,2} \approx 0$), we may average the Hamiltonian (5.104) over time keeping $\Theta_{1,2}$ and $P_{1,2}$ fixed which yields[24]

$$H_{\text{eff}} = \frac{P_1^2 + P_2^2}{2} + V_{\text{eff}}(\Theta_1 - \Theta_2), \tag{5.105}$$

where

$$V_{\text{eff}} = \lambda \sum_k f_{-2k} e^{ik(\Theta_1 - \Theta_2)} = \frac{\lambda}{\sqrt{k_0}} \sum_{\substack{k=-k_0/2 \\ k \neq 0}}^{k_0/2} e^{ik(\Theta_1 - \Theta_2) + i\varphi_k}, \tag{5.106}$$

with a constant term neglected. All used canonical transformations are linear and thus the classical effective Hamiltonian (5.105) is identical to the quantum effective Hamiltonian that can be obtained by means of the quantum secular approximation method [14].

The total momentum of atoms, $P = P_1 + P_2$, is conserved and in the quantum description the Hamiltonian (5.105) can be diagonalized in each subspace of different quantized values of P separately, e.g. for $P = 0$. The relative position

---

[24] In order to have the secular approximation applicable we have to introduce a cut-off n_c in the expansion of the Dirac-delta potential, i.e. $\delta(\theta_1 - \theta_2) \approx \sum_{n \le n_c} e^{in(\theta_1 - \theta_2)}/(2\pi)$, because the exact Dirac-delta potential is an infinite barrier for classical particles which they are not able to overcome. In other words, the exact Dirac-delta potential would never be a small perturbation for classical particles regardless how large momenta $|p_{1,2}| = \omega$ we choose.

degree of freedom, $\Theta_1 - \Theta_2$, experiences the effective potential $V_{\text{eff}}(\Theta_1 - \Theta_2)$ that is a disordered potential characterized by the correlation length $\sqrt{2}/k_0$, zero mean and standard deviation λ. Anderson localization in the relative position of atoms is possible and will be observed if the localization length ξ is much smaller than the length of a ring, i.e. $\xi \ll 2\pi$. If so, we can see that in the moving frame (5.103), two atoms keep together at a certain distance $\Theta_1 - \Theta_2 \approx \Theta_0$ which is exponentially well defined. That is, the probability density,

$$|\psi(\Theta_1, \Theta_2)|^2 \propto e^{-|\Theta_1 - \Theta_2 - \Theta_0|/\xi}, \tag{5.107}$$

shows that two atoms form an exponentially localized bound state. The localization length ξ depends on an eigenenergy of the Hamiltonian (5.105). For similar eigenenergies the localization length ξ is similar but the corresponding eigenstates can have very different localization positions Θ_0. Different Θ_0 means that the bound states of two atoms can have very different sizes. When we excite such a molecule, the localization length ξ increases and the relative distance between atoms becomes worse and worse defined and finally ξ becomes comparable to the length of the ring and signatures of a bound state disappear.

When we return to the laboratory frame we can see that both atoms are moving in the opposite directions but there are spatial and temporal correlations between them. If we ask how the probability for detection of both atoms at the same position, $\theta_1 = \theta_2$, changes in time, it turns out that this probability is exponentially localized around a certain moment of time, see Fig. 5.23.

Anderson localization is a result of destructive interference between different scattering paths [78]. Thus, Anderson molecules we have just described correspond to formation of bound states of atoms not due to attractive interactions between them but because of the destructive interference. It occurs when translational motion of atoms is resonantly coupled to periodic repetition of random modulations of the interaction potential. Experimental realization of Anderson molecules seems attainable in ultra-cold atoms prepared, e.g. in a toroidal trap.

References

1. Abrahams, E., Anderson, P.W., Licciardello, D.C., Ramakrishnan, T.V.: Scaling theory of localization: absence of quantum diffusion in two dimensions. Phys. Rev. Lett. **42**(10), 673–676 (1979). https://doi.org/10.1103/PhysRevLett.42.673. http://link.aps.org/abstract/PRL/v42/p673
2. Agarwal, K., Gopalakrishnan, S., Knap, M., Müller, M., Demler, E.: Anomalous diffusion and Griffiths effects near the many-body localization transition. Phys. Rev. Lett. **114**, 160401 (2015). http://link.aps.org/doi/10.1103/PhysRevLett.114.160401
3. Altman, E., Vosk, R.: Universal dynamics and renormalization in many-body-localized systems. Annu. Rev. Condens. Matter Phys. **6**, 383 (2015). http://dx.doi.org/10.1146/annurev-conmatphys-031214-014701
4. Anderson, P.W.: Absence of diffusion in certain random lattices. Phys. Rev. **109**, 1492–1505 (1958). http://link.aps.org/doi/10.1103/PhysRev.109.1492

5. Anisimovas, E., Žlabys, G., Anderson, B.M., Juzeliūnas, G., Eckardt, A.: Role of real-space micromotion for bosonic and fermionic Floquet fractional Chern insulators. Phys. Rev. B **91**, 245135 (2015). https://link.aps.org/doi/10.1103/PhysRevB.91.245135
6. Asbóth, J., Oroszlány, L., Pályi, A.: A Short Course on Topological Insulators: Band Structure and Edge States in One and Two Dimensions. Lecture Notes in Physics. Springer International Publishing, Cham (2016). https://books.google.pl/books?id=RWKhCwAAQBAJ
7. Bar Lev, Y., Reichman, D.R.: Dynamics of many-body localization. Phys. Rev. B **89**, 220201 (2014). http://link.aps.org/doi/10.1103/PhysRevB.89.220201
8. Bar Lev, Y., Cohen, G., Reichman, D.R.: Absence of diffusion in an interacting system of spinless fermions on a one-dimensional disordered lattice. Phys. Rev. Lett. **114**, 100601 (2015). http://link.aps.org/doi/10.1103/PhysRevLett.114.100601
9. Bardarson, J.H., Pollmann, F., Moore, J.E.: Unbounded growth of entanglement in models of many-body localization. Phys. Rev. Lett. **109**, 017202 (2012). http://dx.doi.org/10.1103/PhysRevLett.109.017202
10. Barišić, O.S., Prelovšek, P.: Conductivity in a disordered one-dimensional system of interacting fermions. Phys. Rev. B **82**, 161106 (2010). http://link.aps.org/doi/10.1103/PhysRevB.82.161106
11. Barišić, O.S., Kokalj, J., Balog, I., Prelovšek, P.: Dynamical conductivity and its fluctuations along the crossover to many-body localization. Phys. Rev. B **94**, 045126 (2016). http://link.aps.org/doi/10.1103/PhysRevB.94.045126
12. Basko, D., Aleiner, I., Altschuler, B.: Metal-insulator transition in a weakly interacting many-electron system with localized single-particle states. Ann. Phys. **321**, 1126 (2006)
13. Berkelbach, T.C., Reichman, D.R.: Conductivity of disordered quantum lattice models at infinite temperature: many-body localization. Phys. Rev. B **81**, 224429 (2010). http://link.aps.org/doi/10.1103/PhysRevB.81.224429
14. Berman, G., Zaslavsky, G.: Theory of quantum nonlinear resonance. Phys. Lett. A **61**(5), 295–296 (1977). https://doi.org/10.1016/0375-9601(77)90618-1. http://www.sciencedirect.com/science/article/pii/0375960177906181
15. Biedroń, K., Dutta, O., Zakrzewski, J.: Topological Rice-Mele model in an emergent lattice: exact diagonalization approach. Phys. Rev. A **93**, 033631 (2016). https://link.aps.org/doi/10.1103/PhysRevA.93.033631
16. Biedroń, K., Łacki, M., Zakrzewski, J.: Extended Bose-Hubbard model with dipolar and contact interactions. Phys. Rev. B **97**, 245102 (2018). https://link.aps.org/doi/10.1103/PhysRevB.97.245102
17. Bomantara, R.W., Gong, J.: Simulation of non-Abelian Braiding in Majorana time crystals. Phys. Rev. Lett. **120**, 230405 (2018). https://link.aps.org/doi/10.1103/PhysRevLett.120.230405
18. Cardano, F., D'Errico, A., Dauphin, A., Maffei, M., Piccirillo, B., de Lisio, C., De Filippis, G., Cataudella, V., Santamato, E., Marrucci, L., Lewenstein, M., Massignan, P.: Detection of Zak phases and topological invariants in a chiral quantum walk of twisted photons. Nat. Commun. **8**, 15516 (2017). http://dx.doi.org/10.1038/ncomms1551610.1038/ncomms15516. https://www.nature.com/articles/ncomms15516#supplementary-information
19. Casati, G., Chirikov, B.V., Izraelev, F.M., Ford, J.: Stochastic behavior of a quantum pendulum under a periodic perturbation. In: G. Casati, J. Ford (eds.) Stochastic Behavior in Classical and Quantum Hamiltonian Systems, pp. 334–352. Springer, Berlin (1979)
20. Chabé, J., Lemarié, G., Grémaud, B., Delande, D., Szriftgiser, P., Garreau, J.C.: Experimental observation of the Anderson metal-insulator transition with atomic matter waves. Phys. Rev. Lett. **101**, 255702 (2008). https://link.aps.org/doi/10.1103/PhysRevLett.101.255702
21. Chandran, A., Kim, I.H., Vidal, G., Abanin, D.A.: Constructing local integrals of motion in the many-body localized phase. Phys. Rev. B **91**, 085425 (2015). http://link.aps.org/doi/10.1103/PhysRevB.91.085425
22. Cooper, N.R., Dalibard, J., Spielman, I.B.: Topological bands for ultracold atoms. Rev. Mod. Phys. **91**, 015005 (2019). https://link.aps.org/doi/10.1103/RevModPhys.91.015005

23. D'Alessio, L., Kafri, Y., Polkovnikov, A., Rigol, M.: From quantum chaos and eigenstate thermalization to statistical mechanics and thermodynamics. Adv. Phys. **65**(3), 239–362 (2016). https://doi.org/10.1080/00018732.2016.1198134
24. Dalla Torre, E.G., Berg, E., Altman, E.: Hidden order in 1D Bose insulators. Phys. Rev. Lett. **97**, 260401 (2006). https://link.aps.org/doi/10.1103/PhysRevLett.97.260401
25. De Luca, A., Scardicchio, A.: Ergodicity breaking in a model showing many-body localization. Europhys. Lett. **101**, 37003 (2013). http://stacks.iop.org/0295-5075/101/i=3/a=37003
26. De Luca, A., Altshuler, B.L., Kravtsov, V.E., Scardicchio, A.: Anderson localization on the Bethe lattice: nonergodicity of extended states. Phys. Rev. Lett. **113**, 046806 (2014). http://link.aps.org/doi/10.1103/PhysRevLett.113.046806
27. Delande, D., Orso, G.: Mobility edge for cold atoms in laser speckle potentials. Phys. Rev. Lett. **113**, 060601 (2014). https://link.aps.org/doi/10.1103/PhysRevLett.113.060601
28. Delande, D., Morales-Molina, L., Sacha, K.: Three-dimensional localized-delocalized Anderson transition in the time domain. Phys. Rev. Lett. **119**, 230404 (2017). https://link.aps.org/doi/10.1103/PhysRevLett.119.230404
29. Deutsch, J.M.: Quantum statistical mechanics in a closed system. Phys. Rev. A **43**, 2046–2049 (1991). https://link.aps.org/doi/10.1103/PhysRevA.43.2046
30. Dutta, O., Gajda, M., Hauke, P., Lewenstein, M., Lühmann, D.-S., Malomed, B.A., Sowiński, T., Zakrzewski, J.: Non-standard Hubbard models in optical lattices: a review. Rep. Prog. Phys. **78**(6), 066001 (2015). http://stacks.iop.org/0034-4885/78/i=6/a=066001
31. Eisert, J., Friesdorf, M., Gogolin, C.: Quantum many-body systems out of equilibrium. Nat. Phys. **11**(2), 124–130 (2015). http://dx.doi.org/10.1038/nphys3215
32. Ejima, S., Fehske, H.: Entanglement properties of the nontrivial Haldane insulator in the 1D extended Bose-Hubbard model. J. Phys. Conf. Ser. **592**(1), 012134 (2015). http://stacks.iop.org/1742-6596/592/i=1/a=012134
33. Else, D.V., Nayak, C.: Classification of topological phases in periodically driven interacting systems. Phys. Rev. B **93**, 201103 (2016)
34. Fishman, S., Grempel, D.R., Prange, R.E.: Chaos, quantum recurrences, and Anderson localization. Phys. Rev. Lett. **49**(8), 509–512 (1982). http://link.aps.org/doi/10.1103/PhysRevLett.49.509
35. Garreau, J.C.: Quantum simulation of disordered systems with cold atoms. C. R. Phys. **18**(1), 31–46 (2017). https://doi.org/10.1016/j.crhy.2016.09.002. http://www.sciencedirect.com/science/article/pii/S1631070516301426. Prizes of the French Academy of Sciences 2015/Prix de l'Académie des sciences 2015
36. Giergiel, K., Sacha, K.: Anderson localization of a Rydberg electron along a classical orbit. Phys. Rev. A **95**, 063402 (2017). https://link.aps.org/doi/10.1103/PhysRevA.95.063402
37. Giergiel, K., Kosior, A., Hannaford, P., Sacha, K.: Time crystals: analysis of experimental conditions. Phys. Rev. A **98**, 013613 (2018). https://link.aps.org/doi/10.1103/PhysRevA.98.013613
38. Giergiel, K., Miroszewski, A., Sacha, K.: Time crystal platform: from quasicrystal structures in time to systems with exotic interactions. Phys. Rev. Lett. **120**, 140401 (2018). https://link.aps.org/doi/10.1103/PhysRevLett.120.140401
39. Giergiel, K., Dauphin, A., Lewenstein, M., Zakrzewski, J., Sacha, K.: Topological time crystals. New J. Phys. **21**(5), 052003 (2019). https://doi.org/10.1088/1367-2630/ab1e5f
40. Goldman, N., Juzeliunas, G., Öhberg, P., Spielman, I.B.: Light-induced gauge fields for ultracold atoms. Rep. Prog. Phys. **77**(12), 126401 (2014). http://stacks.iop.org/0034-4885/77/i=12/a=126401
41. Gornyi, I.V., Mirlin, A.D., Polyakov, D.G.: Interacting electrons in disordered wires: Anderson localization and low-t transport. Phys. Rev. Lett. **95**, 206603 (2005). http://link.aps.org/doi/10.1103/PhysRevLett.95.206603
42. Gramsch, C., Rigol, M.: Quenches in a quasidisordered integrable lattice system: dynamics and statistical description of observables after relaxation. Phys. Rev. A **86**, 053615 (2012). http://link.aps.org/doi/10.1103/PhysRevA.86.053615

43. Grempel, D.R., Prange, R.E., Fishman, S.: Quantum dynamics of a nonintegrable system. Phys. Rev. A **29**, 1639–1647 (1984). https://link.aps.org/doi/10.1103/PhysRevA.29.1639
44. Guo, L., Marthaler, M.: Synthesizing lattice structures in phase space. New J. Phys. **18**(2), 023006 (2016). https://doi.org/10.1088/1367-2630/18/2/023006
45. Guo, L., Marthaler, M., Schön, G.: Phase space crystals: a new way to create a quasienergy band structure. Phys. Rev. Lett. **111**, 205303 (2013). https://link.aps.org/doi/10.1103/PhysRevLett.111.205303
46. Guo, L., Liu, M., Marthaler, M.: Effective long-distance interaction from short-distance interaction in a periodically driven one-dimensional classical system. Phys. Rev. A **93**, 053616 (2016). https://link.aps.org/doi/10.1103/PhysRevA.93.053616
47. Guo, L., Liang, P.: Condensed matter physics in time crystals. New Journal of Physics. IOP Publishing. **22**(7), 075003 (2020). https://doi.org/10.1088/1367-2630/ab9d54
48. Haake, F.: Quantum Signatures of Chaos. Physics and Astronomy Online Library. Springer, Berlin (2001). https://books.google.pl/books?id=Orv0BXoorFEC
49. Haldane, F.D.M.: Continuum dynamics of the 1-D Heisenberg antiferromagnet: identification with the o(3) nonlinear sigma model. Phys. Lett. A **93**, 464 (1983)
50. Haldane, F.D.M.: Nonlinear field theory of large-spin Heisenberg antiferromagnets: semiclassically quantized solitons of the one-dimensional easy-Axis Néel state. Phys. Rev. Lett. **50**, 1153–1156 (1983). https://link.aps.org/doi/10.1103/PhysRevLett.50.1153
51. Hannaford, P., Sacha, K.: Time crystals enter the real world of condensed matter. Phys. World **33**, 42 (2020)
52. Hasan, M.Z., Kane, C.L.: Colloquium. Rev. Mod. Phys. **82**, 3045–3067 (2010). https://link.aps.org/doi/10.1103/RevModPhys.82.3045
53. Huse, D.A., Nandkishore, R., Oganesyan, V., Pal, A., Sondhi, S.L.: Localization-protected quantum order. Phys. Rev. B **88**, 014206 (2013). http://link.aps.org/doi/10.1103/PhysRevB.88.014206
54. Huse, D.A., Nandkishore, R., Oganesyan, V.: Phenomenology of fully many-body-localized systems. Phys. Rev. B **90**, 174202 (2014). http://link.aps.org/doi/10.1103/PhysRevB.90.174202
55. Janot, C.: Quasicrystals: A Primer. Oxford University Press, Oxford (1994)
56. Kennedy, T., Tasaki, H.: Hidden $z_2 \times z_2$ symmetry breaking in Haldane-gap antiferromagnets. Phys. Rev. B **45**, 304–307 (1992). https://link.aps.org/doi/10.1103/PhysRevB.45.304
57. Khemani, V., Nandkishore, R., Sondhi, S.L.: Nonlocal adiabatic response of a localized system to local manipulations. Nat. Phys. **11**, 560 (2015). http://dx.doi.org/10.1038/nphys3344
58. Khemani, V., Moessner, R., Sondhi, S.L.: A brief history of time crystals. arXiv e-prints, arXiv:1910.10745 (2019)
59. Kitagawa, T., Berg, E., Rudner, M., Demler, E.: Topological characterization of periodically driven quantum systems. Phys. Rev. B **82**, 235114 (2010). https://link.aps.org/doi/10.1103/PhysRevB.82.235114
60. Kjäll, J.A., Bardarson, J.H., Pollmann, F.: Many-body localization in a disordered quantum Ising chain. Phys. Rev. Lett. **113**, 107204 (2014). http://dx.doi.org/10.1103/PhysRevLett.113.107204
61. Kozarzewski, M., Prelovšek, P., Mierzejewski, M.: Distinctive response of many-body localized systems to a strong electric field. Phys. Rev. B **93**, 235151 (2016). http://link.aps.org/doi/10.1103/PhysRevB.93.235151
62. Kuhn, R.C., Sigwarth, O., Miniatura, C., Delande, D., Müller, C.A.: Coherent matter wave transport in speckle potentials. New J. Phys. **9**(6), 161 (2007). http://stacks.iop.org/1367-2630/9/i=6/a=161
63. Lemarié, G., Chabé, J., Szriftgiser, P., Garreau, J.C., Grémaud, B., Delande, D.: Observation of the Anderson metal-insulator transition with atomic matter waves: theory and experiment. Phys. Rev. A **80**, 043626 (2009). http://link.aps.org/doi/10.1103/PhysRevA.80.043626
64. Lichtenberg, A., Lieberman, M.: Regular and Chaotic Dynamics. Applied Mathematical Sciences. Springer, Berlin (1992). https://books.google.pl/books?id=2ssPAQAAMAAJ

65. Luitz, D.J., Laflorencie, N., Alet, F.: Extended slow dynamical regime prefiguring the many-body localization transition. Phys. Rev. B **93**, 060201 (2015). http://dx.doi.org/10.1103/PhysRevB.93.060201
66. Lustig, E., Sharabi, Y., Segev, M.: Topological aspects of photonic time crystals. Optica **5**(11), 1390–1395 (2018). https://doi.org/10.1364/OPTICA.5.001390. http://www.osapublishing.org/optica/abstract.cfm?URI=optica-5-11-1390
67. Maccone, L., Sacha, K.: Quantum measurements of time. Phys. Rev. Lett. **124**, 110402 (2020). https://link.aps.org/doi/10.1103/PhysRevLett.124.110402
68. MacKinnon, A., Kramer, B.: The scaling theory of electrons in disordered solids: additional numerical results. Z. Physik B Condens. Matter **53**(1), 1–13 (1983). https://doi.org/10.1007/BF01578242
69. Maeda, H., Gallagher, T.F.: Nondispersing wave packets. Phys. Rev. Lett. **92**, 133004 (2004). http://link.aps.org/doi/10.1103/PhysRevLett.92.133004
70. Maeda, H., Gallagher, T.F.: Nondispersing wave packets. Phys. Rev. A **75**, 033410 (2007). https://link.aps.org/doi/10.1103/PhysRevA.75.033410
71. Maeda, H., Gurian, J.H., Gallagher, T.F.: Nondispersing Bohr wave packets. Phys. Rev. Lett. **102**, 103001 (2009). https://link.aps.org/doi/10.1103/PhysRevLett.102.103001
72. Maffei, M., Dauphin, A., Cardano, F., Lewenstein, M., Massignan, P.: Topological characterization of chiral models through their long time dynamics. New J. Phys. **20**(1), 013023 (2018). http://stacks.iop.org/1367-2630/20/i=1/a=013023
73. Meier, E.J., An, F.A., Dauphin, A., Maffei, M., Massignan, P., Hughes, T.L., Gadway, B.: Observation of the topological Anderson insulator in disordered atomic wires. Science **362**(6417), 929–933 (2018). https://doi.org/10.1126/science.aat3406. https://science.sciencemag.org/content/362/6417/929
74. Mierzejewski, M., Herbrych, J., Prelovšek, P.: Universal dynamics of density correlations at the transition to many-body localized state. Phys. Rev. B **94**, 224207 (2016). https://doi.org/10.1103/PhysRevB.94.224207. http://arxiv.org/abs/1607.04992
75. Mierzejewski, M., Giergiel, K., Sacha, K.: Many-body localization caused by temporal disorder. Phys. Rev. B **96**, 140201 (2017). https://link.aps.org/doi/10.1103/PhysRevB.96.140201
76. Mondaini, R., Rigol, M.: Many-body localization and thermalization in disordered Hubbard chains. Phys. Rev. A **92**, 041601(R) (2015). http://dx.doi.org/10.1103/PhysRevA.92.041601
77. Monthus, C., Garel, T.: Many-body localization transition in a lattice model of interacting fermions: statistics of renormalized hoppings in configuration space. Phys. Rev. B **81**, 134202 (2010). http://link.aps.org/doi/10.1103/PhysRevB.81.134202
78. Müller, C.A., Delande, D.: Disorder and Interference: Localization Phenomena, vol. 91, chap. 9. Oxford Scholarship (2011). https://doi.org/10.1093/acprof:oso/9780199603657.003.0009
79. Nandkishore, R., Huse, D.A.: Many-body-localization and thermalization in quantum statistical mechanics. Annu. Rev. Condens. Matter Phys. **6**, 15 (2015)
80. Oganesyan, V., Huse, D.A.: Localization of interacting fermions at high temperature. Phys. Rev. B **75**, 155111 (2007). http://link.aps.org/doi/10.1103/PhysRevB.75.155111
81. Ozawa, T., Price, H.M., Amo, A., Goldman, N., Hafezi, M., Lu, L., Rechtsman, M.C., Schuster, D., Simon, J., Zilberberg, O., Carusotto, I.: Topological photonics. Rev. Mod. Phys. **91**, 015006 (2019). https://link.aps.org/doi/10.1103/RevModPhys.91.015006
82. Pal, A., Huse, D.A.: Many-body localization phase transition. Phys. Rev. B **82**, 174411 (2010). http://dx.doi.org/10.1103/PhysRevB.82.174411
83. Pengfei, L., Michael, M., Guo, L.: Floquet many-body engineering: topology and many-body physics in phase space lattices. New J. Phys. **20**(2), 023043 (2018). http://stacks.iop.org/1367-2630/20/i=2/a=023043
84. Pethick, C., Smith, H.: Bose-Einstein Condensation in Dilute Gases. Cambridge University Press, Cambridge (2002)
85. Pichard, J.L., Sarma, G.: Finite size scaling approach to Anderson localisation. J. Phys. C Solid State Phys. **14**(6), L127–L132 (1981). https://doi.org/10.1088/0022-3719/14/6/003

86. Potter, A.C., Morimoto, T., Vishwanath, A.: Classification of interacting topological Floquet phases in one dimension. Phys. Rev. X **6**, 041001 (2016). http://link.aps.org/doi/10.1103/PhysRevX.6.041001
87. Prelovšek, P., Herbrych, J.: Self-consistent approach to many-body localization and subdiffusion. Phys. Rev. B **96**(3), 035130 (2017). https://doi.org/10.1103/PhysRevB.96.035130
88. Press, W.H., Teukolsky, S.A., Vetterling, W.T., Flannery, B.P.: Numerical Recipes: The Art of Scientific Computing, 3 edn. Cambridge University Press, Cambridge (2007)
89. Przysiężna, A., Dutta, O., Zakrzewski, J.: Rice–Mele model with topological solitons in an optical lattice. New J. Phys. **17**(1), 013018 (2015). http://stacks.iop.org/1367-2630/17/i=1/a=013018
90. Rademaker, L., Ortuño, M.: Explicit local integrals of motion for the many-body localized state. Phys. Rev. Lett. **116**, 010404 (2016). http://link.aps.org/doi/10.1103/PhysRevLett.116.010404
91. Rigol, M., Dunjko, V., Olshanii, M.: Thermalization and its mechanism for generic isolated quantum systems. Nature **452**, 854–858 (2008). https://doi.org/10.1038/nature06838
92. Ros, V., Müller, M., Scardicchio, A.: Integrals of motion in the many-body localized phase. Nucl. Phys. B **891**, 420 (2015). http://dx.doi.org/10.1016/j.nuclphysb.2014.12.014. http://www.sciencedirect.com/science/article/pii/S0550321314003836
93. Rossini, D., Fazio, R.: Phase diagram of the extended Bose Hubbard model. New J. Phys. **14**(6), 065012 (2012). http://stacks.iop.org/1367-2630/14/i=6/a=065012
94. Roy, R., Harper, F.: Abelian Floquet symmetry-protected topological phases in one dimension. Phys. Rev. B **94**, 125105 (2016). http://link.aps.org/doi/10.1103/PhysRevB.94.125105
95. Rudner, M.S., Lindner, N.H.: The Floquet Engineer's Handbook. arXiv e-prints, arXiv:2003.08252 (2020)
96. Sacha, K.: Anderson localization and Mott insulator phase in the time domain. Sci. Rep. **5**, 10787 (2015). http://dx.doi.org/10.1038/srep1078710.1038/srep10787. https://www.nature.com/articles/srep10787
97. Sacha, K., Delande, D.: Anderson localization in the time domain. Phys. Rev. A **94**, 023633 (2016). http://link.aps.org/doi/10.1103/PhysRevA.94.023633
98. Sacha, K., Zakrzewski, J.: Time crystals: a review. Rep. Prog. Phys. **81**(1), 016401 (2017). https://doi.org/10.1088/1361-6633/aa8b38
99. Sacha, K., Zakrzewski, J., Delande, D.: Controlling nonspreading wavepackets. Eur. Phys. J. D **1**(3), 231–234 (1998). https://doi.org/10.1007/s100530050086
100. Schreiber, M., Hodgman, S.S., Bordia, P., Lüschen, H.P., Fischer, M.H., Vosk, R., Altman, E., Schneider, U., Bloch, I.: Observation of many-body localization of interacting fermions in a quasi-random optical lattice. Science **349**, 842 (2015). http://dx.doi.org/10.1126/science.aaa7432
101. Senthil, T.: Symmetry-protected topological phases of quantum matter. Annu. Rev. Condens. Matter Phys. **6**(1), 299 (2015). https://doi.org/10.1146/annurev-conmatphys-031214-014740
102. Serbyn, M., Papić, Z., Abanin, D.A.: Local conservation laws and the structure of the many-body localized states. Phys. Rev. Lett. **111**, 127201 (2013). https://link.aps.org/doi/10.1103/PhysRevLett.111.127201
103. Serbyn, M., Knap, M., Gopalakrishnan, S., Papić, Z., Yao, N.Y., Laumann, C.R., Abanin, D.A., Lukin, M.D., Demler, E.A.: Interferometric probes of many-body localization. Phys. Rev. Lett. **113**, 147204 (2014). http://link.aps.org/doi/10.1103/PhysRevLett.113.147204
104. Serbyn, M., Papić, Z., Abanin, D.A.: Criterion for many-body localization-delocalization phase transition. Phys. Rev. X **5**, 041047 (2015). http://link.aps.org/doi/10.1103/PhysRevX.5.041047
105. Sierant, P., Delande, D., Zakrzewski, J.: Many-body localization due to random interactions. Phys. Rev. A **95**, 021601 (2017). https://link.aps.org/doi/10.1103/PhysRevA.95.021601
106. Slevin, K., Ohtsuki, T.: Critical exponent for the Anderson transition in the three-dimensional orthogonal universality class. New J. Phys. **16**(1), 015012 (2014). https://doi.org/10.1088/1367-2630/16/1/015012

107. Srednicki, M.: Chaos and quantum thermalization. Phys. Rev. E **50**, 888–901 (1994). http://link.aps.org/doi/10.1103/PhysRevE.50.888
108. Steinigeweg, R., Herbrych, J., Pollmann, F., Brenig, W.: Scaling of the optical conductivity in the transition from thermal to many-body localized phases. Phys. Rev. B **94**, 180401(R) (2016). https://doi.org/10.1103/PhysRevB.94.180401
109. Su, W.P., Schrieffer, J.R., Heeger, A.J.: Solitons in polyacetylene. Phys. Rev. Lett. **42**, 1698–1701 (1979). https://link.aps.org/doi/10.1103/PhysRevLett.42.1698
110. von Keyserlingk, C.W., Sondhi, S.L.: Phase structure of one-dimensional interacting Floquet systems. I. Abelian symmetry-protected topological phases. Phys. Rev. B **93**, 245145 (2016). http://link.aps.org/doi/10.1103/PhysRevB.93.245145
111. Wyker, B., Ye, S., Dunning, F.B., Yoshida, S., Reinhold, C.O., Burgdörfer, J.: Creating and transporting Trojan wave packets. Phys. Rev. Lett. **108**, 043001 (2012). https://link.aps.org/doi/10.1103/PhysRevLett.108.043001
112. Xu, S., Wu, C.: Space-time crystal and space-time group. Phys. Rev. Lett. **120**, 096401 (2018). https://link.aps.org/doi/10.1103/PhysRevLett.120.096401
113. Yablonovitch, E.: Photonic band-gap crystals. J. Phys. Condens. Matter **5**(16), 2443–2460 (1993). https://doi.org/10.1088/0953-8984/5/16/004
114. Žnidarič, M., Prosen, T., Prelovšek, P.: Many-body localization in the Heisenberg xxz magnet in a random field. Phys. Rev. B **77**, 064426 (2008). http://link.aps.org/doi/10.1103/PhysRevB.77.064426
115. Zwerger, W.: Mott-Hubbard transition of cold atoms in optical lattices. J. Opt. B Quantum Semiclassical Opt. **5**(2), S9 (2003). http://stacks.iop.org/1464-4266/5/i=2/a=352

Chapter 6
Phase Space Crystals

Abstract A particle trapped in a harmonic potential and resonantly driven by a time-periodic perturbation can possess a band structure in the quasi-energy spectrum. The corresponding Floquet states reveal crystalline structures in the phase space. If a particle is appropriately driven, such a phase space crystal can possess properties of topological insulators. In the many-body case the effective interactions between particles in the phase space are long range despite the fact that the original interactions between particles in the configuration space are short range.

6.1 Idea of Phase Space Crystals

Guo et al. showed that if a particle trapped in the harmonic potential is resonantly driven, energy bands form in the quasi-energy spectrum of the system which is accompanied by the presence of a crystalline structure in the phase space [6]. It takes place despite the fact that the Hamiltonian of the driven system does not possess any potential energy that is spatially periodic. The work [6] by Guo and coworkers was the first publication where it was shown that the resonantly driven single-particle system can reveal a quasi-energy band structure known from the condensed matter physics. In contrast to what we have described in Chap. 5, the authors focus on the formation of a crystalline structure not in the time dimension but in the phase space. In the subsequent publications, the concept of the phase space crystals was developed and extended to many-body resonantly driven systems [5, 7, 8], (see also a review article [9]).

Let us consider the following single-particle quantum Hamiltonian:

$$H = \frac{p^2}{2} + \frac{\Omega^2 x^2}{2} + \frac{\nu}{2}x^4 + 2V_0 x^s \cos\omega t, \tag{6.1}$$

where the third and fourth terms are considered as weak perturbations of a particle moving in the harmonic potential [6]. The frequency of the time-periodic perturbation is assumed to fulfill $\omega \approx s\Omega$ where s is a large integer number. In other words the $s : 1$ resonance condition is satisfied. When we switch to the moving

K. Sacha, *Time Crystals*, Springer Series on Atomic, Optical, and Plasma Physics 114, https://doi.org/10.1007/978-3-030-52523-1_6

frame by means of the time-dependent unitary transformation $U = e^{i\omega t\hat{a}^\dagger\hat{a}/s}$, where $\hat{a} = (x + ip)/\sqrt{2}$, and drop all terms in the Hamiltonian which oscillate in time, we arrive at the effective Hamiltonian

$$H_{\mathrm{RWA}} = \left(\Omega - \frac{\omega}{s}\right)\hat{a}^\dagger\hat{a} + \frac{3\nu}{4\Omega^2}\hat{a}^\dagger\hat{a}(\hat{a}^\dagger\hat{a} + 1) + \frac{V_0}{(2\Omega)^{s/2}}(\hat{a}^{\dagger s} + \hat{a}^s). \quad (6.2)$$

The approximation we have just applied is the so-called rotating wave approximation (RWA) [6]. The Hamiltonian (6.2) possesses a symmetry which is not displayed by the original Hamiltonian, i.e. it commutes with the unitary operator $e^{-i2\pi\hat{a}^\dagger\hat{a}/s}$. As we will see this symmetry implies that eigenstates of H_{RWA} reveal a crystalline structure in the phase space. In order to show it let us define the radial $\hat{r}$ and the angular $\hat{\theta}$ operators,

$$\hat{a} = \frac{e^{-i\hat{\theta}}\hat{r}}{\sqrt{2\lambda}}, \qquad \hat{a}^\dagger = \frac{\hat{r}e^{i\hat{\theta}}}{\sqrt{2\lambda}}, \quad (6.3)$$

which fulfill the commutation relation $[\hat{r}^2, e^{i\hat{\theta}}] = 2\lambda e^{i\hat{\theta}}$ where $\lambda = 3\nu/(4\Omega^2|\omega/s - \Omega|)$ can be considered as an effective Planck constant. In terms of these operators, the effective Hamiltonian (6.2) takes the form

$$H_{\mathrm{RWA}} = \frac{1}{\lambda}\left|\frac{\omega}{s} - \Omega\right| g(\hat{\theta}, \hat{r}), \quad (6.4)$$

where

$$g = \frac{1}{4}(\hat{r}^2 + \lambda - 1)^2 + \frac{\mu}{2}\left[\left(\hat{r}e^{i\hat{\theta}}\right)^s + \left(\hat{r}e^{-i\hat{\theta}}\right)^s\right], \quad (6.5)$$

$$\mu = \frac{2\lambda V_0}{|\omega/s - \Omega|}\left(\frac{\Omega|\omega/s - \Omega|}{3\nu}\right)^{s/2}. \quad (6.6)$$

The symmetry of the Hamiltonian H_{RWA} is related to the fact that $g(\hat{\theta}, \hat{r})$ does not change under the rotation $\hat{\theta} \to \hat{\theta} + 2\pi/s$. Eigenstates of H_{RWA}, in the angular representation,

$$g(\hat{\theta}, \hat{r})\, \psi_m(\theta) = g(m)\, \psi_m(\theta), \quad (6.7)$$

are Bloch waves, $\psi_m(\theta) = u_m(\theta)e^{im\theta}$ where $u_m(\theta + 2\pi/s) = u_m(\theta)$, and reveal crystalline structures in the phase space. The corresponding energy levels form energy bands. These features are illustrated in Fig. 6.1 where in order to visualize a

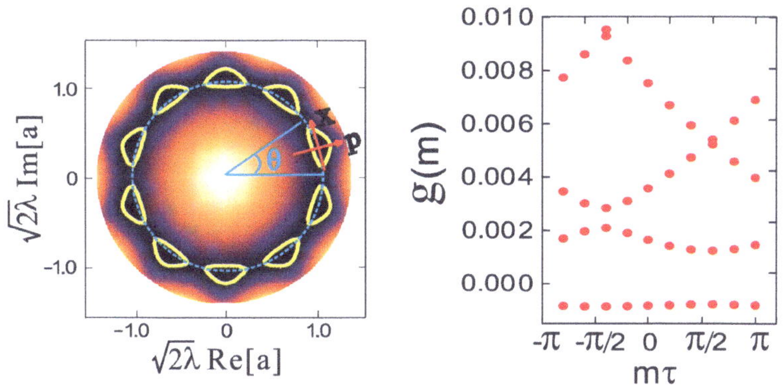

Fig. 6.1 Left panel illustrates a crystalline structure in the phase space. That is, the average value $\langle\alpha|H_{\rm RWA}|\alpha\rangle$ is presented where $\alpha = {\rm Re}(a) + i{\rm Im}(a)$ and $|\alpha\rangle$'s are the coherent states of the harmonic oscillator, $\hat{a}|\alpha\rangle = \alpha|\alpha\rangle$, for $s = 10$ and $\mu = 4.2 \cdot 10^{-3}$. Periodic structure visible in this panel is a picture of a phase space crystal [6]. Right panel shows eigenvalues $g(m)$, cf. (6.7), that form a band structure (presented in the reduced Brillouin zone, i.e. $m\tau = m2\pi/s \in (-\pi, \pi])$ for $s = 10$, $\mu = 3.2 \cdot 10^{-3}$ and $\lambda = 1/205$. Figure adapted with permission from Guo et al. [6]

crystalline structure in the phase space the coherent state representation is used.[1] That is, the average value of the Hamiltonian (6.2), $\langle\alpha|H_{\rm RWA}|\alpha\rangle$, is presented where $\alpha = {\rm Re}(a) + i{\rm Im}(a)$. It corresponds to the substitution of the annihilation and creation operators $\hat{a}$ and $\hat{a}^\dagger$ by complex numbers α and α^*, respectively. The eigenvalues $g(m)$ of $H_{\rm RWA}$, which are actually quasi-energies of the driven system (6.1), form separated groups consisting s energy levels which in the $s \to \infty$ limit form energy bands.

The range of the validity of the effective Hamiltonian $H_{\rm RWA}$ can be estimated if we require that energy associated with (6.4) is smaller than the lowest oscillating frequency of the neglected oscillating terms of the full Hamiltonian. The former corresponds to the coefficient in front of $g(\hat{\theta}, \hat{r})$ in (6.4) because $g(\hat{\theta}, \hat{r})$ is a dimensionless operator while the latter to $2\omega/s$. The resulting condition for the rotating wave approximation to be valid reads

$$\Omega\left(\frac{\omega}{s} - \Omega\right)^2 < \frac{3}{2}\nu, \tag{6.8}$$

and it has been confirmed by the numerical simulations presented in Ref. [6].

[1]Coherent states are defined as eigenstates of the annihilation operator $\hat{a}|\alpha\rangle = \alpha|\alpha\rangle$ where α is an arbitrary complex number. The coherent states $|\alpha\rangle = e^{-|\alpha|^2/2} \sum_{n=0}^{\infty} \alpha^n/\sqrt{n!}|n\rangle$, where $|n\rangle$ are the harmonic oscillator eigenstates, are normalized but form a non-orthogonal and overcomplete basis [3].

6.2 Topological Phase Space Crystals

In the previous section we have seen that a crystalline structure emerges in the phase space if a particle moving in a harmonic potential is periodically driven [6]. In the Hamiltonian (6.1), the time-periodic perturbation corresponds to a power-law potential, x^s, which can be difficult to realize in the laboratory. In ultra-cold atomic gases it is quite easy to perturb trapped atoms by means of an optical standing wave that is periodically modulated in time. It turns out that such a perturbation leads also to the formation of phase space crystals [5, 7, 8]. In the present section we focus on the case when a periodic potential in space (like the potential for atoms created by an optical standing wave) is periodically turned on for very short periods of time [8]. Then, the single-particle quantum Hamiltonian that we want to analyze takes the form of the kicked harmonic oscillator [1, 2],

$$H = \frac{p^2}{2} + \frac{x^2}{2} + KT\cos x \sum_{n=-\infty}^{\infty} \delta(t - nT), \tag{6.9}$$

where T is the period of the kicking and K determines its strength. The position and momentum operators fulfill $[x, p] = i\lambda$ where λ is the effective Planck constant.[2] The period of the harmonic oscillator in (6.9) is 2π and we assume that the period of the kicking T is s times shorter than the harmonic oscillator period, i.e. $T = 2\pi/s$ where s is an integer number. It means we deal with the $s:1$ resonance. Applying the rotating wave approximation, i.e. switching to the moving frame by means of the time-dependent unitary transformation, $U = e^{i\hat{a}^\dagger \hat{a} t/\lambda}$ where $\hat{a} = (x + ip)/\sqrt{2\lambda}$, and next neglecting all oscillating terms, we obtain the effective Hamiltonian

$$H_{\mathrm{RWA}} = \frac{K}{s} \sum_{j=1}^{s} \cos\left[x \cos\left(\frac{2\pi j}{s}\right) + p \sin\left(\frac{2\pi j}{s}\right)\right]. \tag{6.10}$$

The effective Hamiltonian (6.10) possesses different symmetries in the phase space depending which $s:1$ resonance we choose. In quantum mechanics we cannot strictly determine a position in the phase space because the position x and momentum p operators do not commute. In order to illustrate the symmetries of the system in the phase space we plot average values of H_{RWA} in the coherent state representation, $\langle\alpha|H_{\mathrm{RWA}}|\alpha\rangle$, where $\hat{a}|\alpha\rangle = \alpha|\alpha\rangle$ [3]. The real and imaginary parts of a complex number α correspond to a point (X, P) in the phase space, i.e. $X = \langle\alpha|x|\alpha\rangle = \sqrt{2\lambda}\mathrm{Re}(\alpha)$ and $P = \langle\alpha|p|\alpha\rangle = \sqrt{2\lambda}\mathrm{Im}(\alpha)$. We should

[2] The units for the position and momentum coordinates used to derive the Hamiltonian (6.9) are $\sqrt{\hbar/(\lambda m\omega_0)}$ and $\sqrt{m\hbar\omega_0/\lambda}$, respectively, where m is the mass of a particle and ω_0 is the frequency of the harmonic potential. The effective Planck constant $\lambda = \hbar k^2/(m\omega_0)$ where k can be associated with the wave number of an optical standing wave that creates the perturbation in (6.9). The constant λ can be controlled and changed in the experiments with ultra-cold atomic gases.

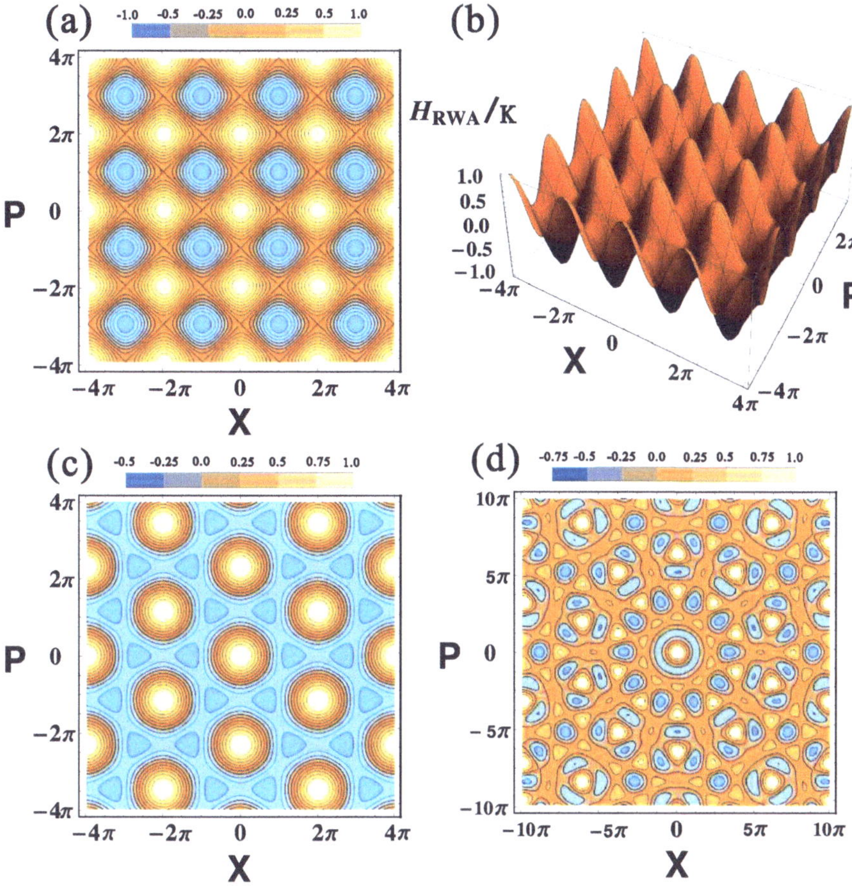

Fig. 6.2 Figure presents average values of the effective Hamiltonian (6.10) in the coherent state representation [3], $\langle\alpha|H_{\text{RWA}}|\alpha\rangle/K$. The real and imaginary parts of a complex number α correspond to a point (X, P) in the phase space, i.e. $X = \langle\alpha|x|\alpha\rangle = \sqrt{2\lambda}\text{Re}(\alpha)$ and $P = \langle\alpha|p|\alpha\rangle = \sqrt{2\lambda}\text{Im}(\alpha)$. Top panels present the $s = 4$ case, in the two-dimensional (**a**) and three-dimensional (**b**) plots, when a squared lattice emerges in the phase space. Panel (**c**) shows a hexagonal lattice that emerges in the $s = 3$ or $s = 6$ cases. Panel (**d**) illustrates a quasi-crystal structure in the phase space corresponding to the $s = 5$ case. Reprinted from [8]

remember that X and P are the mean values of the operators in a coherent state and the variances are determined by the effective Planck constant λ [3]. In other words (X, P) does not have a strict meaning of a point in the phase space.

Plots of $\langle\alpha|H_{\text{RWA}}|\alpha\rangle$ reveal: a squared lattice for $s = 4$, hexagonal lattice for $s = 3$ or $s = 6$, and quasi-crystal structures for $s = 5$ and $s \geq 7$, see Fig. 6.2. For the translationally symmetric lattices, the energy spectra of H_{RWA}, which correspond to quasi-energy spectra of the time-periodic Hamiltonian (6.9), possess band structures [8].

Let us focus on the case when $s = 4$, i.e. when the system reveals a squared lattice in the phase space, and let us show that the non-commutative geometry in the phase space leads to topological properties of the system. For $s = 4$, the effective Hamiltonian (6.10) reduces to a simple form

$$H_{\mathrm{RWA}} = \frac{K}{2}\left(\cos x + \cos p\right). \tag{6.11}$$

The Hamiltonian (6.11) is invariant under translations in the phase space by 2π both along the x and p directions. That is, it commutes with the translational unitary operators $T_1 = e^{i2\pi p/\lambda}$ and $T_2 = e^{i2\pi x/\lambda}$. Hence, the operators $T_{1,2}$ generate an invariance group of the effective Hamiltonian H_{RWA} but because they do not commute with each other the group is non-abelian,

$$\left[T_1^k, T_2^l\right] = T_1 T_2 \left(1 - e^{-i2\pi(2\pi kl/\lambda)}\right), \tag{6.12}$$

where k and l are integer numbers. However, the relation (6.12) indicates that if $2\pi kl/\lambda$ is an integer number, the translation operators to the power of k and l, i.e. T_1^k and T_2^l, generate an invariance subgroup of the system which is abelian. In other words T_1^k and T_2^l commute with H_{RWA} and they also commute with each other. Then, eigenstates of the effective Hamiltonian H_{RWA} are also eigenstates of the translational operators T_1^k and T_2^l.

Let us assume that the effective Planck constant $\lambda/(2\pi) = n/m$, where n and m are coprime integers and let us choose $k = 1$ and $l = n$ which leads to the generators of the abelian subgroup

$$T_X \equiv T_1^k = T_1 = e^{i2\pi p/\lambda}, \qquad T_P \equiv T_2^l = T_2^n = e^{i2\pi nx/\lambda}. \tag{6.13}$$

The eigenvalues of the operators T_X and T_P can be written in the form $e^{i2\pi k_X}$ and $e^{i2\pi nk_P}$, respectively, where $0 \leq k_X \leq 1$ and $0 \leq k_P \leq 1/n$ determine the two-dimensional Brillouin zone. The quantity k_X is a quasi-momentum while k_P, which is related to the periodicity of the system along the momentum direction, is called a quasi-coordinate [8]. Eigenenergies and eigenstates of the effective Hamiltonian (6.11),

$$H_{\mathrm{RWA}}|\psi_{b,\mathbf{k}}\rangle = E_b(\mathbf{k})|\psi_{b,\mathbf{k}}\rangle, \tag{6.14}$$

can be labeled by the quasi-momentum and quasi-coordinate, $\mathbf{k} = (k_X, k_P)$. For each $\mathbf{k}$ we can have a few eigenenergies labeled by b. The index b denotes actually the number of an energy band. The system has also a chiral symmetry (cf. background information in Sect. 5.3.1). That is, the unitary operator $T_{\mathrm{chiral}} = e^{i\pi x/\lambda}e^{i\pi p/\lambda}$ fulfills

$$T_{\mathrm{chiral}} H_{\mathrm{RWA}} T_{\mathrm{chiral}}^{\dagger} = -H_{\mathrm{RWA}}, \tag{6.15}$$

and consequently the energy spectrum is symmetric, i.e. eigenstates appear in pairs $|\psi_{b,\mathbf{k}}\rangle$ and $T^{\dagger}_{\text{chiral}}|\psi_{b,\mathbf{k}}\rangle$ corresponding to eigenvalues $\pm E_b(\mathbf{k})$. Solving the Schrödinger equation (6.14), with the Hamiltonian (6.11), allows us to find eigenenergies which must fulfill

$$\cos(m\lambda k_X) + \cos(m\lambda k_P) = 1 + \frac{1}{2}\mathrm{Tr}\prod_{j=1}^{m}\begin{bmatrix} \frac{E_b(k_X,k_P)}{K} - 2\cos(j\lambda) & -1 \\ 1 & 0 \end{bmatrix}. \tag{6.16}$$

For different rational values of $\lambda/(2\pi) = n/m$, we obtain different numbers of energy bands. Due to the fact that Eq. (6.16) does not change if $\lambda \to \lambda + 2\pi$, it is sufficient to analyze how the spectrum changes in the range $0 \le \lambda/(2\pi) < 1$. Figure 6.3 shows that the energy spectra plotted for different rational values of $\lambda/(2\pi) = n/m$ reveal a Hofstadter's butterfly structure like in quantum Hall systems [4]. The figure presents also the energy bands for a few different choices of $\lambda/(2\pi) = n/m$. The number m determines the number of energy bands while n their degeneracy [8].

Each energy band of the effective H_{RWA} can be characterized by a topological invariant, i.e. an integer-valued Chern number [4, 8] (see background information in Sect. 7.1.2)

$$C_b = \oint_{\mathscr{C}} \langle\psi_{b,\mathbf{k}}|\partial_{\mathbf{k}}|\psi_{b,\mathbf{k}}\rangle \cdot d\mathbf{k}, \tag{6.17}$$

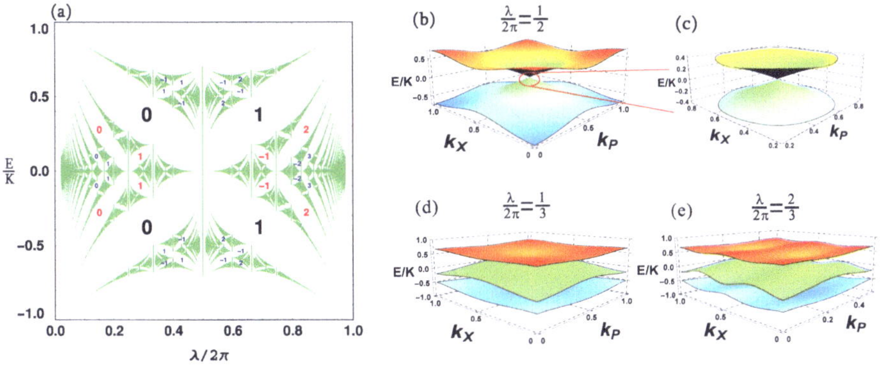

Fig. 6.3 Panel (**a**): energy spectrum of the effective Hamiltonian (6.11) for different rational values of $\lambda/(2\pi) = n/m$ where λ is the effective Planck constant. Due to the chiral symmetry of the effective Hamiltonian, the spectrum is symmetric with respect to the zero energy value. Numbers in the plot denote sums of the Chern numbers of the bands below and above the zero energy value. Panels (**b**) and (**c**) show energy bands of the effective Hamiltonian (6.11), i.e. energy eigenvalues as a function of the quasi-momentum k_X and quasi-coordinate k_P, for $\lambda/(2\pi) = 1/2$. The energy bands for $\lambda/(2\pi) = 1/3$ and $\lambda/(2\pi) = 2/3$ are presented in (**d**) and (**e**), respectively. Reprinted from [8]

where the integration runs over the boundary $\mathscr{C}$ of the Brillouin zone. In Fig. 6.3 a few examples of sums of the Chern numbers for the bands below and above the zero energy value are presented.

The rotating wave approximation used to derive the effective Hamiltonian H_{RWA} requires a weak kicking strength $K \ll 1$ in (6.9). The validity of the approximation was confirmed in numerical integration of the exact Schrödinger equation by Liang and coworkers [8]. They took also into account coupling of the kicked harmonic oscillator to a thermal bath. Assuming that the bath is Markovian and at zero temperature, they investigated time evolution of the density matrix of the kicked harmonic oscillator by solving the quantum master equation

$$\dot{\rho} = -\frac{i}{\lambda}[H, \rho] + \kappa \left(\hat{a}\rho\hat{a}^{\dagger} - \frac{1}{2}\hat{a}^{\dagger}\hat{a}\rho - \frac{1}{2}\rho\hat{a}^{\dagger}\hat{a} \right), \tag{6.18}$$

where H is given by (6.9) and κ characterizes the dissipation rate. In order to illustrate the state of the system in the phase space one can use again the coherent state representation and calculate the so-called Husimi Q-distribution, $Q(\alpha) = \langle\alpha|\rho|\alpha\rangle/\pi$ [3]. Time evolution of the Husimi distribution for the $s:1$ resonant kicking, with $s = 4$ is presented in Fig. 6.4. We can see that starting with the ground state of the harmonic oscillator in (6.9), due to the resonant kicking, a particle explores the phase space and the plots of the Husimi distribution reflect the squared crystalline structure expected for $s = 4$ and depicted in Fig. 6.2.

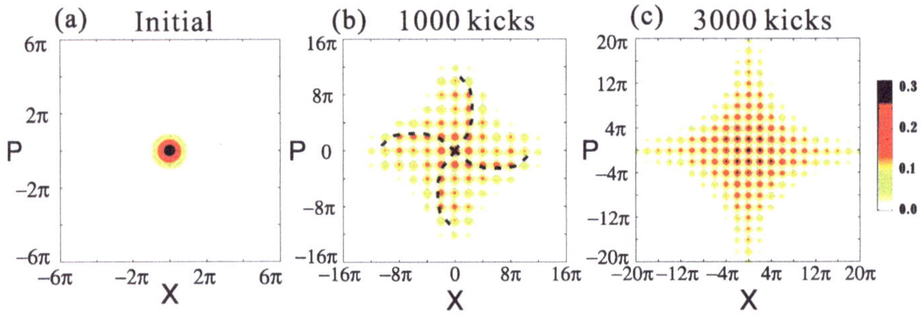

Fig. 6.4 Panel (**a**) presents Husimi Q-distribution [3], $Q(\alpha) = \langle\alpha|\rho|\alpha\rangle/\pi$, for the ground state of the harmonic oscillator in (6.9). The real and imaginary parts of a complex number α correspond to a point (X, P) in the phase space, i.e. $X = \sqrt{2\lambda}\mathrm{Re}(\alpha)$ and $P = \sqrt{2\lambda}\mathrm{Im}(\alpha)$. The ground state is the initial state in the dissipative dynamics described by the quantum master equation (6.18) with the dissipation rate $\kappa = 0.0001$. Panels (**b**) and (**c**) show the Husimi distribution after 1000 and 3000 kicks, respectively. In the Hamiltonian (6.9), the strength of the kicking $K = 0.1$ and the kicking period $T = 2\pi/s$ with $s = 4$ is chosen. For $s = 4$ the squared lattice in the phase space is expected according to the effective Hamiltonian (6.11), cf. Fig. 6.2. The results of the full numerical simulations shown in the present figure confirm predictions of the rotating wave approximation approach, i.e. a kicked quantum particle explores the squared lattice. In the long time evolution a particle reaches a stationary state where the Husimi distribution is localized on a finite area of the phase space due to the dissipative dynamics. The Reprinted from [8]

Without the dissipation a particle would experience unbounded diffusion with the average energy of the harmonic oscillator increasing infinitely. In the presence of the dissipation, the system approaches a stationary state where the Husimi distribution becomes localized on a finite area in the phase space whose size depends on the kicking strength K and the dissipation rate κ. The results presented in Fig. 6.4 confirm the prediction of the rotating wave approximation approach and demonstrate what one can expect in the real experiment.

6.3 Many-Body Physics in Phase Space Crystals

So far we have seen that a resonantly driven single-particle harmonic oscillator can reveal crystalline structures in the phase space. In the present section we consider N interacting particles in a harmonic potential that are resonantly driven and show that two-body interactions described by a potential $V(x_i - x_j)$ result in effective two-body interactions that depend on the relative distance between particles but in the phase space. For simplicity we will present classical calculations only [5] but the quantum approach leads to similar effective interaction potentials if the coherent state representation is used to describe positions of particles in the phase space [8].

Let us consider the classical Hamiltonian

$$H = \sum_{i=1}^{N} \left[\frac{p_i^2}{2} + \frac{x_i^2}{2} + KT \cos x_i \sum_{n=-\infty}^{\infty} \delta(t - nT) \right] + \sum_{i<j=1}^{N} V(x_i - x_j), \tag{6.19}$$

that describes N particles in the one-dimensional harmonic oscillator potential which are kicked periodically (similarly like in the single-particle case investigated in the previous section) and interact via a two-body interaction potential $V(x_i - x_j)$. We assume the resonant driving, i.e. $T = 2\pi/s$, and focus on $s = 4$. In the spirit of the approach used in the previous section we apply a classical version of the rotating wave approximation. We perform the canonical transformation to the frame oscillating with the frequency of the harmonic oscillator potential,

$$\begin{aligned} x_i &= x_i' \cos t + p_i' \sin t, \\ p_i &= -x_i' \sin t + p_i' \cos t, \end{aligned} \tag{6.20}$$

and drop all terms oscillating in time in the Hamiltonian what yields the effective classical Hamiltonian

$$H_{\text{RWA}} = \frac{K}{2} \sum_{i=1}^{N} (\cos x_i + \cos p_i) + \sum_{i<j=1}^{N} U(R_{ij}), \tag{6.21}$$

where we have omitted primes in x_i' and p_i'. The Hamiltonian H_{RWA} is a many-body generalization of the effective single-particle Hamiltonian (6.11). The canonical transformation (6.20) mixes position and momentum coordinates and consequently the effective interaction potential in (6.21) depends on the relative distance between particles but in the phase space,

$$R_{ij} = \sqrt{(x_i - x_j)^2 + (p_i - p_j)^2}. \tag{6.22}$$

That is,

$$U(R_{ij}) = \int_{-\infty}^{+\infty} dq\ \tilde{V}(q)\ J_0(qR_{ij}), \tag{6.23}$$

where $\tilde{V}(q) = (2\pi)^{-1} \int_{-\infty}^{+\infty} dx V(x) e^{-iqx}$ and $J_0(w)$ is the Bessel function of the zeroth order. The effective potential (6.23) has been obtained by neglecting all terms that oscillate in time. The original interaction potential V in the oscillating frame (6.20) reads

$$V\left[(x_i - x_j)\cos t + (p_i - p_j)\sin t\right] = V\left[R_{ij}\cos(t - \theta_{ij})\right], \tag{6.24}$$

where $\tan\theta_{ij} = (p_i - p_j)/(x_i - x_j)$, and can be written in the form

$$\begin{aligned} V\left[R_{ij}\cos(t - \theta_{ij})\right] &= \int_{-\infty}^{+\infty} dq \tilde{V}(q) e^{iqR_{ij}\cos(t-\theta_{ij})} \\ &= \sum_{m=-\infty}^{+\infty} i^m \int_{-\infty}^{+\infty} dq\ \tilde{V}(q)\ J_m(qR_{ij})\ e^{im(t-\theta_{ij})}, \end{aligned} \tag{6.25}$$

where $J_m(w)$'s are the Bessel functions. The time-independent term in (6.25), i.e. the term corresponding to $m = 0$, constitutes the effective interaction potential (6.23).

We have arrived at the effective Hamiltonian (6.21) where particles experience a crystalline structure in the phase space (cf. the previous section) and interact via the potential depending on relative distances between particles in the phase space. Let us analyze possible shapes of the effective interaction potential $U(R_{ij})$.

In ultra-cold atoms the interaction potential can be modeled by the Dirac-delta contact potential $\varepsilon\delta(x_i - x_j)$ where ε is proportional to the atomic s-wave scattering length. Let us represent such a potential with the help of a Lorentz function whose width tends to zero,

$$V(x_i - x_j) = \varepsilon\delta(x_i - x_j) = \lim_{\gamma\to 0} \frac{\varepsilon}{\pi} \frac{\gamma}{(x_i - x_j)^2 + \gamma^2}. \tag{6.26}$$

The Fourier transform $\tilde{V}(q) = \varepsilon e^{-\gamma|q|}/(2\pi)$ and consequently the effective potential (6.23) takes the form

$$U(R_{ij}) = \frac{\varepsilon}{\pi}\frac{1}{\sqrt{R_{ij}^2 + \gamma^2}}. \tag{6.27}$$

For $R_{ij} \gg \gamma$, the effective potential $U(R_{ij})$ behaves like a Coulomb-like potential. It could be surprising that the original contact interactions between particles result in the effective long-range interactions. The reason is similar like in the case of atoms resonantly bouncing on an oscillating atom mirror considered in Sect. 5.4. That is, particles oscillating back and forth meet each other in the configuration space and interact. When such interactions are averaged over time, they become effectively long-range interactions.

If original two-body interactions are described by a hardcore potential, that we model by means of

$$V(x_i - x_j) = \frac{(2a)^n}{(x_i - x_j)^n}, \tag{6.28}$$

with $n \gg 1$ (for example $n = 20$ is chosen in Fig. 6.5), the effective interaction potential $U(R_{ij})$ increases with the phase space distance R_{ij} between particles [8],

$$U(R_{ij}) = \frac{2a}{\pi}R_{ij}. \tag{6.29}$$

Such a potential mimics interactions between quarks in quantum chromodynamics, i.e. the interaction energy increases if the distance between particles becomes larger. Intuitively the form of the effective potential (6.29) can be explained as follows [8]. There is a small hardcore radius a and therefore two particles prefer to oscillate in phase in order not to collide with each other in the configuration space. The in-phase oscillations mean that the distance between them in the phase space does not change significantly. If they oscillate out of phase, they can collide with each other in the configuration space that leads to a large interaction energy. The out of phase oscillations correspond to a large phase space distance R_{ij} between particles and thus the increase of the effective interaction energy with R_{ij} is expected which justifies the confinement-like potential (6.29).

In Fig. 6.5 the exact classical dynamics of three particles with the original interactions corresponding to contact and hardcore potentials is presented and compared with the dynamics described by the effective Hamiltonian (6.21). The results confirm the validity of the rotating wave approximation approach.

As we have already mentioned, in the quantum description the effective interaction potentials take similar form as (6.23) if the coherent state representation is used, i.e. when positions of particles in the phase space are described with the help of coherent states [8].

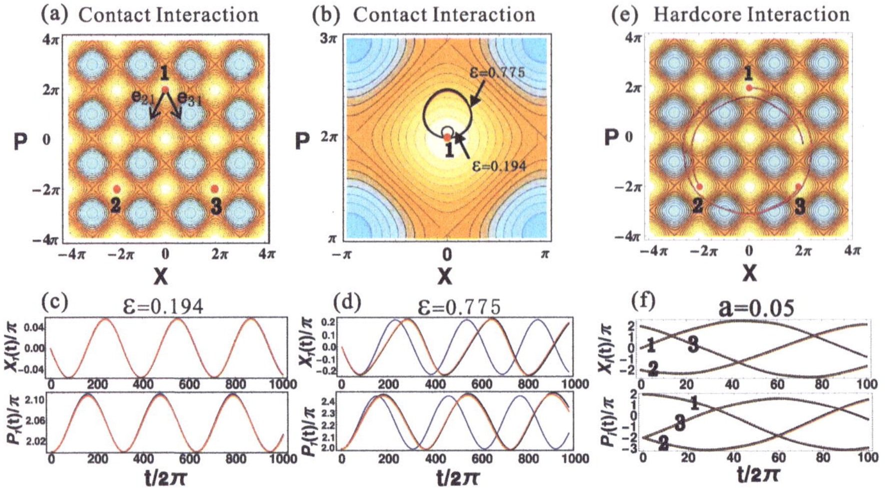

Fig. 6.5 Comparison of the exact classical dynamics with the results based on the effective Hamiltonian (6.21) for three-particle systems ($N = 3$). In panels (**a**), (**b**), and (**e**) the squared lattice created in the phase space by non-interacting resonantly kicked particles is shown. Red filled circles indicate initial positions of three particles. In all panels red dots correspond to exact trajectories obtained for the kicking strength $K = -0.02/\pi$ and sampled stroboscopically every period $T = 2\pi/4$, cf. (6.19), while black dots to the trajectories obtained with the help of the effective Hamiltonian (6.21). Panels (**a**), (**b**), (**c**), and (**d**) correspond to the contact interactions between particles. In (**a**) initial conditions for three particles are presented while in (**b**) trajectories of the first particle are shown for the strength of the contact interactions $\varepsilon = 0.194$ and $\varepsilon = 0.775$. All particles oscillate around the stable configuration in the phase space. Panels (**c**) and (**d**) present how the position and momentum of the first particle evolves in time. Blue curves in (**d**) are related to the trajectories obtained from a linearized version of the effective equations of motion [8]. In panels (**e**) and (**f**) the case of the hardcore interactions is presented for $a = 0.05$ and $n = 20$, cf. (6.28). The interaction energy is so big that the particles are able to overcome potential barriers created by the phase space lattice and explore a large area of the phase space. Reprinted from [8]

References

1. Berman, G.P., Rubaev, V.Y., Zaslavsky, G.M.: The problem of quantum chaos in a kicked harmonic oscillator. Nonlinearity **4**(2), 543–566 (1991). https://doi.org/10.1088/0951-7715/4/2/015
2. Carvalho, A.R.R., Buchleitner, A.: Web-assisted tunneling in the kicked harmonic oscillator. Phys. Rev. Lett. **93**, 204101 (2004). https://link.aps.org/doi/10.1103/PhysRevLett.93.204101
3. Gerry, C., Knight, P.: Introductory Quantum Optics. Cambridge University Press, New York (2005)
4. Goldman, N., Juzeliunas, G., Öhberg, P., Spielman, I.B.: Light-induced gauge fields for ultracold atoms. Rep. Prog. Phys. **77**(12), 126401 (2014). http://stacks.iop.org/0034-4885/77/i=12/a=126401
5. Guo, L., Marthaler, M.: Synthesizing lattice structures in phase space. New J. Phys. **18**(2), 023006 (2016). https://doi.org/10.1088/1367-2630/18/2/023006
6. Guo, L., Marthaler, M., Schön, G.: Phase space crystals: a new way to create a quasienergy band structure. Phys. Rev. Lett. **111**, 205303 (2013). https://link.aps.org/doi/10.1103/PhysRevLett.111.205303

7. Guo, L., Liu, M., Marthaler, M.: Effective long-distance interaction from short-distance interaction in a periodically driven one-dimensional classical system. Phys. Rev. A **93**, 053616 (2016). https://link.aps.org/doi/10.1103/PhysRevA.93.053616
8. Pengfei, L., Michael, M., Guo, L.: Floquet many-body engineering: topology and many-body physics in phase space lattices. New J. Phys. **20**(2), 023043 (2018). http://stacks.iop.org/1367-2630/20/i=2/a=023043
9. Guo, L., Liang, P.: Condensed matter physics in time crystals. New Journal of Physics. IOP Publishing. **22**(7), 075003 (2020). https://doi.org/10.1088/1367-2630/ab9d54

Chapter 7
Photonic Time Crystals

Abstract Particles moving in spatially periodic potentials can model condensed matter systems where electrons interact with ions that form crystalline structures in space. Condensed matter-like behavior can be observed not only for massive particles but also for photons. Photonic crystals are systems where the refractive index of a material changes periodically in space. The corresponding dispersion relation reveals a band structure where certain frequencies of electromagnetic waves are forbidden and the waves cannot propagate in a material. When the refractive index changes periodically not in space but in time, photonic time crystals can be observed. Then, gaps of the dispersion relation are present in the wave number domain instead of the frequency domain. Here we describe photonic time crystals and show that they can reveal topological behavior whose signatures are visible in the time dimension. Introduction to photonic crystals in space is also given.

7.1 Photonic Crystals in Space

We know that a particle moving in a time-independent potential that changes periodically in space mimics a condensed matter system where electrons experience a spatially periodic structure created by regularly distributed ions. It turns out that electromagnetic waves propagating in a material where the refractive index changes periodically in space also possess properties similar to properties of condensed matter systems [10, 21, 22]. These kind of systems are dubbed photonic crystals and they are intensively investigated theoretically and experimentally [9].

7.1.1 Introduction to Photonic Crystals in Space

Electromagnetic waves are described by the Maxwell equations which in the absence of free charges and the current density read

K. Sacha, *Time Crystals*, Springer Series on Atomic, Optical, and Plasma Physics 114, https://doi.org/10.1007/978-3-030-52523-1_7

$$\nabla \cdot [\varepsilon_0 \epsilon(\mathbf{r})\mathbf{E}] = 0, \qquad \nabla \times \mathbf{E} = -\mu_0 \frac{\partial \mathbf{H}}{\partial t}, \tag{7.1}$$

$$\nabla \cdot \mathbf{H} = 0, \qquad \nabla \times \mathbf{H} = \varepsilon_0 \epsilon(\mathbf{r}) \frac{\partial \mathbf{E}}{\partial t}, \tag{7.2}$$

where **E** and **H** are the electric and magnetic fields, ε_0 and μ_0 are the vacuum electric permittivity and magnetic permeability, respectively. We assume non-magnetic linear medium where the magnetic induction $\mathbf{B} = \mu_0 \mathbf{H}$ and the displacement field $\mathbf{D} = \varepsilon_0 \epsilon(\mathbf{r})\mathbf{E}$ with the dielectric constant $\epsilon(\mathbf{r}) \geq 1$ which changes periodically in space. We look for solutions of the Maxwell equations which are changing harmonically in time with a frequency ω, i.e. $\mathbf{E}(\mathbf{r}, t) = \mathbf{E}(\mathbf{r})e^{-i\omega t}$ and $\mathbf{H}(\mathbf{r}, t) = \mathbf{H}(\mathbf{r})e^{-i\omega t}$ in the complex-valued field representation. Taking the curl of the second equation in (7.1) and using the second equation in (7.2), we obtain

$$\nabla \times (\nabla \times \mathbf{E}) - \frac{\omega^2}{c^2} \epsilon(\mathbf{r})\mathbf{E} = 0, \tag{7.3}$$

where $c = 1/\sqrt{\mu_0 \varepsilon_0}$. If we know a solution of (7.3), we also know the corresponding magnetic field $\mathbf{H} = (\nabla \times \mathbf{E})/(i\mu_0\omega)$.

Let us assume that the dielectric constant changes periodically along the z axis, i.e. $\epsilon(z + \Lambda) = \epsilon(z)$, and an electromagnetic wave propagates along the z direction and is polarized along the x axis, i.e. $\mathbf{E}(\mathbf{r}) = E(z)\mathbf{e}_x$. Then, Eq. (7.3) reduces to

$$\left[\frac{d^2}{dz^2} + \frac{\omega^2}{c^2}\epsilon(z)\right] E(z) = 0. \tag{7.4}$$

Due to the fact that $\epsilon(z)$ is a periodic function we may employ the Bloch theorem in order to solve a generalized eigenvalue problem[1] (7.4). The theorem guarantees that

$$E(z) = u_{n,q}(z)e^{iqz}, \tag{7.5}$$

where $u_{n,q}(z + \Lambda) = u_{n,q}(\Lambda)$ and q is a wave number which is an analogue of a quasi-momentum in condensed matter physics. The index n labels frequency bands, $\omega = \omega_n(q)$, that can emerge from the dispersion relation of the system. The range of the wave number q can be restricted to the Brillouin zone, $-\pi/\Lambda \leq q < \pi/\Lambda$.

We are going to analyze a concrete example where the dielectric constant $\epsilon(z)$ changes periodically in space in an alternate way as depicted in Fig. 7.1 [20]. The layers A (of the width $d_A = 0.4\Lambda$) correspond to $\epsilon = \epsilon_A = 4$ while the layers B (of the width $d_B = 0.6\Lambda$) to $\epsilon = \epsilon_B = 1$. In each of the layers A and B, a solution of (7.4) has a form

[1] Equation (7.4) written in the form $-d^2E(z)/dz^2 = \omega\epsilon(z)E(z)$ is the generalized eigenvalue problem for the Hermitian operators: $-d^2/dz^2$ and positive defined $\epsilon(z)$ [9].

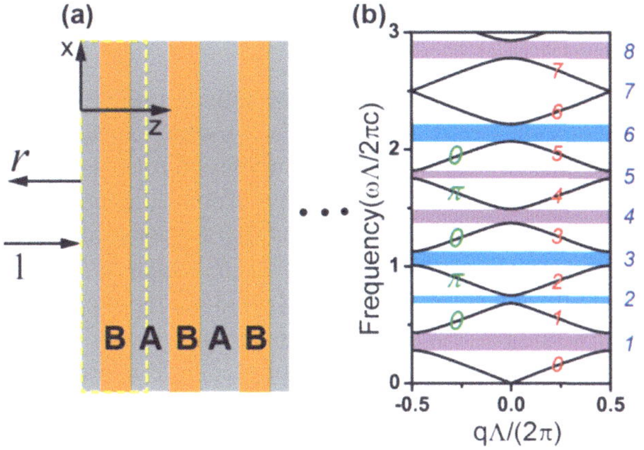

Fig. 7.1 Panel (**a**) illustrates a photonic crystal structure consisting of A and B layers. The dashed yellow line indicates a unit cell. In Sect. 7.1.2 an electromagnetic wave incidents normally on the semi-infinite crystal (located at $z \geq 0$ as shown in the panel) is analyzed—r denotes the reflection coefficient of the electric field. In panel (**b**) the dispersion relation (7.7) is depicted which corresponds to the widths $d_A = 0.4\Lambda$ and $d_B = 0.6\Lambda$ and the dielectric constants $\epsilon = \epsilon_A = 4$ and $\epsilon = \epsilon_B = 1$ of the layers A and B, respectively—the dielectric constant is periodic $\epsilon(z + \Lambda) = \epsilon(z)$. The dispersion relation reveals a band structure. The numbers of the bands and of the gaps are depicted with red and blue labels, respectively. The Zak phases of the bands are labeled in green while the magenta and cyan strips denote the gaps for which the imaginary parts of the impedance are positive and negative, respectively (see Sect. 7.1.2). Reprinted from [20]

$$E(z) = C_A e^{ik_A z} + D_A e^{-ik_A z}, \qquad E(z) = C_B e^{ik_B z} + D_B e^{-ik_B z}, \tag{7.6}$$

respectively, where $k_A = \omega\sqrt{\epsilon_A}/c$ and $k_B = \omega\sqrt{\epsilon_B}/c$. Imposing the continuity conditions of $E(z)$ and $dE(z)/dz$ at boundaries between neighboring layers and periodicity of $u_{n,q}(z)$ and $du_{n,q}(z)/dz$, one obtains the dispersion relation [20]

$$\cos(q\Lambda) = \cos(k_A d_A)\cos(k_B d_B) - \frac{1}{2}\left(\sqrt{\frac{\epsilon_B}{\epsilon_A}} + \sqrt{\frac{\epsilon_A}{\epsilon_B}}\right)\sin(k_A d_A)\sin(k_B d_B), \tag{7.7}$$

which determines how the frequency ω of an electromagnetic wave propagating in the photonic crystal depends on the wave number q. It turns out that not all values of ω are allowed, see Fig. 7.1. There are gaps in the frequency domain and a wave with a frequency in a gap cannot enter the medium and is totally reflected.

The band structure indicates that the system has similar properties like an electron moving in a periodic crystalline structure in a solid state system. We have considered a one-dimensional example but similar behavior can be observed in media where the dielectric constant $\epsilon(\mathbf{r})$ possesses various translational symmetries known from solid state crystallography. The entire field is dubbed photonic crystals and is intensively developed both in the theoretical and experimental research [9].

7.1.2 Topological Properties of Photonic Crystals in Space

Photonic crystals in space can reveal topological properties [15]. The one-dimensional system considered in Fig. 7.1 is a useful example to illustrate how topology can determine results of experimental measurements [20].

Suppose that a linearly polarized plane wave from a vacuum incidents normally on the semi-infinite photonic crystal that consists of a stack of two kinds of layers as depicted in Fig. 7.1. The surface of the crystal, which is located at $z = 0$, is precisely in the middle of a layer A. Thus, if we had not the semi-infinite but infinite photonic crystal it would possess the reflection symmetry, i.e. the dielectric constant $\epsilon(-z) = \epsilon(z)$. Assume that the frequency of the wave incident normally on the crystal corresponds to a gap of the dispersion relation (7.7) and consequently the wave is totally reflected and cannot enter the photonic crystal. It means that the reflection coefficient $r = e^{i\phi_R}$ where ϕ_R is the phase the wave acquires when it is totally reflected from the crystal (i.e. reflected from the right part of the space in Fig. 7.1). The reflection of the wave can be also described in terms of the wave impedances Z_R and Z_0 [20], i.e.

$$r_R = \frac{Z_R - Z_0}{Z_R + Z_0}, \tag{7.8}$$

where

$$Z_R = \frac{E(z = 0^+)}{H(z = 0^+)}, \tag{7.9}$$

with $\mathbf{E}(\mathbf{r}) = E(z)\mathbf{e}_x$ and the corresponding magnetic field $\mathbf{H}(\mathbf{r}) = H(z)\mathbf{e}_y$ and $Z_0 = c = 1/\sqrt{\varepsilon_0\mu_0}$ is the vacuum impedance. In the case of the total reflection, the ratio of the impedances is purely imaginary, i.e. $Z_R/Z_0 = i\zeta_R$ where ζ_R is a real number which determines the reflection phase $\phi_R = \pi - 2\arctan\zeta_R$.

Now suppose that there are two different semi-infinite photonic crystals glued together at $z = 0$ and we ask the question if it is possible to observe a state of the electromagnetic field, that corresponds to the frequency in a common gap of the photonic crystals, which is localized on the interface of them? It would be an electromagnetic analogue of the edge states of the Su–Schrieffer–Heeger model described in Sect. 5.3.1. The semi-infinite crystal on the left side of $z = 0$ is characterized by the impedance Z_L. The appearance of a surface electromagnetic mode requires a pole of the reflection coefficient between these two media [11, 12, 20],

$$r_{LR} = \frac{Z_R - Z_L}{Z_R + Z_L}, \tag{7.10}$$

which happens when

$$Z_R + Z_L = 0. \tag{7.11}$$

We have seen that for the frequency chosen in a gap of the right photonic crystal $Z_R/Z_0 = i\zeta_R$. It means that in order to fulfill $Z_R + Z_L = 0$, the impedance of the left photonic crystal must be $Z_L/Z_0 = i\zeta_L$ and $\zeta_L = -\zeta_R$. Thus, the necessary condition for the existence of a surface mode is the following: ζ_L and ζ_R must have opposite signs. Now topology comes into the play because it turns out that the signs of ζ_L and ζ_R are determined by topological invariants, i.e. by the Zak phases of the bands below the common gap of the photonic crystals [20]. Before we continue our analysis it should be useful for a reader if we provide basic information about the Zak phase [23].

Zak Phase

Consider the time-independent Schrödinger equation,

$$H(\mathbf{X})|\psi_n(\mathbf{X})\rangle = E_n(\mathbf{X})|\psi_n(\mathbf{X})\rangle, \tag{7.12}$$

where $\mathbf{X} = (X_1, \ldots, X_p)$ is a set of parameters of a system. If at $t = 0$ we start with an eigenstate $|\psi_n[\mathbf{X}(0)]\rangle$ and change the parameters $\mathbf{X}(t)$ very slowly in time, the adiabatic approximation applies which says that a system follows an instantaneous eigenstate $|\psi_n[\mathbf{X}(t)]\rangle$ provided $E_n[\mathbf{X}(t)]$ is not degenerate at any t. If at $t = T$, the parameters return to their initial values, $\mathbf{X}(T) = \mathbf{X}(0)$, a system also comes back to the initial state up to an overall phase $e^{i(\gamma_n^{\text{dyn}}+\gamma_n^{\text{geo}})}$. This phase factor consists of the dynamical contribution, $\gamma_n^{\text{dyn}} = -\int_0^T E_n[\mathbf{X}(t)]dt$ (we assume $\hbar = 1$), and the Berry phase [3],

$$\gamma_n^{\text{geo}} = i \oint \langle\psi_n(\mathbf{X})|\partial_{\mathbf{X}}\psi_n(\mathbf{X})\rangle \cdot d\mathbf{X}, \tag{7.13}$$

which is the integral of the Berry connection

$$\mathbf{A}(\mathbf{X}) = i\langle\psi_n(\mathbf{X})|\partial_{\mathbf{X}}\psi_n(\mathbf{X})\rangle, \tag{7.14}$$

along a closed loop in the space of the system parameters. The Berry phase is a geometrical phase and can be calculated even if no parameters of a system have been changed adiabatically in time.

Let us consider a particle in a one-dimensional potential which changes periodically in space $V(z+a) = V(z)$. Eigenstates of such a system are Bloch waves $\psi_{n,k}(z) = e^{ikz}u_{n,k}(z)$ where k is a quasi-momentum, n labels energy bands and $u_{n,k}(z+a) = u_{n,k}(z)$. The corresponding eigenvalues are periodic, $E_n(k + 2\pi/a) = E_n(k)$, and consequently the Hamiltonian is invariant when we perform translation in the reciprocal space by $2\pi/a$. Thus, if we change the quasi-momentum over the Brillouin zone, e.g. from $k = -\pi/a$ to $k = \pi/a$, it actually means that k makes a closed loop because a system is invariant under such a change and we can calculate the geometrical phase (7.13),[2]

(continued)

[2]Note that the scalar product in (7.15) involves integration over a unit cell only, i.e. $\langle u_{n,k}|\partial_k u_{n,k}\rangle = \int_{\text{unit cell}} dz u^*_{n,k}(z)\partial_k u_{n,k}(z)$.

$$\theta_n^{\mathrm{Zak}} = i \int_{-\pi/a}^{\pi/a} \langle u_{n,k} | \partial_k u_{n,k} \rangle dk, \tag{7.15}$$

where k is considered as a parameter $\mathbf{X} = k$ of a system [3]. The phase θ_n^{Zak} is called the Zak phase [23]. Superposing Bloch waves of the band n one can obtain the Wannier functions $w_n(z - ma)$ corresponding to this band which are localized in different elementary cells of a periodic potential $V(z)$ [5]. When we expand Bloch waves in the Wannier function basis,

$$e^{ikz} u_{n,k}(z) = \sum_m e^{ikma} w_n(z - ma), \tag{7.16}$$

we can express the Zak phase in terms of the Wyckoff position z_n of a one-dimensional crystal[3] [23]

$$\theta_n^{\mathrm{Zak}} = \frac{2\pi}{a} z_n = \frac{2\pi}{a} \int_{-\infty}^{\infty} dz z |w_n(z)|^2. \tag{7.17}$$

If there is no symmetry of a one-dimensional potential $V(z)$, the Wyckoff position z_n can take any value as a function of the band number n. However, if there is the reflection symmetry of a potential, then z_n can be equal 0 or $a/2$ only [23] (modulo translations by integer multiples of a). Indeed, if $V(-z) = V(z)$, the probability density of the Wannier function $|w_n(z)|^2$ is symmetric under inversion with respect to 0 or $a/2$. In the former case $z_n = 0$ while in the latter case $z_n = a/2$. Consequently, the Zak phase for reflection symmetric potentials can be either 0 or π only (modulo 2π). The fact that the Zak phase is determined modulo 2π is related to the freedom of choice of global phases of Bloch waves [23]. That is, instead of $u_{n,k}(z)$ we may use

$$\tilde{u}_{n,k}(z) = e^{-i[jka+f(k)]} u_{n,k}(z) = e^{-if(k)} \sum_m e^{-ik(z-ma+ja)} w_n(z - ma), \tag{7.18}$$

where j is an integer number and $f(k)$ is any function that fulfills $f(k + 2\pi/a) = f(k)$. When we substitute $\tilde{u}_{n,k}(z)$ to (7.15) we obtain

(continued)

[3]In order to show (7.17), we use (7.16) and substitute $u_{n,k}(z) = \sum_m e^{-ik(z-ma)} w_n(z - ma)$ to (7.15) which yields $\theta_n^{\mathrm{Zak}} = (2\pi/a) \sum_m \int_{-ma}^{-ma+a} dz z |w_n(z)|^2$ and consequently (7.17).

$$\theta_n^{\mathrm{Zak}} = \frac{2\pi}{a}\int_{-\infty}^{\infty} dz z|w_n(z)|^2 + 2\pi j. \tag{7.19}$$

Thus, different choices of global phases of Bloch waves can lead to the Zak phases which are different by integer multiples of 2π.

In Sect. 5.3.1, the Su–Schrieffer–Heeger (SSH) model is analyzed where the winding number ν is the topological invariant characterizing topologically different regimes of the system parameters. The relation between the Zak phase and the winding number in the SSH model is the following $\theta^{\mathrm{Zak}} = \pi\nu$ [3].

We have considered one-dimensional systems but it is also worth mentioning the case of two-dimensional crystalline structures. If $\mathbf{X}$ in (7.13) is a two-dimensional vector in the reciprocal space of a two-dimensional crystalline structure, $\mathbf{X} = \mathbf{k} = (k_1, k_2)$, then the Berry phase divided by 2π is the Chern number of the band n [3]

$$C_n = \frac{1}{2\pi}\oint_{\mathscr{C}} \mathbf{A}(\mathbf{k}) \cdot d\mathbf{k}, \tag{7.20}$$

where

$$\mathbf{A}(\mathbf{k}) = i\langle u_{n,\mathbf{k}}|\partial_{\mathbf{k}} u_{n,\mathbf{k}}\rangle \tag{7.21}$$

is the Berry connection and $e^{i\mathbf{k}\cdot\mathbf{r}}u_{n,\mathbf{k}}(\mathbf{r})$ is a Bloch wave. The line integral in (7.20) is performed along the boundary $\mathscr{C}$ of the Brillouin zone. One can define the gauge-invariant Berry curvature

$$F(\mathbf{k}) = \mathbf{e}_3\cdot[\nabla_{\mathbf{k}}\times\mathbf{A}(\mathbf{k})] = \partial_{k_1}A_2 - \partial_{k_2}A_1 = i\epsilon_{3ij}\partial_{k_i}\langle u_{n,\mathbf{k}}|\partial_{k_j}u_{n,\mathbf{k}}\rangle, \tag{7.22}$$

and employing the Stokes' theorem exchange the line integral in (7.20) by the flux of the Berry curvature through the Brillouin zone (BZ),

$$C_n = \frac{1}{2\pi}\int_{\mathrm{BZ}} F(\mathbf{k})\, d^2k. \tag{7.23}$$

The Chern number is the topological invariant that characterizes energy bands of topological insulators in two-dimensional space, i.e. systems which are band insulators in the bulk but they can possess conducting surface (edge) states [3].

The Zak phase of the isolated band n of a one-dimensional photonic crystal described by Eq. (7.4) reads [20]

$$\theta_n^{\mathrm{Zak}} = \int_{-\pi/\Lambda}^{\pi/\Lambda} \left[i \int_{\text{unit cell}} dz\, \epsilon(z)\, u_{n,q}^*(z)\, \partial_q u_{n,q}(z) \right] dq. \tag{7.24}$$

We assume that the dielectric constant $\epsilon(z+\Lambda) = \epsilon(z)$ and a crystal has the inversion symmetry, i.e. $\epsilon(-z) = \epsilon(z)$. Figure 7.1 presents an example of a band structure of a photonic crystal and the values of the Zak phases corresponding to each band. If the frequency of a wave incidents on the semi-infinite crystal is located in the nth gap, then the ratio of the impedances has a purely imaginary value, i.e. $Z_R^{(n)}/Z_0 = i\zeta_R^{(n)}$ where $\zeta_R^{(n)}$ is a real number. Xiao et al. showed that the sign of $\zeta_R^{(n)}$ is determined by the Zak phases of the bands below the gap [20]

$$\mathrm{sgn}\left[\zeta_R^{(n)}\right] = (-1)^n(-1)^l \exp\left(i \sum_{m=0}^{n-1} \theta_m^{\mathrm{Zak}}\right), \tag{7.25}$$

where l is the number of band crossings below the nth gap (i.e. when two bands touch each other and there is no gap between them, see the 6th and 7th bands in Fig. 7.1) below the nth gap. The Zak phase of the lowest 0th band is determined in the following way $e^{i\theta_0^{\mathrm{Zak}}} = \mathrm{sgn}[1 - \epsilon_A/\epsilon_B]$, see Fig. 7.1.

According to (7.11) two semi-infinite photonic crystals, which are glued together, can possess a surface state when the signs of ζ_R and ζ_L, corresponding to a common gap, are different. In Fig. 7.2 there is an example of two photonic crystals for which this condition is fulfilled for the 7th gap which is common for both of them [20]. When such two photonic crystals are finite, transmission of electromagnetic waves through the system, with frequencies corresponding to gaps, is stopped unless there is a surface state. In the latter case one observes a resonant transmission due to the presence of an interface state, see the transmission spectrum in Fig. 7.2 for frequencies around the 7th gap.

7.2 Photonic Crystals in Time

In Sect. 7.1 photonic crystals in space were introduced which are related to propagation of electromagnetic waves in media where the dielectric constant changes periodically in space. It was an introduction to the main part of this chapter where we consider photonic time crystals, i.e. propagation of electromagnetic waves in materials where the dielectric constant is changing periodically in time.

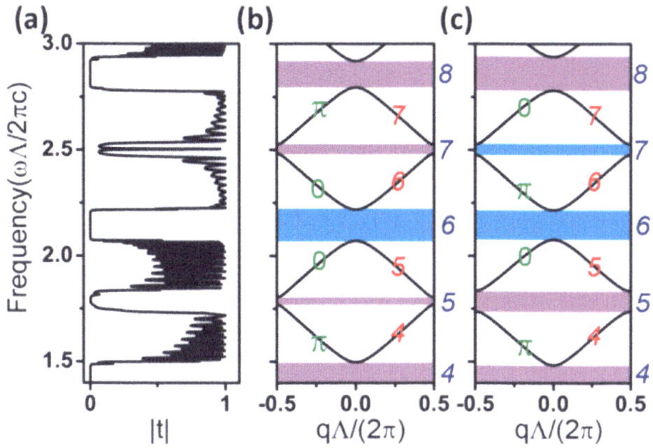

Fig. 7.2 Panel (**a**): transmission spectrum through a system of two finite photonic crystals which are glued together and which are composed of 10 unit cells. The band structures of the corresponding infinite crystals are presented in (**b**) and (**c**). The crystals are stacks of layers A and B, like in Fig. 7.1, with the parameters: $\epsilon_A = 3.8$, $\epsilon_B = 1$, $d_A = 0.42\Lambda$, $d_B = 0.58\Lambda$ (**b**) and $\epsilon_A = 4.2$, $\epsilon_B = 1$, $d_A = 0.38\Lambda$, $d_B = 0.62\Lambda$ (**c**). The numbers of the bands and of the gaps are depicted with red and blue labels, respectively. The Zak phases of the bands are labeled in green, while the magenta and cyan strips denote the gaps for which the imaginary parts of the impedance are positive and negative, respectively. The 7th gaps of the crystals correspond to opposite signs of the imaginary parts of the impedances and therefore fulfill the necessary condition for the existence of an interface state. A signature of the presence of such an interface state is visible in panel (**a**) where the resonance transmission at frequency $\omega\Lambda/(2\pi c) = 2.5$ can be observed. Reprinted from [20]

7.2.1 Introduction to Photonic Time Crystals

A sudden change of the dielectric constant in time results in reflection similar like in the case when it changes suddenly in space [1, 17]. We have seen that electromagnetic waves propagating in media where the dielectric constant varies periodically in space can behave like an electron moving in a solid state crystal and this kind of systems is dubbed photonic crystals. In the present section we concentrate on photonic time crystals which are related to materials where the dielectric constant is modulated periodically in time [2, 4, 6–8, 13, 14, 16, 17, 19, 24–26].

We assume a linear non-magnetic medium with instantaneous response, i.e. the magnetic induction $\mathbf{B} = \mu_0 \mathbf{H}$ and the displacement field $\mathbf{D} = \varepsilon_0 \epsilon(t)\mathbf{E}$, where the dielectric constant is changing periodically in time $\epsilon(t+T) = \epsilon(t)$. In the absence of free charges and the current density, the Maxwell equations read

$$\nabla \cdot \mathbf{E} = 0, \qquad \nabla \times \mathbf{E} = -\mu_0 \frac{\partial \mathbf{H}}{\partial t}, \tag{7.26}$$

$$\nabla \cdot \mathbf{H} = 0, \qquad \nabla \times \mathbf{H} = \frac{\partial}{\partial t}\left[\varepsilon_0 \epsilon(t)\mathbf{E}\right], \tag{7.27}$$

and allow one to show that **E** fulfills the following wave equation

$$\nabla^2 \mathbf{E} - \frac{1}{c^2}\frac{\partial^2}{\partial t^2}\left[\epsilon(t)\mathbf{E}\right] = 0, \tag{7.28}$$

where $c = 1/\sqrt{\mu_0 \varepsilon_0}$. Supposing plane wave solutions propagating along the z direction and polarized along the x axis, i.e. $\mathbf{E}(\mathbf{r}, t) = \mathbf{e}_x E(t) e^{ikz}$, Eq. (7.28) can be reduced to

$$\frac{d^2}{dt^2}[\epsilon(t)E(t)] + k^2c^2E(t) = 0. \tag{7.29}$$

Due to the fact that $\epsilon(t+T) = \epsilon(t)$, the Floquet theorem guarantees that

$$E(t) = u_{n,\Omega}(t)e^{-i\Omega t}, \tag{7.30}$$

where $u_{n,\Omega}(t+T) = u_{n,\Omega}(t)$, the index n labels bands of the wave number k and the frequency Ω can be restricted to the Floquet zone $-\pi/T \leq \Omega < \pi/T$ [25].

Let us consider an example where the dielectric constant changes periodically in time, $\epsilon(t+T) = \epsilon(t)$, in an alternate way like in Fig. 7.3a [13]. That is, $\epsilon(t) = \epsilon_1 = 3$ during the time interval t_1 and $\epsilon(t) = \epsilon_2 = 1$ during the remaining part of the period $t_2 = T - t_1$. The system is spatially homogeneous and consequently the wave number k is a well defined quantity. In the time intervals $t_{1,2}$, solutions of (7.29) are parameterized by $\omega_{1,2} = kc/\sqrt{\epsilon_{1,2}}$. Applying similar continuity conditions as for the photonic crystals in space (see Sect. 7.1.1) but in the time domain one obtains the dispersion relation [13, 19]

$$\cos(\Omega T) = \cos(\omega_1 t_1)\cos(\omega_2 t_2) - \frac{1}{2}\left(\sqrt{\frac{\epsilon_2}{\epsilon_1}} + \sqrt{\frac{\epsilon_1}{\epsilon_2}}\right)\sin(\omega_1 t_1)\sin(\omega_2 t_2), \tag{7.31}$$

which determines how the wave number k depends on the frequency Ω, cf. (7.7). The dispersion relation is shown in Fig. 7.3b where a band structure in the wave number domain is visible.

The wave number k is conserved due to the homogeneity of space. However, modulation of the dielectric constant in time results in partial reflection of waves to their time-reversed pairs. The time-reversed pairs are waves with the same k as in the incident waves but with opposite frequencies. In Fig. 7.3c propagation of an electromagnetic pulse, whose wave number components are located in a band is illustrated. In the case when at $t = t_0$ the dielectric constant starts changing in an alternate way in time, like in Fig. 7.3a, each plane wave component of the pulse is coupled to two Floquet solutions (7.30) with real $\Omega(k)$ and consequently two pulses propagate at $t > t_0$. When the modulation of the dielectric constant ends, the two pulses split again and four pulses are present when the photonic time crystal is over [13].

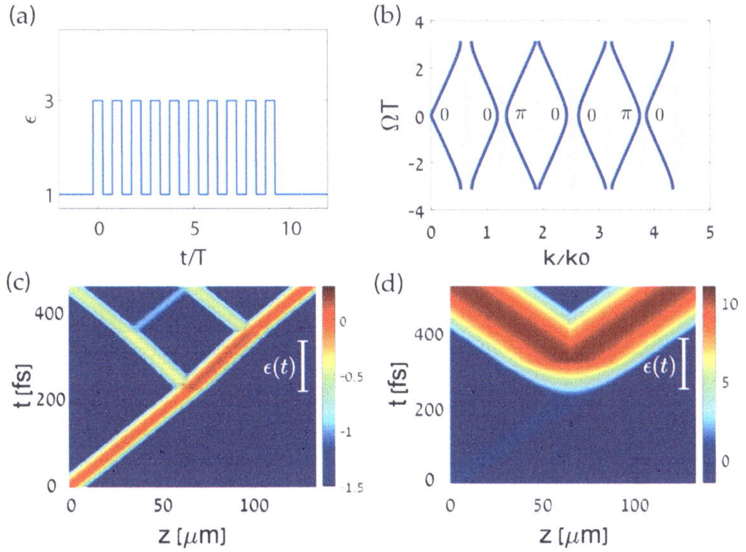

Fig. 7.3 Panel (**a**): photonic time crystal with alternate changes of the dielectric constant, i.e. $\epsilon(t+T) = \epsilon(t)$ where $T = 2$ fs, $\epsilon(t) = 3$ during $t_1 = T/2$ and $\epsilon(t) = 1$ for the remaining part of the period $t_2 = T - t_1$. Panel (**b**) shows the corresponding dispersion relation (7.31). Blue lines present bands separated by gray regions of gaps. Values of the Zak phases (0 or π) are depicted for each band. The wave number k is in the units of $k_0 = 2\pi/(Tc)$. Panels (**c**) and (**d**): simulations of propagation of pulses of the width of 45 fs in the presence of the photonic time crystal with the parameters as in (**a**)—the amplitude of the electric displacement field is presented (color bar in log scale). The pulse in (**c**) has the center wavelength 1.4 μm which is in a band, while the pulse in (**d**) has the center wavelength 0.94 μm which falls in a gap. The time crystal is turned on at $t = 220$ fs and turned off at $t = 340$ fs, i.e. it consists of $N = 60$ periods of the oscillations of $\epsilon(t)$. In (**c**), when the time crystal is turned on, the pulse splits into two pulses which in turn split again into two pairs of pulses when the crystal is turned off. In (**d**) the center wavelength of the pulse belongs to a gap and when the time crystal is turned on, complex values of $\Omega(k)$ are responsible for an exponential increase of the amplitude of the electromagnetic field so that when the time crystal is over two strong pulses propagate in the opposite directions. Adapted with permission from [13] © The Optical Society

If a propagating pulse composed of plane waves with wave numbers k in a gap of the dispersion relation, the frequencies $\Omega(k)$ are complex-valued and the modulation of the dielectric constant generates an exponential increase of an electromagnetic field in time, see Fig. 7.3d. When the time crystal ends, two pulses which are much stronger than the initial pulse propagate in the opposite directions.

7.2.2 Topological Properties of Photonic Time Crystals

In Sect. 7.1.2 we have seen that an electromagnetic wave incident on a semi-infinite photonic crystal in space is totally reflected if the frequency of the wave is chosen in a gap of the dispersion relation. Then, the wave impedance is purely imaginary and for a photonic crystal with the inversion symmetry, the sign of the imaginary part of the impedance is determined by the topological invariants, i.e. the Zak phases of the bands below the gap. In the case of photonic time crystals, even if the wave number k is chosen in a gap of the dispersion relation (7.31), the wave cannot propagate back in time and it enters a photonic time crystal. However, we will see that for a photonic time crystal with the inversion symmetry in time, i.e. $\epsilon(-t) = \epsilon(t)$, values of the Zak phases have still measurable consequences.

Topological photonic time crystals were first considered by Lustig et al. [13] (see also [19] and for similar behavior but of massive particles see Sect. 5.3.1). If the system is symmetric to time inversion, $\epsilon(-t) = \epsilon(t)$, the Zak phase for the band n, given by

$$\theta_n^{\text{Zak}} = \int_{-\pi/T}^{\pi/T} \left[i \int_{\text{temporal period}} dt\ \epsilon(t)\ u_{n,\Omega}^*(t)\ \partial_\Omega u_{n,\Omega}(t) \right] d\Omega, \tag{7.32}$$

takes the value 0 or π [13], compare (7.24). In (7.32), $u_{n,\Omega}(t)$ stands for a solution (7.30). The Zak phases for the photonic time crystal where the dielectric constant changes periodically in time in an alternate way are depicted in Fig. 7.3b. Assume that an electromagnetic wave, with the wave number in the nth gap, experiences N periods of a photonic time crystal. When N is large, the transmitted wave E_t and the reflected wave E_r have asymptotically the same amplitudes, $|E_t| = |E_r|$, cf. Fig. 7.3d. However, the relative phase $E_t/E_r = e^{i\phi^{(n)}}$ is different for different gaps. It turns out that the sign of the phase $\phi^{(n)}$ is determined by the Zak phases of the bands below the gap [13], i.e.

$$\text{sgn}\left[\phi^{(n)}\right] = (-1)^n(-1)^l \exp\left(i \sum_{m=0}^{n-1} \theta_m^{\text{Zak}}\right), \tag{7.33}$$

where $e^{i\theta_0^{\text{Zak}}} = \text{sgn}[1 - \epsilon_1/\epsilon_2]$ and l is the number of band crossings below the nth gap, cf. (7.25). The relative phase between two pulses can have significant measurable consequences, e.g., for coherent control of pump-probe experiments [13].

In Sect. 7.1.2 we have seen that if two photonic crystals in space are glued together, then interface states of an electromagnetic field can be observed for frequencies in common gaps of the crystals provided the imaginary parts of the crystals' impedances have different signs [20]. One can also arrange two different photonic time crystals that follow each other. It turns out that a state of an electromagnetic field (with the wave number corresponding to a common gap of the time crystals) localized on the edge of the crystals can form if the crystals have different topological properties. The edge of photonic time crystals is actually an edge in time. In Fig. 7.4a an example of two photonic time crystals when one follows the other is presented. For both crystals $\epsilon(t+T)=\epsilon(t)$, however, the first crystal has $\epsilon(t)=\epsilon_1=3$ during $t_1=T/2$ and $\epsilon(t)=\epsilon_2=1$ during $t_2=T-t_1$, while in the second crystal the values of ϵ_1 and ϵ_2 are exchanged. Hence, both crystals have the same dispersion relation but different Zak phases of the bands. When a propagating electromagnetic wave experiences the first photonic time crystal and the wave number is in a gap of its dispersion relation, the intensity of the wave increases exponentially up to a moment when the second time crystal appears, see Fig. 7.4b. Then the intensity decreases exponentially which demonstrates that the electromagnetic field is localized around the edge in time between these two photonic time crystals (compare the edge states of the Su–Schrieffer–Heeger model

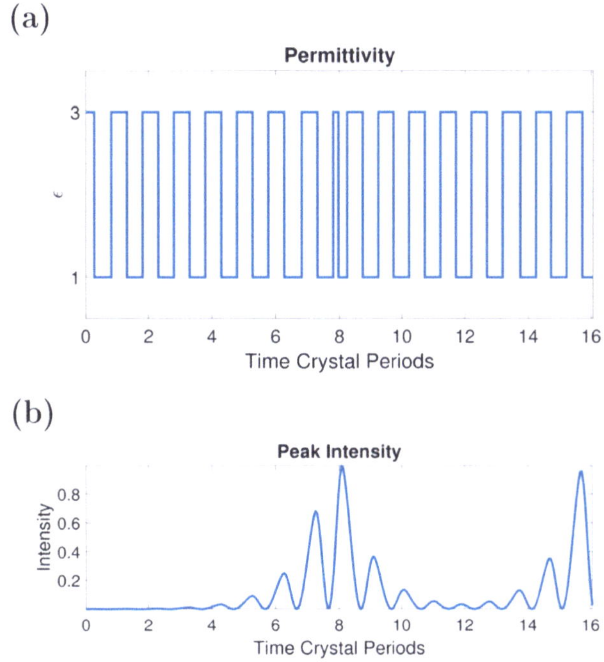

Fig. 7.4 Panel (**a**): two photonic time crystals with different Zak phases of the bands—the first crystal begins at $t=0$ and ends at $t=8T$ and is followed by the second time crystal. Panel (**b**) presents the intensity of an electromagnetic wave with the wavelength belonging to a common gap of the time crystals. When the first time crystal is turned on, the intensity increases exponentially in time up to the moment when the second time crystal is tuned on. Then, the intensity starts decreasing exponentially and consequently we can observe a state of the electromagnetic field localized on the edge between the time crystals which is an edge in time. After the decrease, the intensity starts increasing exponentially again like in the case of a single photonic time crystal presented in Fig. 7.3d. Adapted with permission from [13] © The Optical Society

described in Sect. 5.3.1). After the intensity decreases, it starts to increase again like in the case of a single photonic time crystal presented in Fig. 7.3d.

In order to observe pronounced features of photonic time crystals, the dielectric constant has to be modulated fast in time (on the time scale of a few periods of an electromagnetic wave propagating in a photonic crystal) and with a large amplitude. Then, the gaps in the wave number domain are big, i.e. of the order of the momentum of the wave [13]. Photonic time crystals were observed at the radio frequencies [18], and there is an effort to demonstrate them in optical frequencies [13].

References

1. Bacot, V., Labousse, M., Eddi, A., Fink, M., Fort, E.: Time reversal and holography with spacetime-transformations. Nat. Phys. **12**(10), 972–977 (2016). https://doi.org/10.1038/nphys3810
2. Biancalana, F., Amann, A., Uskov, A.V., O'Reilly, E.P.: Dynamics of light propagation in spatiotemporal dielectric structures. Phys. Rev. E **75**, 046607 (2007). https://link.aps.org/doi/10.1103/PhysRevE.75.046607
3. Cooper, N.R., Dalibard, J., Spielman, I.B.: Topological bands for ultracold atoms. Rev. Mod. Phys. **91**, 015005 (2019). https://link.aps.org/doi/10.1103/RevModPhys.91.015005
4. Deck-Léger, Z.L., Chamanara, N., Skorobogatiy, M., Silveirinha, M.G., Caloz, C.: Uniform-velocity spacetime crystals. Adv. Photonics **1**(5), 1–26 (2019). https://doi.org/10.1117/1.AP.1.5.056002
5. Dutta, O., Gajda, M., Hauke, P., Lewenstein, M., Lühmann, D.-S., Malomed, B.A., Sowiński, T., Zakrzewski, J.: Non-standard Hubbard models in optical lattices: a review. Rep. Prog. Phys. **78**(6), 066001 (2015). http://stacks.iop.org/0034-4885/78/i=6/a=066001
6. Elachi, C.: Waves in active and passive periodic structures: a review. Proc. IEEE **64**(12), 1666–1698 (1976). https://doi.org/10.1109/PROC.1976.10409
7. Harfoush, F., Taflove, A.: Scattering of electromagnetic waves by a material half-space with a time-varying conductivity. IEEE Trans. Antennas Propag. **39**(7), 898–906 (1991). https://doi.org/10.1109/8.86907
8. Holberg, D., Kunz, K.: Parametric properties of fields in a slab of time-varying permittivity. IEEE Trans. Antennas Propag. **14**(2), 183–194 (1966). https://doi.org/10.1109/TAP.1966.1138637
9. Joannopoulos, J.D., Johnson, S.G., Winn, J.N., Meade, R.D.: Photonic Crystals: Molding the Flow of Light, 2 edn. Princeton University Press, Princeton (2008)
10. John, S.: Strong localization of photons in certain disordered dielectric superlattices. Phys. Rev. Lett. **58**, 2486–2489 (1987). https://link.aps.org/doi/10.1103/PhysRevLett.58.2486
11. Lawrence, F.J., Botten, L.C., Dossou, K.B., de Sterke, C.M., McPhedran, R.C.: Impedance of square and triangular lattice photonic crystals. Phys. Rev. A **80**, 023826 (2009). https://link.aps.org/doi/10.1103/PhysRevA.80.023826
12. Lawrence, F.J., Botten, L.C., Dossou, K.B., McPhedran, R.C., de Sterke, C.M.: Photonic-crystal surface modes found from impedances. Phys. Rev. A **82**, 053840 (2010). https://link.aps.org/doi/10.1103/PhysRevA.82.053840
13. Lustig, E., Sharabi, Y., Segev, M.: Topological aspects of photonic time crystals. Optica **5**(11), 1390–1395 (2018). https://doi.org/10.1364/OPTICA.5.001390. http://www.osapublishing.org/optica/abstract.cfm?URI=optica-5-11-1390
14. Ma, J., Wang, Z.G.: Band structure and topological phase transition of photonic time crystals. Opt. Express **27**(9), 12914–12922 (2019). https://doi.org/10.1364/OE.27.012914. http://www.opticsexpress.org/abstract.cfm?URI=oe-27-9-12914

15. Malkova, N., Hromada, I., Wang, X., Bryant, G., Chen, Z.: Observation of optical Shockley-like surface states in photonic superlattices. Opt. Lett. **34**(11), 1633–1635 (2009). https://doi.org/10.1364/OL.34.001633. http://ol.osa.org/abstract.cfm?URI=ol-34-11-1633
16. Martínez-Romero, J.S., Halevi, P.: Parametric resonances in a temporal photonic crystal slab. Phys. Rev. A **98**, 053852 (2018). https://link.aps.org/doi/10.1103/PhysRevA.98.053852
17. Morgenthaler, F.R.: Velocity modulation of electromagnetic waves. IRE Trans. Microwave Theory Tech. **6**(2), 167–172 (1958). https://doi.org/10.1109/TMTT.1958.1124533
18. Reyes-Ayona, J.R., Halevi, P.: Observation of genuine wave vector (k or ?) gap in a dynamic transmission line and temporal photonic crystals. Appl. Phys. Lett. **107**(7), 074101 (2015). https://doi.org/10.1063/1.4928659
19. Wang, N., Zhang, Z.Q., Chan, C.T.: Photonic Floquet media with a complex time-periodic permittivity. Phys. Rev. B **98**, 085142 (2018). https://link.aps.org/doi/10.1103/PhysRevB.98.085142
20. Xiao, M., Zhang, Z.Q., Chan, C.T.: Surface impedance and bulk band geometric phases in one-dimensional systems. Phys. Rev. X **4**, 021017 (2014). https://link.aps.org/doi/10.1103/PhysRevX.4.021017
21. Yablonovitch, E.: Inhibited spontaneous emission in solid-state physics and electronics. Phys. Rev. Lett. **58**, 2059–2062 (1987). https://link.aps.org/doi/10.1103/PhysRevLett.58.2059
22. Yablonovitch, E.: Photonic band-gap crystals. J. Phys. Condens. Matter **5**(16), 2443–2460 (1993). https://doi.org/10.1088/0953-8984/5/16/004
23. Zak, J.: Berry's phase for energy bands in solids. Phys. Rev. Lett. **62**, 2747–2750 (1989). https://link.aps.org/doi/10.1103/PhysRevLett.62.2747
24. Zhang, Z., Zhang, W., Zeng, L.: Analyzing photonic space-time crystal with FDTD. Mod. Phys. Lett. B **34**(06), 2050082 (2020). https://doi.org/10.1142/S0217984920500827
25. Zurita-Sánchez, J.R., Halevi, P., Cervantes-González, J.C.: Reflection and transmission of a wave incident on a slab with a time-periodic dielectric function $\epsilon(t)$. Phys. Rev. A **79**, 053821 (2009). https://link.aps.org/doi/10.1103/PhysRevA.79.053821
26. Zurita-Sánchez, J.R., Abundis-Patiño, J.H., Halevi, P.: Pulse propagation through a slab with time-periodic dielectric function ϵ(t). Opt. Express **20**(5), 5586–5600 (2012). https://doi.org/10.1364/OE.20.005586. http://www.opticsexpress.org/abstract.cfm?URI=oe-20-5-5586

Index

A

Action-angle variables, 48–53, 101, 102, 174, 175, 188, 210, 217, 226
Aharonov-Bohm ring, 21

B

Berry phase, 255, 257
Bloch wave, 40, 176–178, 182, 212, 238, 252, 255–257
Bogoliubov approach, 67–70, 75, 80
Bose–Einstein condensate, 15, 21, 24, 33, 35, 36, 58, 59, 63, 67–72, 74, 76, 77, 80, 103, 112, 113, 128–131, 133, 201, 208, 214
Bose–Hubbard model, 111–114, 213–216, 218–225

C

Canonical ensemble, 81, 92, 217
Canonical transformation, 47–49, 51, 52, 174, 182, 194, 223, 226, 228, 245, 246
Center of mass, 8–10, 22, 23, 25, 26, 80
Chemical potential, 24, 35
Chern number, 243, 244, 257
Circular orthogonal ensemble, 85, 87
Coherent state, 239–241, 244, 245, 247
Contact interactions, 21–23, 56, 68, 80, 91, 103, 111, 216, 219, 228, 246–248

D

Depletion of a condensate, 67, 69–72
Dicke model, 134, 135, 137–140
Dispersion relation, 251–254, 260–263

E

Edge states, 201, 202, 205, 208, 209, 254, 263
Effective mass, 32, 51, 73, 175, 177, 188, 190, 206, 210
Eigenstate thermalization hypothesis, 81, 92, 93, 95, 96, 99, 217
Energy gap, 31, 55, 56, 58, 73, 77, 112, 201, 203–205, 208, 213, 214, 217, 220, 222

F

Feshbach resonance, 215, 219, 221, 227
Fibonacci quasi-crystal, 109, 110, 112, 114, 115, 210–212
Field operator, 11, 12, 15, 34, 35, 45, 57, 67, 73, 102, 111, 213
Fock states, 11–13, 15, 16, 35, 45, 58, 74, 102, 213, 214, 225
Fractional time crystals, 99, 104, 105, 107

G

Gravitational units, 47, 56, 110, 184, 208
Gross–Pitaevskii equation, 22–24, 58, 59, 63, 64, 66–70, 75, 76, 78, 129, 130, 132
Ground canonical ensemble, 35

H

Haldane insulator, 209, 222
Hofstadter's butterfly structure, 243

I

Impedance, 253, 254, 258, 259, 262

K. Sacha, *Time Crystals*, Springer Series on Atomic, Optical, and Plasma Physics 114, https://doi.org/10.1007/978-3-030-52523-1

J
Jaynes–Cummings model, 41
Josephson effect, 34

K
Kac normalization, 105
Kicked harmonic oscillator, 240, 244
Kicked rotor, 180, 181

L
Lindblad generator, 136, 141, 142, 144–146, 151, 153
Lipkin–Meshkov–Glick model, 62, 63, 66, 79, 91, 92, 99, 105, 108, 109
Long-range interactions, 91, 94–99, 105–108, 112, 116, 119, 121, 145, 215, 218–220, 224, 225, 237, 247

M
Magnon condensate, 36
Many-body localization, 81, 84, 85, 87, 93, 116, 118, 119, 122, 127, 134, 173, 217, 218
Master equation, 136, 141–146, 149–152, 154, 244
Mean field, 24, 62, 63, 74–79, 91, 120, 129, 133, 136–139, 141, 146, 153
Mott insulator, 173, 214, 215, 222, 224, 225

N
n-tupling time crystal, 72, 92, 106, 108, 138, 139
Nuclear magnetic resonance (NMR), 36, 123, 124

O
Optical lattice, 179, 201

P
Phase diagram, 65, 66, 70, 85–87, 114, 117, 118, 138, 146
Poisson distribution, 85, 87, 90
Prethermal phase, 87, 93, 95–99, 108

R
Rotating wave approximation (RWA), 41, 42, 131, 238–240, 244, 245, 247

S
Scaling symmetry, 52, 53
Scaling theory of localization, 180, 192, 197
Scattering length, 56, 130, 131, 213, 215, 219–222, 227, 246
Schrödinger cat, 59, 60, 76, 77, 83, 84, 103, 113, 116, 125
Second quantization, 10, 12–14, 20, 23, 24, 43, 45, 57, 74, 76, 213
Secular approximation, 48, 50–55, 73, 100, 174–177, 182, 183, 188, 190, 194, 206, 217, 220, 222, 224, 228
Semiclassical approach, 53–55, 58, 101, 102, 108, 124, 190
Soliton, 22, 23, 25, 26, 80
Superfluid, 36, 159, 214, 215, 222
Superradiance, 135, 137
Su–Schrieffer–Heeger (SSH) model, 201, 202, 254, 257, 263

T
Transfer matrix, 194, 196, 197

U
U(1) symmetry, 35, 57, 67, 75, 133

W
Wannier states, 177, 185, 207, 224, 256
Winding number, 203–205, 208, 209, 257
Wyckoff position, 256

Z
Zak phase, 205, 253, 255–259, 261–263
Z_2-symmetry, 62, 63, 66, 79, 90–92, 95, 96, 133, 136–138, 140, 144, 146–150